Zhihua Zhang

Environmental Data Analysis

Also of Interest

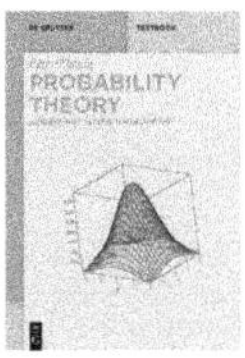

Probability Theory and Statistical Applications.
A Profound Treatise for Self-Study
Peter Zörnig, 2016
ISBN 978-3-11-036319-7, e-ISBN 978-3-11-040283-4

An Introduction to Nonlinear Optimization Theory
Marius Durea, Radu Strugariu, 2014
ISBN 978-3-11-042603-8, e-ISBN 978-3-11-042735-6

Asymptotic Statistics.
With a View to Stochastic Processes
Reinhard Höpfner, 2014
ISBN 978-3-11-025024-4, e-ISBN 978-3-11-036778-2

Compressive Sensing.
Applications to Sensor Systems and Image Processing
Joachim Ender, 2017
ISBN 978-3-11-033531-6, e-ISBN 978-3-11-039027-8

Scientific Computing.
For Scientists and Engineers
Timo Heister, Leo G. Rebholz, 2015
ISBN 978-3-11-035940-4, e-ISBN 978-3-11-038680-6

Zhihua Zhang

Environmental Data Analysis

Methods and Applications

DE GRUYTER

Author
Prof. Zhihua Zhang
College of Global Change &
Earth System Science
Beijing Normal University
19 Xinjiekou Wai St.
100875 Beijing
People's Republic of China
zhangzh@bnu.edu.cn

ISBN 978-3-11-043001-1
e-ISBN (PDF) 978-3-11-042490-4
e-ISBN (EPUB) 978-3-11-042498-0
Set-ISBN 978-3-11-042491-1

Library of Congress Cataloging-in-Publication Data
A CIP catalog record for this book has been applied for at the Library of Congress.

Bibliographic information published by the Deutsche Nationalbibliothek
The Deutsche Nationalbibliothek lists this publication in the Deutsche Nationalbibliografie; detailed bibliographic data are available on the Internet at http://dnb.dnb.de.

Cover image: Merve Sarac/iStock/thinkstock
Typesetting: PTP-Berlin, Protago-TEX-Production GmbH, Berlin
Printing and binding: CPI books GmbH, Leck
♾ Printed on acid-free paper
Printed in Germany

www.degruyter.com

Preface

Environmental data provide huge amounts of information, but it is complex to process due to the size, variety, and dynamic nature of the data. In order to develop solutions to many environmental issues and make predictions to determine how resources are best allocated, environmental researchers have spent considerable time ensuring well-conducted data collection, analyzing and interpreting environmental data, and describing environmental changes with sound and validated models. Therefore, researchers in environmental science need to be familiar with various advanced techniques for exploration, identification and analysis of patterns in data.

This book covers the comprehensive range of topics in data analysis in space, time and spectral domains which are necessary knowledge for environmental research. Main topics include Models for Linear and Nonlinear Environmental Systems, Statistical and Numerical Methods, Data Envelopment Analysis, Risk Assessments, and Life Cycle Assessments. It is a concise and accessible book suitable for anyone interested in learning and understanding advanced methods and applications in environmental data analysis.

DOI 10.1515/9783110424904-001

Contents

1 Time series analysis

The main objectives of environmental time series analysis are to describe environmental change, explain the mechanisms underlying these changes, predict future environmental change under given perturbation and avoid undesired environmental impacts. In environmental time series analysis, autoregressive moving average (ARMA) models and state-space models are two most useful tools to discover the dynamical characteristics of environmental change through fitting them to environmental time series with complex dynamics.

1.1 Stationary time series

A time series is a set of observations $\{x_t\}$, where each t is a specific time. Time series are encountered in a variety of fields such as temperature, rainfall, pollution, emissions and population.

To draw inferences from time series, involving prediction and interrelationships, and to understand the mechanism generating the series, one needs to establish a probability model. The observation is postulated to be a realization of the probability model. For a complete probability model $\{X_t\}$, all of the joint distributions of $(X_1, \dots, X_n)^T$ $(n \in \mathbb{Z}_+)$ would be specified, i.e., for any $n \in \mathbb{Z}_+$, the probabilities $P(X_1 \le x_1, \dots, X_n \le x_n)$ are specified. If all the joint distributions are multivariate normal, the distribution properties of the probability model would be determined completely by the means EX_t and covariances $\mathrm{Cov}(X_{t+h}, X_t)$ $(t \in \mathbb{Z}_+,\ h = 0, 1, \dots)$. Generally speaking, a complete probability model cannot be established using one data observation. In most practical problems involving time series, only one realization is seen. Moreover, to obtain all of the joint distributions, one needs to estimate too many parameters. However, from a linear prediction viewpoint, the minimum mean squared error linear prediction depends only on means and covariances. Therefore, one characterizes time series models using the second-order properties.

A time series model for observed data is a sequence of random variables $\{X_t\}$ with mean function EX_t and covariance function $\mathrm{Cov}(X_{t+h}, X_t)$. Denote

$$\mu_X(t) = EX_t,$$
$$\gamma_X(t+h, t) = \mathrm{Cov}(X_{t+h}, X_t).$$

We say a time series $\{X_t\}$ is *stationary* if its mean function $\mu_X(t)$ is independent of t and the autocovariance function $\gamma_X(t+h, t)$ is independent of t for each lag h.

DOI 10.1515/9783110424904-002

1.1.1 Autocovariance functions

If a time series $\{X_t\}$ is stationary, denote $\mu_X = \mu_X(t)$, the autocovariance function of $\{X_t\}$ at lag h is $\gamma_X(h) = \gamma_X(t+h, t)$ and the autocorrelation function at lag h is $\rho_X(h) = \gamma_X(h)/\gamma_X(0)$.

The basic properties of autocovariance function $\gamma_X(h)$ are as follows:

- $\gamma_X(0) \geq 0$ and $\gamma_X(h)$ is even, and $|\gamma_X(h)| \leq \gamma_X(0)$ for all h;
- the matrix $(\gamma_X(i-j))_{i,j=1,\dots,n}$ is nonnegative definite, i.e., for any real-valued vector $\boldsymbol{a} = (a_1, \dots, a_n)^{\mathrm{T}}$,

$$\sum_{i,j=1}^{n} a_i \gamma_X(i-j) a_j \geq 0 \quad (n \in \mathbb{Z}).$$

In fact, it is clear that

$$\gamma_X(0) = \mathrm{Cov}(X_0, X_0) = \mathrm{Var}(X_0) \geq 0,$$
$$\gamma_X(-h) = \mathrm{Cov}(X_{-h}, 0) = \mathrm{Cov}(X_0, X_h) = \mathrm{Cov}(X_h, X_0) = \gamma_X(h).$$

By the Schwarz inequality,

$$\begin{aligned} 0 \leq |\gamma_X(h)| &= |\mathrm{Cov}(X_h, X_0)| = |E[(X_h - EX_h)(X_0 - EX_0)]| \\ &\leq (E[(X_h - EX_h)^2])^{\frac{1}{2}} \cdot (E[(X_0 - EX_0)^2])^{\frac{1}{2}} \\ &= (\mathrm{Var}(X_h)\,\mathrm{Var}(X_0))^{\frac{1}{2}} = \mathrm{Var}(X_0) = \gamma_X(0). \end{aligned}$$

Without loss of generality, assume $\mu_X = 0$. Let $\boldsymbol{X}_n = (X_n, \dots, X_1)^{\mathrm{T}}$. Then

$$0 \leq \mathrm{Var}(a^{\mathrm{T}}\boldsymbol{X}_n) = E[(a^{\mathrm{T}}\boldsymbol{X}_n)^2] = E[a^{\mathrm{T}}\boldsymbol{X}_n\boldsymbol{X}_n^{\mathrm{T}}a] = a^{\mathrm{T}}E[\boldsymbol{X}_n\boldsymbol{X}_n^{\mathrm{T}}]a = \sum_{i,j=1}^{n} a_i\gamma_X(i-j)a_j,$$

where $\mathrm{Var}(X)$ means the variance of X.

A time series $\{X_t\}$ is *strictly stationary* if $(X_1, \dots, X_n)$ and $(X_{1+h}, \dots, X_{n+h})$ $(h \in \mathbb{Z}, n \in \mathbb{Z}_+)$ have the same joint distributions. If $\{X_t\}$ is strictly stationary, then $\{X_t\}$ must be stationary. Conversely, it is not true.

For example, consider a time series $\{X_t\}$ which is obtained by tossing a penny repeatedly and scoring $+1$ for each head and -1 for each tail. Its time series model $\{X_t\}$ is an independent random variable sequence with

$$P(X_t = 1) = \tfrac{1}{2},$$
$$P(X_t = -1) = \tfrac{1}{2}, \quad (t \in \mathbb{Z}_+).$$

Its mean function $EX_t = 0$ and variance $\mathrm{Var}(X_t) = 1$, and autocovariance function $\gamma_X(t+h, t) = 0$ $(h \neq 0)$. Clearly, both $\mu_X(t)$ and $\gamma_X(t+h, t)$ are independent of t. So $\{X_t\}$ is a stationary time series with mean 0 and autocovariance function $\gamma_X(h) = \delta_{0,h}$, where δ_{ij} is the Kronecker delta. Take

$$S_t = X_1 + X_2 + \cdots + X_t \quad (t \in \mathbb{Z}_+),$$
$$S_0 = 0,$$

which is called the *random walk*. Clearly, $\mu_S(t) = ES_t = 0$ and

$$\begin{aligned}\gamma_X(t+h,t) &= \mathrm{Cov}(S_{t+h}, S_t)\\ &= \mathrm{Cov}(S_t + X_{t+1} + \cdots + X_{t+h}, S_t) = \mathrm{Cov}(S_t, S_t)\\ &= E[(X_1 + \cdots + X_t)^2] = EX_1^2 + \cdots + EX_t^2 = t.\end{aligned}$$

Since $\gamma_X(t+h,t)$ depends on t, S_t is not a stationary time series. Let $Y_t = X_t + \frac{1}{2}X_{t-1}$. Then $\mu_Y(t) = 0$ and

$$\gamma_X(t+h,t) = \begin{cases} \frac{5}{4}, & h = 0,\\ \frac{1}{2}, & h = \pm 1,\\ 0, & |h| > 1,\end{cases}$$

and so the random walk is stationary with mean 0 and

$$\gamma_X(h) = \begin{cases} \frac{5}{4}, & h = 0,\\ \frac{1}{2}, & h = \pm 1,\\ 0, & |h| > 1.\end{cases}$$

For example, consider a time series

$$X_t = A\cos(\omega t) + B\sin(\omega t) \quad (t \in \mathbb{Z}),$$

where A and B are uncorrelated random variables with mean 0 and variance 1, and ω is a constant. The mean and covariance are, respectively,

$$\begin{aligned}EX_t &= (EA)\cos(\omega t) + (EB)\sin(\omega t) = 0,\\ \mathrm{Cov}(X_{t+h}, X_t) &= E[(A\cos\omega(t+h) + B\sin\omega(t+h))(A\cos(\omega t) + B\sin(\omega t))]\\ &= E[A^2\cos\omega(t+h)\cos(\omega t) + AB\cos\omega(t+h)\sin(\omega t)\\ &\qquad + AB\sin\omega(t+h)\cos(\omega t) + B^2\sin\omega(t+h)\sin(\omega t)].\end{aligned}$$

By the assumption that $EA^2 = EB^2 = 1$ and $E[AB] = 0$,

$$\mathrm{Cov}(X_{t+h}, X_t) = \cos\omega(t+h)\cos(\omega t) + \sin\omega(t+h)\sin(\omega t) = \cos(\omega h).$$

Therefore, $\{X_t\}$ is a stationary time series with the mean $\mu = 0$ and autocovariance function $\gamma(h) = \cos(\omega h)$.

1.1.2 White noise and linear process

The simplest model for a time series is white noise which is a sequence of uncorrelated random variables $\{Z_t\}$ with mean 0 and variance σ^2. Denote $\{Z_t\} \sim \mathrm{WN}(0, \sigma^2)$.

Since $EZ_t = 0$ $(t \in \mathbb{Z})$ and $\mathrm{Cov}(Z_{t+h}, Z_t) = \delta_{h,0}\sigma^2$ $(t \in \mathbb{Z})$, the time series $\{Z_t\}$ is stationary. Its mean is $\mu = 0$ and the covariance matrix of $Z_1, \ldots, Z_n$ is $\sigma^2 I_n$, where I_n is the identity matrix of order n.

White noise plays an important role as a building block for more complicated time series. For example, consider the time series

$$X_t = Z_t + \frac{1}{3} Z_{t-1},$$
$$\{Z_t\} \sim \mathrm{WN}(0,1) \quad (t \in \mathbb{Z}).$$

Its mean is $EX_t = 0$ and for all t,

$$E[X_{t+h}, X_t] = \begin{cases} \frac{10}{9} & \text{if } h = 0, \\ \frac{1}{3} & \text{if } h = \pm 1, \\ 0 & \text{if } |h| > 1. \end{cases}$$

So X_t is a stationary time series.

We say that a time series $\{X_t\}$ is a *linear process* if

$$X_t = \sum\nolimits_{j \in \mathbb{Z}} \psi_j Z_{t-j},$$
$$\{Z_t\} \sim \mathrm{WN}(0, \sigma^2),$$

where each ψ_j is constant and $\sum_{j\in\mathbb{Z}} |\psi_j| < \infty$. Since $\sum_{j\in\mathbb{Z}} |\psi_j| < \infty$, clearly, $\sum_{j\in\mathbb{Z}} \psi_j^2 < \infty$. This further deduces that $\sum_{j\in\mathbb{Z}} \psi_j Z_{t-j}$ is convergent in the square mean. So the linear process X_t is stationary.

In fact, from

$$E[X_t] = \sum\nolimits_{j\in\mathbb{Z}} \psi_j E[Z_{t-j}] = 0,$$
$$E[X_{t+h}X_t] = \sum\nolimits_{k\in\mathbb{Z}} \sum\nolimits_{j\in\mathbb{Z}} \psi_j \psi_k E[Z_{t+h-j} Z_{t-k}],$$
$$E[Z_{t+h-j} Z_{t-k}] = \delta_{j-h,k} \sigma^2,$$

it follows that

$$\gamma(h) = E[X_{t+h}X_t] = \sum_{k\in\mathbb{Z}} \psi_{k+h} \psi_k \sigma^2, \tag{1.1.1}$$

and so $\{X_t\}$ is stationary.

Assume that $\{X_t\}$ is a stationary time series satisfying the following:

$$\begin{aligned} X_t &= \varphi X_{t-1} + Z_t, \\ \{Z_t\} &\sim \mathrm{WN}(0, \sigma^2) \quad (t \in \mathbb{Z}), \end{aligned} \tag{1.1.2}$$

where $|\varphi| < 1$ and Z_t is uncorrelated with X_s. Then

$$X_t = \varphi X_{t-1} + Z_t = \varphi(\varphi X_{t-2} + Z_{t-1}) + Z_t = \varphi^2 X_{t-2} + \varphi Z_{t-1} + Z_t$$
$$= \sum_0^k \varphi^j Z_{t-j} + \varphi^{k+1} X_{t-k-1}.$$

Since $\{X_t\}$ is stationary,

$$E[(X_t - \sum_{j=0}^{k} \varphi^j Z_{t-j})^2] = \varphi^{2k+2} E[X_{t-k-1}^2] \to 0 \quad (k \to \infty),$$

and so $X_t = \sum_{j=0}^{\infty} \varphi^j Z_{t-j}$ in the mean square sense. This is a linear process and $\sum_{j=0}^{\infty} \varphi^j < \infty$. By (1.1.1),

$$\gamma_X(h) = \sum_0^{\infty} \varphi^{k+h} \varphi^k \sigma^2 = \frac{\sigma^2 \varphi^h}{1 - \varphi^2} \quad (h \geq 0). \tag{1.1.3}$$

1.1.3 Sample autocorrelation function

In practical problems, we do not start with a model but with observed data $\{x_1, x_2, \ldots, x_n\}$. Using these observation data, we estimate the mean μ, autocovariance function $\gamma(h)$, and autocorrelation function $\rho(h)$. The *sample mean* of $x_1, \ldots, x_n$ is $\bar{x} = \frac{1}{n} \sum_{t=1}^{n} x_t$. The *sample autocovariance function* is

$$\hat{\gamma}(h) = \frac{1}{n} \sum_{1}^{n-|h|} (x_{t+|h|} - \bar{x})(x_t - \bar{x}) \quad (-n < h < n).$$

1.1.4 Classical decomposition model

When we analyze any time series, we first plot data on a sheet of graph paper. If there are discontinuities, we break it into homogeneous segments. If there are outlying observations, we discard them. Inspection of a graph may also suggest the possibility of representing the data as a realization of a process. The classical decomposition model is

$$X_t = m_t + s_t + Y_t,$$

where m_t is a trend component, s_t is a seasonal component, and Y_t is a stationary random noise component.

We hope to estimate m_t and s_t such that the residual Y_t is stationary. First we consider the nonseasonal model with trend $X_t = m_t + Y_t$ $(t = 1, \ldots, n)$, where $E[Y_t] = 0$. The moving average is

$$\hat{m}_t = \frac{1}{2p+1} \sum_{-p}^{p} X_{t-j} \quad (p - 1 \leq t \leq n - p).$$

It provides an estimate of the trend m_t.

A general classical decomposition model is

$$X_t = m_t + s_t + Y_t \quad (t = 1, \ldots, n),$$

where $E[Y_t] = 0$, $s_{t+d} = s_t$, and $\sum_{j=1}^{d} s_j = 0$. For observations $\{x_1, \dots, x_n\}$, if the period $d = 2p$ is even, then the estimate of trend

$$\hat{m}_t = \frac{1}{2p}\left(\frac{1}{2}x_{t-p} + x_{t-p+1} + \cdots + x_{t+p-1} + x_{t+p}\right) \quad (p < t \le n-p);$$

if the period $d = 2p + 1$ is odd, then the estimate of trend

$$\hat{m}_t = \frac{1}{2p+1}\sum_{-p}^{p} x_{t-j} \quad (p+1 \le t \le n-p).$$

For each $k = 1, \dots, d$, let ω_k be the average of the deviations $\{x_{k+jd} - \hat{m}_{k+jd}\}$ $(p < k + jd \le n - p)$ and let

$$\hat{s}_k = \omega_k - \frac{1}{d}\sum_{1}^{d} \omega_i \quad (k = 1, \dots, d), \quad \hat{s}_k = s_{k-d} \quad (k > d).$$

Define $d_t = x_t - \hat{s}_t$ $(t = 1, \dots, n)$. Reestimating the trend from $\{d_t\}$ as above, the estimated noise series is

$$\hat{Y}_t = x_t - \hat{m}_t - \hat{s}_t \quad (t = 1, \dots, n).$$

1.2 Prediction of time series

If a time series $\{X_t\}$ is independent for different time t, we cannot make predictions. If it is correlated, and its mean and covariance are known, we may predict X_{t+h} from $X_0, \dots, X_t$. In fact, this requires finding a linear combination τ^* of $X_0, \dots, X_t$ such that the mean square error $E[(X_{t+h} - \tau^*)^2]$ is minimal. The τ^* is called the *best linear approximation* and is used as the *forecast* of X_{t+h}.

1.2.1 The best linear approximation

If a random variable Y is approximated by random variable X, define the mean square error as $E[(Y - X)^2]$. The simplest case is that if a random variable Y is approximated by a constant, then EY is its best approximation and the minimum mean square error is $\mathrm{Var}(Y)$.

Orthogonality principle

Let Y be a random variable. Denote by τ_n all linear combinations of $1, X_1, \dots, X_n$, i.e.,

$$\tau_n = \left\{ a_0 + \sum_{1}^{n} a_k X_k, \quad \text{where each } a_k \text{ is constant} \right\}.$$

For $X \in \tau_n$, define the approximation error as $E[(Y - X)^2]$. If $X^* \in \tau_n$ is such that $E[(Y - X^*)X_k] = 0$ $(k = 0, 1, \dots, n)$, then X^* is the best linear approximation of Y in τ_n, i.e., for all $X \in \tau_n$,

$$E[(Y - X^*)^2] \le E[(Y - X)^2]$$

and the minimum mean error is

$$E[(Y - X^*)^2] = E[Y^2] - E[X^{*2}].$$

Conversely, if $X^* \in \tau_n$ is the best linear approximation of Y in τ_n, then

$$E[(Y - X^*)X_k] = 0 \quad (k = 0, 1, \dots, n).$$

1.2.2 Prediction of stationary time series

Let $\{X_t\}$ be a stationary time series with mean μ and autocovariance function γ. We estimate X_{n+h} $(h > 0)$ by the linear combination of $1, X_1, \dots, X_n$ such that the mean square error attains the minimal value. Denote by P_nX_{n+h} the best linear predictor and

$$P_nX_{n+h} = a_0^* + a_1^*X_n + \cdots + a_n^*X_1, \tag{1.2.1}$$

where $a_0^*, a_1^*, \dots, a_n^*$ are undetermined coefficients.

By the orthogonality principle, the best approximation P_nX_{n+h} of X_{n+h} in all linear combinations of $\{1, X_1, \dots, X_n\}$ satisfies

$$E[(X_{n+h} - P_nX_{n+h})] = 0,$$
$$E[(X_{n+h} - P_nX_{n+h})X_{n+1-j}] = 0 \quad (j = 1, \dots, n).$$

This implies the following proposition.

Let $\{X_t\}$ be a stationary time series with mean μ and autocovariance function γ. Then

(a) the best predictor P_nX_{n+h} of X_{n+h} $(h > 0)$ by the linear combinations of $1, X_1, \dots, X_n$ is

$$P_nX_{n+h} = \mu + \sum_1^n a_i^*(X_{n+1-i} - \mu)$$

and the coefficient $\boldsymbol{a}_n^* = (a_1^*, \dots, a_n^*)^{\mathrm{T}}$ satisfies the system of linear equations

$$\begin{cases} a_0^* = \mu\left(1 - \sum_{i=1}^n a_i^*\right), \\ \Gamma_n\boldsymbol{a}_n^* = \boldsymbol{R}_n(h), \end{cases} \tag{1.2.2}$$

where $\Gamma_n = (\gamma(i-j))_{n\times n}$, and $\boldsymbol{R}_n(h) = (\gamma(h), \gamma(h+1), \dots, \gamma(h+n-1))^{\mathrm{T}}$;

(b) the minimum mean square error $E[(X_{n+h} - P_nX_{n+h})^2] = \gamma(0) - \boldsymbol{a}_n^{*T}\boldsymbol{R}_n(h)$.

Consider a time series $\{X_t\}$ satisfying

$$X_t = \frac{1}{2}X_{t-1} + Z_t,$$
$$\{Z_t\} \sim \mathrm{WN}(0,1) \quad (t \in \mathbb{Z}).$$

The best predictor is

$$P_n X_{n+1} = \sum_{i=1}^{n} a_i^* X_{n+1-i} \tag{1.2.3}$$

and the coefficient $\boldsymbol{a}_n^* = (a_1^*, \dots, a_n^*)^{\mathrm{T}}$ satisfies the equation

$$(\gamma(i-j))_{n\times n}(a_1^*, \dots, a_n^*)^{\mathrm{T}} = (\gamma(1), \dots, \gamma(n))^{\mathrm{T}}.$$

From (1.1.3), it follows that $\gamma(h) = \frac{4}{3}2^{-h}$. So

$$\begin{pmatrix} 1 & 2^{-1} & 2^{-2} & \cdots & 2^{-n+1} \\ 2^{-1} & 1 & 2^{-1} & \cdots & 2^{-n+2} \\ \vdots & \vdots & \vdots & \vdots & \vdots \\ 2^{-n+1} & 2^{-n+2} & 2^{-n+3} & \cdots & 1 \end{pmatrix} \begin{pmatrix} a_1^* \\ a_2^* \\ \vdots \\ a_n^* \end{pmatrix} = \begin{pmatrix} 2^{-1} \\ 2^{-2} \\ \vdots \\ 2^{-n} \end{pmatrix}$$

and the solution is $a_1^* = \frac{1}{2}$, $a_2^* = \cdots = a_n^* = 0$. By (1.2.3),

$$P_n X_{n+1} = \frac{1}{2}X_n.$$

So the minimum mean square error is

$$E[(X_{n+1} - P_n X_{n+1})^2] = \gamma(0) - a_1^*\gamma(1) = \frac{4}{3} - \frac{2}{3}a_1^* = 1.$$

1.2.3 Durbin–Levinson and innovation algorithms

Generally speaking, to obtain $a_1^*, \dots, a_n^*$, we need to solve the system (1.2.2) of linear equations. This is very difficult for large n. The following Durbin–Levinson algorithm gives a simple recursive formula of coefficients of the best linear prediction for a stationary time series $\{X_t\}$ with mean μ and covariance function $\gamma(h)$. For convenience, we assume that the mean $\mu = 0$ and $\gamma(h) \to 0\,(h \to \infty)$.

Durbin–Levinson algorithm

Let $P_n X_{n+1} = a_{n1}^* X_n + \cdots + a_{nn}^* X_1$. The minimum mean square error is

$$U_n := \gamma(0) - (a_{n1}^*\gamma(n) + \cdots + a_{nn}^*\gamma(1)).$$

The coefficients $a_{n1}^*, \dots, a_{nn}^*$ can be computed recursively as follows:

$$a_{nn}^* = \left(\gamma(n) - \sum_{j=1}^{n-1} a_{n-1,j}^* \gamma(n-j) \right) U_{n-1}^{-1}, \quad U_n = U_{n-1}(1-(a_{nn}^*)^2),$$

$$\begin{pmatrix} a_{n1}^* \\ \vdots \\ a_{n,n-1}^* \end{pmatrix} = \begin{pmatrix} a_{n-1,1}^* \\ \vdots \\ a_{n-1,n-1}^* \end{pmatrix} - a_{nn}^* \begin{pmatrix} a_{n-1,n-1}^* \\ \vdots \\ a_{n-1,1}^* \end{pmatrix},$$

where $a_{11}^* = \gamma(1)/\gamma(0)$ and $U_0 = \gamma(0)$.

Now we do not consider stability. For a general time series $\{X_t\}$, if both its mean function $\mu_X(t) = E[X_t]$ and covariance function $\gamma_X(i,j) = \mathrm{Cov}(X_i, X_j)$ are known, we may predict X_{n+h} by $1, X_1, \dots, X_n$. Similar to the stationary case that the best linear predictor of X_{n+h} is

$$P_n X_{n+h} = E[X_{n+h}] + \sum_{i=1}^{n} a_i (X_{n+1-i} - E[X_{n+1-i}])$$

and the coefficient $\boldsymbol{a}_n^* = (a_1^*, \dots, a_n^*)^{\mathrm{T}}$ satisfy the following system of linear equations

$$\widetilde{\Gamma}_n \boldsymbol{a}_n^* = \widetilde{\boldsymbol{R}}_n(h), \tag{1.2.4}$$

where

$$\widetilde{\Gamma} = (\mathrm{Cov}(X_{n+1-i}, X_{n+1-j}))_{i,j=1,\dots,n},$$
$$\widetilde{\boldsymbol{R}}_n(h) = (\mathrm{Cov}(X_{n+h}, X_n), \dots, \mathrm{Cov}(X_{n+h}, X_1))^{\mathrm{T}}.$$

The mean square error is

$$E[(X_{n+h} - P_n X_{n+h})^2] = \mathrm{Var}(X_{n+h}) - \boldsymbol{a}_n^{*T} \boldsymbol{R}_n(h).$$

Consider the time series $\{X_t\}$ satisfying

$$X_t = \tfrac{1}{2} X_{t-1} + Z_t,$$
$$\{Z_t\} \sim \mathrm{WN}(0,1).$$

Formula (1.1.3) has given the autocovariance function

$$\gamma(h) = \frac{2^{-h+2}}{3}.$$

Suppose that the observations X_1 and X_3 are known but we miss the observation value X_2. We want to estimate X_2 by the linear combination of X_1 and X_3. Now

$$\widetilde{\Gamma}_2 = \begin{pmatrix} \mathrm{Cov}(X_3, X_3) & \mathrm{Cov}(X_3, X_1) \\ \mathrm{Cov}(X_1, X_3) & \mathrm{Cov}(X_1, X_1) \end{pmatrix} = \begin{pmatrix} \gamma(0) & \gamma(2) \\ \gamma(2) & \gamma(0) \end{pmatrix} = \begin{pmatrix} \frac{4}{3} & \frac{1}{3} \\ \frac{1}{3} & \frac{4}{3} \end{pmatrix},$$
$$\widetilde{\boldsymbol{R}}_2(1) = (\mathrm{Cov}(X_2, X_3), \mathrm{Cov}(X_2, X_1))^{\mathrm{T}} = (\gamma(1), \gamma(1))^{\mathrm{T}} = \left(\tfrac{2}{3}, \tfrac{2}{3}\right)^{\mathrm{T}}.$$

This gives a system of equations

$$\begin{pmatrix} \frac{4}{3} & \frac{1}{3} \\ \frac{1}{3} & \frac{4}{3} \end{pmatrix} \begin{pmatrix} a_1 \\ a_2 \end{pmatrix} = \begin{pmatrix} \frac{2}{3} \\ \frac{2}{3} \end{pmatrix}$$

with solution $a_1 = a_2 = \frac{2}{5}$. Therefore, the best estimator of X_2 is

$$P(X_2 \mid 1, X_1, X_3) = E[X_2] + \tfrac{2}{5}(X_3 + X_1) = \tfrac{2}{5}(X_1 + X_3).$$

The minimum mean square error is

$$E[X \mid 1, X_1, X_3] = \operatorname{Var}(X_2) - (a_1, a_2)\begin{pmatrix} \frac{2}{3} \\ \frac{2}{3} \end{pmatrix} = \frac{4}{3} - \frac{8}{15} = \frac{4}{5}.$$

From this example, we see that even if a time series is stationary, to estimate its missing values we will use (1.2.4) but not use the Durbin–Levinson method.

To avoid solving the system of linear equations (1.2.4) for large n, one gives the innovation algorithm. It is a simple recurrence algorithm for the best linear prediction, but allows the time series to not be stationary.

Innovation algorithm

Assume that the mean $EX_t = 0\,(t \in \mathbb{Z}_+)$ and the covariance matrix $(h_{ij})_{n\times n}$ is nonsingular, where $h_{ij} = \operatorname{Cov}(X_i, X_j)$. Then the best linear prediction is

$$X^o_{n+1} = \sum_{j=1}^{n} \alpha_{nj}(X_{n+1-j} - X^o_{n+1-j}),$$

where $X^o_{n+1} = P_n X_{n+1}$ $(n \in \mathbb{Z}_+)$ and $X^o_0 = 0$, and the coefficients α_{nj} and mean square errors λ_n satisfy

$$\begin{aligned} \lambda_0 &= h_{11}, \quad \lambda_{nn} = \frac{1}{\lambda_0} h_{n+1,1}, \\ \lambda_{n,n-k} &= \frac{1}{\lambda_k} h_{n+1,k+1} - \frac{1}{\lambda_k}\sum_{j=0}^{k-1} \alpha_{k,k-j}\alpha_{n,n-j}\lambda_j \quad (k = 1, \ldots, n-1), \\ \lambda_n &= h_{n+1,n+1} - \sum_{j=0}^{n-1} \alpha^2_{n,n-j}\lambda_j, \\ E[(X_{n+1} &- P_n X_{n+1})^2] = \lambda_n. \end{aligned} \tag{1.2.5}$$

The use of this recursive formula is in the order $\lambda_0, \alpha_{11}, \lambda_1, \alpha_{22}, \alpha_{21}, \lambda_2, \alpha_{33}, \alpha_{32}, \alpha_{31}, \lambda_3, \ldots$.

For example, consider the time series

$$X_t = Z_t + \tfrac{1}{2} Z_{t-1},$$
$$\{Z_t\} \sim \mathrm{WN}(0, 1).$$

Note that

$$h_{ij} = E[X_i X_j] = E[(Z_i + \tfrac{1}{2}Z_{i-1})(Z_j + \tfrac{1}{2}Z_{j-i})],$$
$$E[Z_i Z_j] = 0 \quad (i \neq j),$$
$$E[Z_i^2] = 1.$$

Then

$$h_{ii} = \tfrac{5}{4}, \quad h_{i,i+1} = \tfrac{1}{2}, \quad h_{ij} = 0 \qquad (|i-j| \geq 2).$$

Applying the innovation algorithm gives

$$\alpha_{nj} = 0 \quad (2 \leq j \leq n), \qquad \alpha_{n1} = \frac{1}{2\lambda_{n-1}},$$
$$\lambda_0 = \frac{5}{4}, \qquad \lambda_n = \frac{5}{4} - \frac{1}{4\lambda_{n-1}}.$$

So the best linear prediction is

$$X_{n+1}^o = \sum_1^n \alpha_{nj}(X_{n+1-j} - X_{n+1-j}^o) = \alpha_{n1}(X_n - X_n^o).$$

For the recursive calculation of the h-step predictors, we may use the following formula:

$$P_n X_{n+h} = \sum_{j=h}^{n+h-1} \alpha_{n+h-1}(X_{n+h-j} - X_{n+h-j}^o),$$

where the coefficients α_j are determined as before by the innovation algorithm and the minimum mean square error is

$$E[(X_{n+h} - P_n X_{n+h})^2] = K(n+h, n+h) - \sum_{j=h}^{n+h-1} \alpha_{n+h-1}^2 U_{n+h-j-1}.$$

Compare these two algorithms. In the Durbin–Levinson algorithm, the best prediction is represented into a linear combination of $X_n, \ldots, X_1$

$$P_n X_{n+h} = \sum_{j=1}^{n} a_{nj} X_{n+1-j},$$

where $\{\varphi_{nj}\}$ are given recursively. The Durbin–Levinson algorithm is suited to autoregressive process

$$X_t - a_1 X_{t-1} - \cdots - a_p X_{t-p} = Z_t$$

since $\varphi_{nj} = 0$ for $j < n - p$. While in the innovation algorithm, the best prediction is represented by another linear combination

$$X_{n+1}^o = \sum_{j=1}^{n} \alpha_{nj}(X_{n+1-j} - X_{n+1-j}^o).$$

When X_t is a moving average of order q, $\alpha_{nj} = 0$ $(j < n - q)$. The innovation algorithm is suited to moving average process

$$X_t = Z_t + \alpha_1 Z_{t-1} + \cdots + \alpha_q Z_{t-q}.$$

1.2.4 Wold decomposition

In order to introduce the Wold decomposition of a stationary process, we explain concepts of convergence for a sequence of random variables.

Let $\{X_n\}_{n=0,1,...}$ be a sequence of random variables and X be a random variable. We say that $X_n \to X$ in *mean square* if $\lim_{n\to\infty} E[(X_n - X)^2] = 0$.

Now we consider the prediction of a stationary process in terms of infinitely many past values. $P_{mn}X_{n+h}$ is the best linear prediction of X_{n+h} $(h > 0)$ by the linear combination of $1, X_m, \ldots, X_0, \ldots, X_n$. Define the best prediction based on the infinite past values $\{X_t\ (-\infty < t \le n)\}$ as

$$\tilde{P}_n X_{n+h} = \lim_{m\to-\infty} P_{mn}X_{n+h}$$

in the mean square sense. A time series $\{X_t\}$ is called *deterministic* if $X_n = \tilde{P}_{n-1}X_n$ for all n.

For example, $X_t = A\cos(\omega t) + B\sin(\omega t)$, where ω is constant and A, B are uncorrelated random variables with mean 0 and variance σ^2. Note that

$$2\cos\omega X_{n-1} = A(\cos(\omega t) + \cos\omega(t-2)) + B(\sin(\omega t) + \sin\omega(t-2)) = X_n + X_{n-2}.$$

Then

$$X_n = 2\cos\omega X_{n-1} - X_{n-2} = \tilde{P}_{n-1}X_n \quad (n \in \mathbb{Z}_+).$$

So $\{X_t\}$ is deterministic.

Wold decomposition

If $\{X_t\}$ is a nondeterministic stationary time series, then

$$X_t = \sum_{j=0}^{\infty} \psi_j Z_{t-j} + V_t,$$

$$\{Z_t\} \sim \mathrm{WN}(0, 1),$$

where $\psi_0 = 1$, $\sum_{j=0}^{\infty} \psi_j^2 < \infty$, $\{V_t\}$ is deterministic, and $\mathrm{Cov}(Z_t, V_{t'}) = 0$ for all t, t'.

For all causal ARMA processes (see Section 1.4), the deterministic component V_t is zero in the Wold decomposition, and so

$$X_t = \sum_{j=0}^{\infty} \psi_j Z_{t-j}.$$

Let $\{X_t\}$ be a stationary time series with mean 0 and autocovariance function $\gamma(h)$ such that $\gamma(k) = 0$ $(k > q)$ and $\gamma(q) \ne 0$. Then

$$X_t = \sum_{j=0}^{q} \psi_j Z_{t-j},$$

$$\{Z_t\} \sim \mathrm{WN}(0, 1),$$

i.e., $\{X_t\}$ is an MA process of order q, where $\psi_0 = 1$.

1.3 Spectral analysis

The spectral representation of a stationary time series $\{X_t\}$ shows that $\{X_t\}$ can be decomposed into a sum of sinusoidal components with uncorrelated random coefficients. Such spectral representation is called *spectral analysis*.

1.3.1 Spectral density

Suppose that $\{X_t\}$ is a stationary time series with mean 0 and autocovariance function $\gamma(h)$ satisfying $\sum_{h\in\mathbb{Z}}|\gamma(h)| < \infty$. The *spectral density* of $\{X_t\}$ is defined as

$$f(\alpha) = \sum_{h\in\mathbb{Z}} \gamma(h)\,\mathrm{e}^{-\mathrm{i}h\alpha} = \gamma(0) + 2\sum_{1}^{\infty} \gamma(h)\cos(h\alpha). \tag{1.3.1}$$

The spectral density $f(\alpha)$ is a nonnegative even function. Using termwise integration, the autocovariance can be expressed into the following integral:

$$\gamma(h) = \frac{1}{2\pi}\int_{-\pi}^{\pi} f(\alpha)\,\mathrm{e}^{\mathrm{i}h\alpha}\,\mathrm{d}\alpha = \int_{-\pi}^{\pi} \cos(h\alpha)f(\alpha)\,\mathrm{d}\alpha.$$

Since $f(\alpha)$ is an even function,

$$\gamma(h) = \frac{1}{2\pi}\int_{-\pi}^{\pi} f(-\alpha)\,\mathrm{e}^{-\mathrm{i}h\alpha}\,\mathrm{d}\alpha = \frac{1}{2\pi}\int_{-\pi}^{\pi} f(\alpha)\,\mathrm{e}^{-\mathrm{i}h\alpha}\,\mathrm{d}\alpha.$$

This implies that the autocovariance function $\gamma(h)$ is Fourier coefficient of the spectral density $f(\alpha)$.

The definition of spectral density is generalized as follows. For a stationary time series $\{X_t\}$ with autocovariance function $\gamma(h)$, if $f(\alpha)$ is a 2π-periodic nonnegative function and $\gamma(h)$ is its Fourier coefficient, then $f(\alpha)$ is called the *spectral density* of $\{X_t\}$.

For example, suppose that $\{X_t\}$ is a stationary time series satisfying

$$X_t = \varphi X_{t-1} + Z_t,$$
$$\{Z_t\} \sim \mathrm{WN}(0, 1) \quad (t \in \mathbb{Z}),$$

where $|\varphi| < 1$. By (1.1.3),

$$\gamma(h) = \frac{1}{1-\varphi^2}\varphi^{|h|}.$$

So $\{X_t\}$ has the spectral density

$$\begin{aligned}
f(\alpha) &= \sum_h \mathrm{e}^{-\mathrm{i}h\alpha}\gamma(h) = \frac{1}{1-\varphi^2}\sum_h \varphi^{|h|}\mathrm{e}^{-\mathrm{i}\alpha h} \\
&= \frac{1}{1-\varphi^2}\left(1+\sum_{h=1}^{\infty}\varphi^h\mathrm{e}^{-\mathrm{i}\alpha h}+\sum_{h=1}^{\infty}\varphi^h\mathrm{e}^{\mathrm{i}\alpha h}\right) \\
&= \frac{1}{1-\varphi^2}\left(1+\frac{\varphi\mathrm{e}^{-\mathrm{i}\alpha}}{1-\varphi\mathrm{e}^{-\mathrm{i}\alpha}}+\frac{\varphi\mathrm{e}^{\mathrm{i}\alpha}}{1-\varphi\mathrm{e}^{\mathrm{i}\alpha}}\right) = \frac{1}{1-2\varphi\cos\alpha+\varphi^2}.
\end{aligned}$$

For another time series

$$\begin{aligned}
X_t &= Z_t + \theta Z_{t-1}, \\
Z_t &\sim WN(0,1)
\end{aligned}$$

with mean $\mu = 0$ and autocovariance function

$$\gamma(h) = \begin{cases} 1+\theta^2, & h = 0, \\ \theta, & h = \pm 1, \\ 0, & |h| > 1, \end{cases}$$

the spectral density is

$$f(\alpha) = \gamma(0) + \gamma(-1)\,\mathrm{e}^{\mathrm{i}\alpha} + \gamma(1)\,\mathrm{e}^{-\mathrm{i}\alpha} = 1+\theta^2+\theta(\mathrm{e}^{\mathrm{i}\alpha}+\mathrm{e}^{-\mathrm{i}\alpha}) = 1+2\theta\cos\alpha+\alpha^2.$$

1.3.2 Spectral estimation

Consider a stationary time series $\{X_t\}$ with mean 0 and autocovariance function $\gamma(h)$. In applications, we only know finitely many observations $x_0, \dots, x_{N-1}$. Since $\bar{x} = \frac{1}{N}\sum_{k=0}^{N-1} x_k \approx 0$, the $\gamma(h)$ is estimated by

$$\hat{\gamma}(h) = \frac{1}{N}\sum_{n=0}^{N-1} x_n x_{n+h}. \tag{1.3.2}$$

The corresponding spectral density $f(\alpha)$ is estimated by $\hat{f}(\alpha) = \sum_{k\in\mathbb{Z}}\hat{\gamma}(k)\,\mathrm{e}^{-\mathrm{i}k\alpha}$. Denote

$$x(n) = \begin{cases} x_n, & 0 < n \le N-1, \\ 0, & \text{otherwise.} \end{cases}$$

So

$$\hat{\gamma}(h) = \frac{1}{N}\sum\nolimits_{n\in\mathbb{Z}} x(n+h)x(n)$$

and the spectral density is estimated by

$$\begin{aligned}
\hat{f}(\alpha) &= \sum\nolimits_{h\in\mathbb{Z}}\hat{\gamma}(h)\,\mathrm{e}^{-\mathrm{i}h\alpha} = \frac{1}{N}\sum\nolimits_{n\in\mathbb{Z}} x(n)\left(\sum\nolimits_{h\in\mathbb{Z}} x(n+h)\,\mathrm{e}^{-\mathrm{i}h\alpha}\right) \\
&= \frac{1}{N}\left(\sum\nolimits_{n\in\mathbb{Z}} x(n)\,\mathrm{e}^{\mathrm{i}n\alpha}\right)\left(\sum\nolimits_{m\in\mathbb{Z}} x(m)\,\mathrm{e}^{-\mathrm{i}m\alpha}\right) = \frac{1}{N}\left|\sum_{0}^{N-1} x_n\mathrm{e}^{-\mathrm{i}n\alpha}\right|^2.
\end{aligned}$$

For each $N \in \mathbb{Z}_+$, define

$$f_N(\alpha) = \frac{1}{N} E\left[\left|\sum_0^{N-1} X_n e^{-in\alpha}\right|^2\right] = \frac{1}{N} E\left[\sum_{n=0}^{N-1} X_n e^{-in\alpha} \sum_{m=0}^{N-1} X_m e^{im\alpha}\right]$$
$$= \frac{1}{N} \sum\nolimits_{|h|<N} (N - |h|)\, e^{-ih\alpha} \gamma(h).$$

This implies that

$$f_N(\alpha) \to f(\alpha) \quad (N \to \infty).$$

So an asymptotic unbiased estimator of the spectral density $f(\alpha)$ is

$$\frac{1}{N} E[|u(\alpha)|^2], \quad \text{where } u(\alpha) = \sum_{n=0}^{N-1} X_n e^{-in\alpha}.$$

1.3.3 Fourier power spectra

Suppose that $\{X_t\}$ is a time series. Consider the discrete Fourier transform of $X = \{X_k\}_{k=0,\dots,N-1}$

$$a_k = \frac{1}{N} \sum_0^{N-1} X_n e^{-in\frac{2\pi k}{N}} \quad (k = 0, \dots, N-1).$$

The inverse discrete Fourier transform is

$$X_n = \sum_0^{N-1} a_k e^{ik\frac{2\pi n}{N}} \quad (n = 0, \dots, N-1).$$

The *Fourier power spectrum* of X is defined as $|a_k|^2$ in Fourier frequent $\frac{2\pi k}{N}$.

Now we consider a simple model for red noise AR (1)

$$\begin{aligned} X_0 &= 0, \\ X_n &= \alpha X_{n-1} + Z_n \quad (n \in \mathbb{Z}_+,\ 0 < \alpha < 1), \end{aligned} \tag{1.3.3}$$

which is often used in applications, where $\{Z_n\}$ is a Gaussian white noise

$$\begin{aligned} E[Z_0] &= 0, \\ \mathrm{Var}(Z_n) &= \sigma^2, \\ E[Z_n Z_m] &= 0 \quad (n \neq m) \end{aligned}$$

and each Z_n is a Gaussian random variable.

Suppose that a red noise model $\{X_n\}$ satisfies (1.3.3) with parameters α and σ^2, and the discrete Fourier transforms of $X_0, \dots, X_{N-1}$ are $a_0, \dots, a_{N-1}$. Then the Fourier power spectrum $|a_k|^2$ in Fourier frequency $\frac{2\pi k}{N}$ satisfies the following condition:

$$\frac{N|a_k|^2}{\sigma^2} \quad \text{is distributed as} \quad \frac{1}{2(1 - 2\alpha \cos\frac{2\pi k}{N} + \alpha^2)} \chi_2^2,$$

where χ_2^2 is the chi-square distribution with two degrees of freedom.

1.3.4 Time-invariant linear filters

Suppose that $\{X_t\}$ is a time series. We say the time series $\{Y_t\}$ is the *output* of a *time-invariant linear filter* $\{\psi_k\}$ applied to the input $\{X_t\}$ if

$$Y_t = \sum_{k\in\mathbb{Z}} \psi_{t-k} X_k = \sum_{k\in\mathbb{Z}} \psi_k X_{t-k}.$$

We say the time-invariant filter $\{\psi_k\}$ is *causal* if $\psi_j = 0$ for $j < 0$. In this case,

$$Y_t = \sum_0^\infty \psi_k X_{t-k},$$

i.e., Y_t is represented only by X_l $(l \le t)$.

Suppose that X_t is a stationary time series with mean 0, autocovariance function γ and $\sum_{k\in\mathbb{Z}} |\psi_k| < \infty$. Then the time series $Y_t = \sum_{k\in\mathbb{Z}} \psi_k X_{t-k}$ is stationary with mean 0 and

(a) its autocovariance function is $\gamma_Y(h) = \sum_{k\in\mathbb{Z}} \sum_{l\in\mathbb{Z}} \psi_k \psi_l \gamma_X(h-k+l)$;

(b) its spectral density is

$$f_Y(\alpha) = |\psi(e^{-\alpha})|^2 f_X(\alpha), \tag{1.3.4}$$

where $\psi(\mathrm{e}^{-\mathrm{i}\alpha}) = \sum_{k\in\mathbb{Z}} \psi_k \mathrm{e}^{-\mathrm{i}k\alpha}$.

In fact, from $E[X_t] = 0$, it follows that $E[Y_t] = \sum_k \psi_k E[X_{t-k}] = 0$ and

$$\begin{aligned}\gamma_Y(h) &= \mathrm{Cov}(Y_{t+h}, Y_t) = E[Y_{t+h} Y_t] = E\left[\left(\textstyle\sum_{k\in\mathbb{Z}} \psi_k X_{t+h-k}\right)\left(\textstyle\sum_{l\in\mathbb{Z}} \psi_l X_{t-l}\right)\right] \\ &= \sum_{k\in\mathbb{Z}} \sum_{l\in\mathbb{Z}} \psi_k \psi_l E[X_{t+h-k} X_{t-l}] = \sum_{k\in\mathbb{Z}} \sum_{l\in\mathbb{Z}} \psi_k \psi_l \gamma_X(h-k+l).\end{aligned} \tag{1.3.5}$$

Therefore, the time series $\{Y_t\}$ is stationary with mean 0. Its autocovariance function can be expressed as the Fourier coefficient of the spectral density

$$\gamma_X(h-k+l) = \frac{1}{2\pi} \int_{-\pi}^{\pi} f_X(\alpha)\, \mathrm{e}^{\mathrm{i}(h-k+l)\alpha}\, \mathrm{d}\alpha.$$

From this and (1.3.5), it follows that

$$\begin{aligned}\gamma_Y(h) &= \frac{1}{2\pi} \sum_{k\in\mathbb{Z}} \sum_{l\in\mathbb{Z}} \psi_k \psi_l \int_{-\pi}^{\pi} f_X(\alpha)\, \mathrm{e}^{\mathrm{i}(h-k+l)\alpha}\, \mathrm{d}\alpha \\ &= \frac{1}{2\pi} \int_{-\pi}^{\pi} \left(\textstyle\sum_{k\in\mathbb{Z}} \psi_k \mathrm{e}^{-\mathrm{i}k\alpha}\right)\left(\textstyle\sum_{l\in\mathbb{Z}} \psi_l \mathrm{e}^{\mathrm{i}l\alpha}\right) f_X(\alpha)\, \mathrm{e}^{\mathrm{i}h\alpha}\, \mathrm{d}\alpha \\ &= \frac{1}{2\pi} \int_{-\pi}^{\pi} \left|\textstyle\sum_{k\in\mathbb{Z}} \psi_k \mathrm{e}^{-\mathrm{i}k\alpha}\right|^2 f_X(\alpha)\, \mathrm{e}^{\mathrm{i}h\alpha}\, \mathrm{d}\alpha.\end{aligned}$$

Note that

$$\gamma_Y(h) = \frac{1}{2\pi}\int_{-\pi}^{\pi} f_Y(\alpha)\,\mathrm{e}^{\mathrm{i}h\alpha}\,\mathrm{d}\alpha.$$

Comparing these two formulas, the spectral density of Y is

$$f_Y(\alpha) = \left|\sum_k \psi_k \mathrm{e}^{-\mathrm{i}k\alpha}\right|^2 f_X(\alpha).$$

Let $\psi_k = \frac{1}{2m+1}$ ($|k| \le m$) and $\psi_k = 0$ ($|k| > m$), then Y_t is a simple moving average and

$$Y_t = \frac{1}{2m+1}\sum_{-m}^{m} \psi_k X_{t-k},$$

$$\psi(\mathrm{e}^{-\mathrm{i}\alpha}) = \frac{1}{2m+1}\sum_{-m}^{m} \mathrm{e}^{-\mathrm{i}k\alpha}.$$

By the sum formula of geometric series,

$$\psi(\mathrm{e}^{-\mathrm{i}\alpha}) = \frac{\mathrm{e}^{\mathrm{i}m\alpha} - \mathrm{e}^{-\mathrm{i}(m+1)\alpha}}{(2m+1)(1-\mathrm{e}^{-\mathrm{i}\alpha})} = \frac{\mathrm{e}^{\mathrm{i}(m+\frac{1}{2})\alpha} - \mathrm{e}^{-\mathrm{i}(m+\frac{1}{2})\alpha}}{(2m+1)(\mathrm{e}^{\mathrm{i}\frac{\alpha}{2}} - \mathrm{e}^{-\mathrm{i}\frac{\alpha}{2}})} = \frac{\sin(m+\frac{1}{2})\alpha}{(2m+1)\sin\frac{\alpha}{2}} \quad (\alpha \neq 0),$$

$$\psi(\mathrm{e}^{-\mathrm{i}\alpha}) = 1 \quad (\alpha = 0).$$

So, by (a) and (b)

$$\gamma_Y(h) = \frac{1}{(2m+1)^2}\sum_{k=-m}^{m}\sum_{l=-m}^{m} \gamma_X(h-k+l),$$

$$f_Y(\alpha) = \frac{\sin^2(m+\frac{1}{2})\alpha}{(2m+1)^2\sin^2\frac{\alpha}{2}} f_X(\alpha) \quad (\alpha \neq 0),$$

$$f_Y(\alpha) = f_X(\alpha), \quad (\alpha = 0).$$

1.4 Autoregressive moving average models

Autoregressive moving average (ARMA) models are the most important class of stationary time series. Here we discuss their basic properties and give Yule–Walker equations. We also introduce partial autocorrelation function and spectral densities of ARMA.

1.4.1 ARMA Models

Let $\{Z_t\} \sim \mathrm{WN}(0, \sigma^2)$. If $\{X_t\}$ is stationary and for each t,

$$X_t - \sum_{1}^{p} \varphi_k X_{t-k} = Z_t + \sum_{1}^{q} \theta_l Z_{t-l}, \tag{1.4.1}$$

where φ_k, θ_l are constants, then $\{X_t\}$ is called an *ARMA* (p, q) *process* and (1.4.1) is called an *ARMA equation*. If $\varphi_1 = \cdots = \varphi_p = 0$, then $\{X_t\}$ is called an *MA* (q) *process*. If $\theta_1 = \cdots = \theta_q = 0$, then $\{X_t\}$ is called an *AR* (p) *process*.

Let $\{X_t\}$ be a stationary time series. Define a *time shift operator* B as $BX_t = X_{t-1}$. So, for any $l \in \mathbb{Z}_+$,

$$B^l X_t = X_{t-l}.$$

Denote

$$\begin{aligned}\varphi(z) &= 1 - \sum_1^p \varphi_k z^k,\\ \theta(z) &= 1 + \sum_1^q \theta_l z^l.\end{aligned} \tag{1.4.2}$$

They are polynomials of complex variable z of degree $\le p$ and degree $\le q$, respectively. Assume that $\varphi(z)$ and $\theta(z)$ have no common root. Denote the operator polynomials

$$\begin{aligned}\varphi(B) &= 1 - \sum_1^p \varphi_k B^k,\\ \theta(B) &= 1 + \sum_1^q \theta_l B^l.\end{aligned}$$

Since

$$\begin{aligned}\varphi(B)X_t &= X_t - \sum_1^p \varphi_k B^k X_t = X_t - \sum_1^p \varphi_k X_{t-k},\\ \theta(B)Z_t &= Z_t + \sum_1^q \theta_l B^l X_t = Z_t + \sum_1^q \theta_l X_{t-l},\end{aligned}$$

the ARMA equation is written into $\varphi(B)X_t = \theta(B)Z_t$. The AR equation is written into $\Phi(B)X_t = Z_t$ and the MA equation is written into $X_t = \theta(B)Z_t$.

Suppose that the polynomial $\varphi(z) = 1 - \sum_{k=1}^p \varphi_k z^k \neq 0$ ($|z| = 1$), where $|z| = 1$ is the unit circle. Then ARMA equation (1.4.1) has a unique solution

$$X_t = \sum_{j\in\mathbb{Z}} \psi_j z_{t-j} \quad (t \in \mathbb{Z}) \tag{1.4.3}$$

in the mean square sense, where the coefficients ψ_j ($j \in \mathbb{Z}$) are determined by

$$\psi(z) = \frac{\theta(z)}{\varphi(z)} = \sum_{j\in\mathbb{Z}} \psi_j z^j \quad (|z| = 1), \tag{1.4.4}$$

where $\theta(z) = 1 + \sum_{l=1}^q \theta_l z^l$.

If the unique solution is $X_t = \sum_{j=0}^\infty \psi_j Z_{t-j}$, then the ARMA (p, q) process is called *causal*.

For example, consider ARMA equation

$$X_t - \frac{10}{3}X_{t-1} + X_{t-2} = Z_t \quad \text{or} \quad \varphi(B)X_t = Z_t,$$

where $\varphi(z) = 1 - \frac{10}{3}z + z^2$ and $\theta(z) = 1$. Since $\varphi(z) = 0$ has two roots $z = 3$ and $z = \frac{1}{3}$ and $\varphi(z) \neq 0$ on $|z| = 1$, the ARMA equation has a unique solution. Note that

$$\psi(z) = \frac{1}{\varphi(z)} = \frac{3}{8}\left(\frac{1}{z-3} - \frac{1}{z-\frac{1}{3}}\right) = -\frac{1}{8}\sum_0^{\infty}\left(\frac{z}{3}\right)^j + \frac{9}{8}\sum_{-\infty}^{-1}(3z)^j \quad (|z| = 1),$$

i.e., $\psi(z) = \sum_{j\in\mathbb{Z}} \psi_j^o z^j$, where $\psi_j^o = -\frac{1}{3^j 8}$ $(j \geq 0)$ and $\psi_j^o = \frac{3^j 9}{8}$ $(j \leq -1)$. By (1.4.3), the solution is

$$X_t = \sum_{j\in\mathbb{Z}} \psi_j^o Z_{t-j}.$$

Generally, the solution is $\psi(z) = \frac{\theta(z)}{\varphi(z)}$ which is a rational function. When $\varphi(z) \neq 0$ $(|z| = 1)$, the function $\psi(z)$ can be expanded into (1.4.4). In fact, if $\varphi(z)$ has p different roots $\alpha_1, \ldots, \alpha_p$, then

$$\psi(z) = \frac{\beta_1}{z-\alpha_1} + \frac{\beta_2}{z-\alpha_2} + \cdots + \frac{\beta_p}{z-\alpha_p} + Q(z), \tag{1.4.5}$$

where $Q(z)$ is a polynomial of z and $\beta_\mu = \lim_{z\to\alpha_\mu} \psi(z)(z-\alpha_\mu)$ $(\mu = 1, \ldots, p)$. If $|\alpha_\mu| > 1$, then $\frac{\beta_\mu}{z-\alpha_\mu}$ can be expanded into a power series on $|z| = 1$; if $|\alpha_\mu| < 1$, then it can be expanded into a negative power series. From this and(1.4.5), we get $\psi(z) = \sum_{j\in\mathbb{Z}} \psi_j z^j$ $(|z| = 1)$.

Causal solution

An ARMA (p, q) process is causal if and only if $\varphi(z) \neq 0$ $(|z| \leq 1)$, where $\varphi(z)$ is stated in (1.4.2).

In fact, if $\varphi(z) = 1 - \varphi_1 z - \cdots - \varphi_p z^p \neq 0$ $(|z| \leq 1)$, then all roots of $\varphi(z)$ lie outside the unit disk. So each fraction $\frac{\beta_j}{z-\alpha_j}$ in (1.4.5) is expanded into a power series. This implies that

$$\psi(z) = \frac{\theta(z)}{\varphi(z)} = \sum_0^{\infty} \psi_j z^j,$$

i.e., the ARMA (p, q) process is causal. Conversely, it is also true.

Another method for finding the coefficients $\{\psi_j\}$ is as follows.

From $\psi(z) = \frac{\theta(z)}{\varphi(z)}$, it follows that

$$(1 - \varphi_1 z - \cdots - \varphi_p z^p)(\psi_0 + \psi_1 z + \cdots) = 1 + \theta_1 z + \cdots + \theta_q z^q.$$

Equating the coefficients of z^j $(j = 0, 1, \ldots)$, we get

$$\psi_0 = \theta_0 + 1,$$
$$\psi_1 = \theta_1 + \varphi_1,$$
$$\psi_2 = \theta_2 + \varphi_2 + \theta_1\varphi_1 + \varphi_1^2,$$

In general,

$$\psi_l = \theta_l + \sum_{k=1}^{l} \varphi_k \psi_{l-k} = \theta_l \quad (l \in \mathbb{Z}_+),$$

where $\theta_l = 0$ $(l > q)$ and $\varphi_k = 0$ $(k > p)$.

Using the above similar argument, we get the following:

Invertible solution

An ARMA (p, q) process is invertible if and only if $\theta(z) \neq 0$ $(|z| \leq 1)$, where $\theta(z)$ is stated in (1.4.2).

For example, consider an ARMA $(1, 1)$ process

$$X_t - \varphi X_{t-1} = Z_t + \theta Z_{t-1} \quad (|\varphi| < 1,\ |\theta| < 1,\ \varphi + \theta \neq 0).$$

It is clear that for all $|z| \leq 1$,

$$\varphi(z) = 1 - \varphi z \neq 0,$$
$$\theta(z) = 1 + \theta z \neq 0.$$

Since $|\varphi z| < 1$ on $|z| = 1$, the ARMA process is causal. Note that

$$\psi(z) = \frac{1 + \theta z}{1 - \varphi z} = 1 + \sum_{1}^{\infty} \varphi^{j-1}(\varphi + \theta) z^j.$$

So $\psi_0 = 1$ and $\psi_j = \varphi^{j-1}(\varphi + \theta)$ $(j = 1, 2, \ldots)$, and

$$X_t = Z_t + (\varphi + \theta) \sum_{1}^{\infty} \varphi^{j-1} Z_{t-j}.$$

Since $|\theta z| < 1$ $(|z| = 1)$, the ARMA process is invertible. Note that

$$\zeta(z) = \frac{1 - \varphi z}{1 + \theta z} = (1 - \varphi z) \sum_{0}^{\infty} (-1)^j \theta^j z^j = 1 + \sum_{j=1}^{\infty} (-1)^j \theta^{j-1}(\varphi + \theta) z^j.$$

So

$$Z_t = X_t + (\varphi + \theta) \sum_{1}^{\infty} (-1)^j \theta^{j-1} X_{t-j},$$

i.e., the ARMA process is invertible.

1.4.2 Yule–Walker equation

The Yule–Walker equation explains the relation between the parameters of an ARMA process $\varphi_1, \ldots, \varphi_p; \theta_1, \ldots, \theta_q; \sigma^2$, and its autocovariance function $\gamma(t)$.

Suppose that $\{X_t\}$ is a causal AR (p) process, $X_t - \sum_{v=1}^{p} \varphi_v X_{t-v} = Z_t$. Multiplying both sides by X_{t-k} and then taking the mean on both sides,

$$E[X_t X_{t-k}] - \sum_{1}^{p} \varphi_v E[X_{t-k} X_{t-v}] = E[Z_t X_{t-k}].$$

Since $\{X_t\}$ is causal, $X_t = \sum_{j=0}^{\infty} \psi_j Z_{t-j}$ $(t \in \mathbb{Z})$. So, for $k = 0, 1, \ldots,$

$$E[Z_t X_{t-k}] = \sum_{0}^{\infty} \psi_j E[Z_t Z_{t-k-j}] = \sigma^2 \sum_{0}^{\infty} \psi_j \delta_{0,k+j} = \sigma^2 \sum_{j=k}^{\infty} \psi_{j-k} \delta_{0,j} = \begin{cases} \sigma^2 & (k = 0), \\ 0 & (k \neq 0). \end{cases}$$

From this and $E[X_t X_{t-k}] = \gamma(k)$, and $E[X_{t-k} X_{t-v}] = \gamma(k - v)$, it follows that the *Yule–Walker equation*

$$\gamma(k) - \sum_{1}^{p} \varphi_v \gamma(k - v) = \begin{cases} \sigma^2 & (k = 0), \\ 0 & (k \neq 0). \end{cases}$$

Its matrix form is

$$\gamma(0) - \boldsymbol{\varphi}^{\mathrm{T}} \gamma_p = \sigma^2,$$

$$\Gamma_p \boldsymbol{\varphi} = \boldsymbol{\gamma}_p,$$

where $\Gamma_p = (\gamma(i - j))_{i,j=1,\ldots,p}$, $\boldsymbol{\gamma}_p = (\gamma(1), \ldots, \gamma(p))^{\mathrm{T}}$, and $\boldsymbol{\varphi} = (\varphi_1, \ldots, \varphi_p)^{\mathrm{T}}$.

For example, consider an AR (2) process

$$X_t - \frac{3}{4} X_{t-1} + \frac{1}{8} X_{t-2} = Z_t,$$

$$\{Z_t\} \sim \mathrm{WN}(0, 1).$$

By Yule–Walker equation, the autocovariance function $\gamma(h)$ satisfies

(a) $\gamma(0) - \frac{3}{4}\gamma(1) + \frac{1}{8}\gamma(2) = 1$,

(b) $-\frac{3}{4}\gamma(0) + \frac{9}{8}\gamma(1) = 0$,

(c) $\gamma(k) - \frac{3}{4}\gamma(k - 1) + \frac{1}{8}\gamma(k - 2) = 0$ $(k \geq 2)$.

Note that the polynomial $\varphi(z) = 1 - \frac{3}{4}z + \frac{1}{8}z^2$ has two zero points $z_1 = 2$ and $z_2 = 4$, the homogeneous linear difference equation (c) has a general solution $\gamma(h) = a_1 2^{-h} + a_2 4^{-h}$ $(h \geq 0)$, where a_1 and a_2 are arbitrary constants. Substituting it into (a) and (b), we get

$$84a_1 + 105a_2 = 128,$$

$$42a_1 + 105a_2 = 0.$$

So $a_1 = \frac{64}{21}$ and $a_2 = -\frac{128}{105}$. This implies $\gamma(h) = 128(\frac{2^{-h}}{42} - \frac{4^{-h}}{105})$ $(h \geq 0)$.

More conveniently, take $k = 2$ in (c). The solution is

$$\gamma(0) = \frac{64}{35}, \quad \gamma(1) = \frac{128}{105}, \quad \gamma(2) = \frac{24}{35},$$

and then $\gamma(3), \gamma(4), \ldots$ can be found from (c) successively. This is an especially convenient method for numerical determination of autocovariances $\gamma(h)$.

For an AR (p) process, from the first $p+1$ Yule–Walker equations

$$\gamma(0) - \sum_1^p \varphi_\nu \gamma(\nu) = \sigma^2,$$

$$\gamma(k) - \sum_1^p \varphi_\nu \gamma(k-\nu) = 0 \quad (k = 1, \ldots, p),$$

we may find out $\gamma(0), \gamma(1), \ldots, \gamma(p)$ and then find out $\gamma(p+1), \gamma(p+2), \ldots$ successively using equations

$$\gamma(k) - \sum_1^p \varphi_\nu \gamma(k-\nu) = 0 \quad (k = p+1, \ldots).$$

More generally, if $\{X_t\}$ is a causal ARMA process

$$X_t - \sum_1^p \varphi_k X_{t-k} = Z_t + \sum_1^q \theta_l Z_{t-l},$$

$$X_t = \sum_0^\infty \psi_j Z_{t-j},$$

the relation between $\varphi_k, \theta_l, \sigma^2$ and $\gamma(t)$ is given by using a similar method as follows:

$$\gamma(k) - \sum_1^p \varphi_\nu \gamma(k-\nu) = \sigma^2 \sum_{j=k}^q \theta_j \psi_{j-k} \quad (0 \le k < m), \tag{1.4.6}$$

$$\gamma(k) - \sum_1^p \varphi_\nu \gamma(k-\nu) = 0 \quad (k \ge m), \tag{1.4.7}$$

where $m = \max\{p, q+1\}$ and $\theta_j = 0 \ (j \ge q+1)$.

The homogeneous linear difference equation (1.4.7) with constant coefficients has a solution

$$\gamma(h) = \sum_1^p a_l \xi_l^{-h} \quad (h \ge m-p), \tag{1.4.8}$$

where $a_1, \ldots, a_p$ are arbitrary constants and $\xi_1, \ldots, \xi_p$ are different roots of the polynomials

$$\varphi(z) = 1 - \sum_{k=1}^p \varphi_k z^k.$$

In fact, substituting $\gamma(h) = \xi_l^{-h}$ $(l = 1, \ldots, p; h > m-p)$ into the left-hand side of the equation (1.4.7),

$$\gamma(k) - \sum_1^p \varphi_\nu \gamma(k-\nu) = \xi_l^{-k}\left(1 - \textstyle\sum_1^p \varphi_\nu \xi_l^\nu\right) = \xi_l^{-k}\varphi(\xi_l) = 0$$

since ξ_l is a root of $\varphi(z)$. Note that (1.4.7) is a homogeneous linear difference equation. The linear combination (1.4.8) of $\xi_1^{-h}, \dots, \xi_p^{-h}$ is the general solution of (1.4.7). Substituting (1.4.8) into (1.4.7), we obtain a system of m linear equations that determine the constants $a_1, \dots, a_p$ and autocovariances $\gamma(0), \dots, \gamma(m-p-1)$.

For example, consider the ARMA (1, 1) process

$$\begin{aligned} X_t - \varphi X_{t-1} &= Z_t + \theta Z_{t-1} \quad (|\varphi| < 1), \\ \{Z_t\} &\sim WN(0, \sigma^2). \end{aligned}$$

Since $|\varphi| < 1$, it has a causal solution

$$X_t = 1 + \sum_1^{\infty} \psi_n Z_{t-n},$$

where $\psi_n = (\varphi + \theta)\varphi^{n-1}$ $(n \geq 1)$. By (1.4.7) and (1.4.8), we get

$$\begin{aligned} \gamma(0) - \varphi\gamma(1) &= \sigma^2(1 + (\varphi + \theta)\theta), \\ \gamma(1) - \varphi\gamma(0) &= \sigma^2\theta. \end{aligned}$$

By (1.4.7), the homogeneous equation

$$\gamma(k) - \varphi\gamma(k-1) = 0 \quad (k \geq 2)$$

has the solution $\gamma(h) = a\varphi^h$ $(h \geq 1)$. Substituting this solution into the above equations, we get

$$\begin{aligned} \gamma(0) - a\varphi^2 &= \sigma^2(1 + \theta(\varphi + \theta)), \\ -\varphi\gamma(0) + a\varphi &= \sigma^2\theta. \end{aligned}$$

Their solution is

$$a = \frac{(\varphi + \theta)(1 + \varphi\theta)}{\varphi(1 - \varphi^2)}\sigma^2, \quad \gamma(0) = \left(1 + \frac{(\varphi + \theta)^2}{1 - \varphi^2}\right)\sigma^2. \tag{1.4.9}$$

So

$$\gamma(h) = a\varphi^h = \frac{(\varphi + \theta)(1 + \varphi\theta)}{1 - \varphi^2}\varphi^{h-1}\sigma^2 \quad (h \geq 1). \tag{1.4.10}$$

1.4.3 Partial autocorrelation function and spectral densities

Suppose that $\{X_t\}$ is a stationary time series with mean 0 and autocorrelation function $\gamma(t)$. Denote the matrix $\Gamma_h = (\gamma(i-j))_{i,j=1,\dots,h}$ and the vector $\gamma_h = (\gamma(1), \dots, \gamma(h))^{\mathrm{T}}$. Let

$$\begin{aligned} \boldsymbol{\Phi}_h &= \Gamma_h^{-1}\gamma_h, \\ \boldsymbol{\Phi}_h &= (\Phi_{i1}, \dots, \Phi_{hh})^{\mathrm{T}}. \end{aligned}$$

The *partial autocorrelation function* $\beta(h)$ is defined as

$$\beta(0) = 1,$$
$$\beta(h) = \Phi_{hh} \quad (h \geq 1).$$

Consider the MA (1) process

$$X_t = Z_t + \theta Z_{t-1},$$
$$\{Z_t\} \sim \mathrm{WN}(0, \sigma^2).$$

By $\gamma(0) = (1+\theta^2)\sigma^2$ and $\gamma(1) = \theta\sigma^2$, it follows that

$$\boldsymbol{\Phi}_1 = \Gamma_1^{-1}\gamma_1 = \frac{\gamma(1)}{\gamma(0)} = \frac{\theta}{1+\theta^2}.$$

So $\beta(1) = \Phi_{11} = \frac{\theta}{1+\theta^2}$.

By $\gamma(0) = (1+\theta^2)\sigma^2$, $\gamma(1) = \theta\sigma^2$, and $\gamma(2) = 0$, it follows that

$$\begin{aligned}\boldsymbol{\Phi}_2 = \Gamma_2^{-1}\gamma_2 &= \begin{pmatrix}\gamma(0) & \gamma(1)\\ \gamma(1) & \gamma(0)\end{pmatrix}^{-1}\begin{pmatrix}\gamma(1)\\ \gamma(2)\end{pmatrix}\\ &= \frac{1}{\gamma^2(0)-\gamma^2(1)}\begin{pmatrix}\gamma(0) & -\gamma(1)\\ -\gamma(1) & \gamma(0)\end{pmatrix}\begin{pmatrix}\gamma(1)\\ 0\end{pmatrix}\\ &= \frac{1}{\gamma^2(0)-\gamma^2(1)}\begin{pmatrix}\gamma(1)\gamma(0)\\ -\gamma^2(1)\end{pmatrix}.\end{aligned}$$

So $\beta(2) = \Phi_{22} = \frac{-\gamma^2(1)}{\gamma^2(0)-\gamma^2(1)} = \frac{-\theta^2}{1+\theta^2+\theta^4}$.

By $\gamma(0) = (1+\theta^2)\sigma^2$, $\gamma(1) = \theta\sigma^2$, $\gamma(2) = \gamma(3) = 0$, and

$$\Gamma_3 = \begin{pmatrix}\gamma(0) & \gamma(1) & 0\\ \gamma(1) & \gamma(0) & \gamma(1)\\ 0 & \gamma(1) & \gamma(0)\end{pmatrix},$$

and $\gamma_3 = (\gamma(1), 0, 0)^{\mathrm{T}}$, it follows that

$$\Gamma_3^{-1} = (\gamma^3(0) - 2\gamma^2(1)\gamma(0))^{-1}\begin{pmatrix}\gamma^2(0)-\gamma^2(1) & -\gamma(0)\gamma(1) & \gamma^2(1)\\ -\gamma(0)\gamma(1) & \gamma^2(0) & -\gamma(0)\gamma(1)\\ \gamma^2(1) & -\gamma(0)\gamma(1) & \gamma^2(0)-\gamma^2(1)\end{pmatrix},$$

$$\boldsymbol{\Phi}_3 = \Gamma_3^{-1}\gamma_3 = \frac{\gamma(1)}{\gamma^3(0)-2\gamma^2(1)\gamma(0)}\begin{pmatrix}\gamma^2(0)-\gamma^2(1)\\ -\gamma(0)\gamma(1)\\ \gamma^2(1)\end{pmatrix}.$$

So $\beta(3) = \Phi_{33} = \frac{\gamma^3(1)}{\gamma^3(0)-2\gamma^2(1)\gamma(0)} = \frac{\theta^3}{1+\theta^2+\theta^4+\theta^6}$.

In general,

$$\beta(h) = \Phi_{hh} = \frac{(-1)^{h+1}\theta^h}{1+\theta^2+\cdots+\theta^{2h}} \quad (h \geq 1).$$

Consider the AR (p) process

$$X_t - \varphi_1 X_{t-1} - \cdots - \varphi_p X_{t-p} = Z_t,$$
$$\{Z_t\} \sim \mathrm{WN}(0, \sigma^2).$$

The Yule–Walker equation is $\boldsymbol{\varphi} = \Gamma_p^{-1}\gamma_p$, where

$$\boldsymbol{\varphi} = (\varphi_1, \dots, \varphi_p),$$
$$\Gamma_p = (\gamma(i-j))_{i,j=1,\dots,p},$$
$$\boldsymbol{\gamma}_p = (\gamma_1, \dots, \gamma_p)^{\mathrm{T}}.$$

Comparing it with the definition of the partial autocorrelation function,

$$\boldsymbol{\Phi}_p = \boldsymbol{\varphi},$$
$$\beta(p) = \Phi_{pp} = \varphi_p,$$
$$\beta(h) = 0 \quad (h > p).$$

For $h < p$, $\beta(h)$ can easily be computed by the definition.

Now we turn to introducing the spectral density.

Suppose that $\{X_t\}$ is a causal ARMA (p, q) process

$$\varphi(B)X_t = \theta(B)Z_t,$$
$$\{Z_t\} \sim \mathrm{WN}(0, \sigma^2).$$

So

$$X_t = \sum_0^\infty \psi_j Z_{t-j}, \quad \varphi(z) \neq 0 \quad (|z| \leq 1),$$

where $\{\psi_j\}$ satisfy $\psi(z) = \frac{\theta(z)}{\varphi(z)} = \sum_0^\infty \psi_j z^j$ and $\sum_0^\infty |\psi_j| < \infty$. By (1.3.4), the spectral density is

$$f_X(\alpha) = |\psi(\mathrm{e}^{\mathrm{i}\alpha})|^2 f_Z(\alpha) = \left|\frac{\theta(\mathrm{e}^{\mathrm{i}\alpha})}{\varphi(\mathrm{e}^{\mathrm{i}\alpha})}\right|^2 f_Z(\alpha).$$

However, for white noise $\{Z_t\}$,

$$E[Z_t] = 0,$$
$$\gamma(0) = E[Z_t^2] = \sigma^2,$$
$$\gamma(\alpha) = E[Z_{t+|\alpha|}Z_t] = 0 \quad (\alpha \in \mathbb{Z}_+),$$

and its spectral density is

$$f_Z(\alpha) = \sum_h \gamma(h)\,\mathrm{e}^{-\mathrm{i}h\alpha} = \gamma(0) = \sigma^2.$$

Thus the spectral density of the causal ARMA (p, q) process $\{X_t\}$ is

$$f_X(\alpha) = \sigma^2 \frac{|\theta(\mathrm{e}^{\mathrm{i}\alpha})|^2}{|\varphi(\mathrm{e}^{-\mathrm{i}\alpha})|^2}.$$

For example, in the causal ARMA (1, 2) process

$$X_t - \varphi_1 X_{t-1} = Z_t + \theta_1 z_{t-1} + \theta_2 z_{t-2},$$

where

$$\theta(z) = 1 + \theta_1 z + \theta_2 z^2,$$
$$\varphi(z) = 1 - \varphi_1 z.$$

So the spectral density of the causal ARMA (1, 2) process is

$$f_X(\alpha) = \sigma^2 \frac{|\theta(e^{i\alpha})|^2}{|\varphi(e^{-i\alpha})|^2} = \sigma^2 \frac{1 + \theta_1^2 - 2\theta_2 + \theta_2^2 + 2(\theta_1\theta_2 + \theta_1)\cos\alpha + 4\theta_2\cos^2\alpha}{1 + \varphi_1^2 - 2\varphi_1\cos\alpha}.$$

1.5 Prediction and modeling of ARMA processes

The innovation algorithm of the linear prediction of time series has been discussed in Section 1.2. For ARMA processes, one can simplify the innovation algorithm. Based on an observed stationary time series, one can determine an appropriate ARMA model fitting it.

1.5.1 Prediction

For a causal ARMA process

$$X_t - \varphi_1 X_{t-1} - \cdots - \varphi_p X_{t-p} = Z_t + \theta_1 Z_{t-1} + \cdots + \theta_q Z_{t-q},$$
$$\{Z_t\} \sim \mathrm{WN}(0, \sigma^2),$$

we use the innovation algorithm to predict X_{n+1}^o from X_{n+1}.

To simplify the algorithm, let $m = \max\{p, q\}$ and

$$\begin{aligned} S_t &= \sigma^{-1} X_t \quad (t = 1, \ldots, m), \\ S_t &= \sigma^{-1}(X_t - \varphi_1 X_{t-1} - \cdots - \varphi_p X_{t-p}) \quad (t > m). \end{aligned} \tag{1.5.1}$$

The autocovariances $h_{ij} = E[S_i S_j]$. A direct computation shows that for $1 \le i, j \le m$, since $\{X_t\}$ is stationary,

$$h_{ij} = \sigma^{-2} E[X_i X_j] = \sigma^{-2}\gamma(i - j),$$

where γ is autocovariance function of $\{X_t\}$ which can be calculated as in Section 1.4.2. For $1 \le i \le m$ and $m \le j \le 2m$,

$$\begin{aligned} h_{ij} &= \sigma^{-2} E[X_i(X_j - \varphi_1 X_{j-1} - \cdots - \varphi_p X_{j-p})] \\ &= \sigma^{-2}(\gamma(j - i) - \varphi_1\gamma(j - 1 - i) - \cdots - \varphi_p\gamma(j - p - i)). \end{aligned}$$

For $1 \le i \le m$ and $j > 2m$, $h_{ij} = 0$.
For $1 \le j \le m$ and $m < i \le 2m$, $h_{ij} = \sigma^{-2}(\gamma(i-j) - \varphi_1\gamma(i-1-j) - \cdots - \varphi_p\gamma(i-p-j))$.
For $1 \le j \le m$ and $i > 2m$, $h_{ij} = 0$.
For $i > m$ and $j > m$,

$$\begin{aligned} h_{ij} &= \sigma^{-2}E[(X_i - \varphi_1 X_{i-1} - \cdots - \varphi_p X_{i-p})(X_j - \varphi_1 X_{j-1} - \cdots - \varphi_p X_{j-p})] \\ &= \sigma^{-2}E[(Z_i - \theta_1 Z_{i-1} + \cdots + \theta_q Z_{i-q})(Z_j + \theta_1 Z_{j-1} + \cdots + \theta_q Z_{j-q})]. \end{aligned}$$

By the autocovariance function formula (1.1.1) of a linear process,

$$h_{ij} = \begin{cases} \theta_0\theta_{j-i} + \theta_1\theta_{1+j-i} + \cdots + \theta_q\theta_{q+j-i} & (j > i), \\ \theta_0\theta_{i-j} + \theta_1\theta_{1+i-j} + \cdots + \theta_q\theta_{q+i-j} & (i > j), \end{cases}$$

where $\theta_0 = 1$. Therefore, the autocovariances $\{h_{ij}\}$ are as follows:

$$h_{ij} = \begin{cases} \sigma^{-2}\gamma(i-j) & (1 \le i, j \le m), \\ \sigma^{-2}\left(\gamma(i-j) - \sum_{l=1}^{p} \varphi_l \gamma(l - |i-j|)\right) & (\min\{i,j\} \le m < \max\{i,j\} \le 2m), \\ \sum_{l=0}^{q} \theta_l \theta_{l+|i-j|} & (\min\{i,j\} > m), \\ 0, & \text{otherwise}. \end{cases} \tag{1.5.2}$$

By using the innovation algorithm to $\{S_t\}$, the best linear predictor of S_{n+1} in terms of $\{1, S_1, \ldots, S_n\}$ is

$$S_{n+1}^o = \begin{cases} \sum_1^n \alpha_{nj}(S_{n+1-j} - S_{n+1-j}^o) & (1 \le n < m), \\ \sum_1^q \alpha_{nj}(S_{n+1-j} - S_{n+1-j}^o) & (n \ge m), \end{cases} \tag{1.5.3}$$

where the coefficients $\{\alpha_{nj}\}$ are found from recurrence formula (1.2.5) in the innovation algorithm.

Denote by X_k^o the best linear predictor of X_k in terms of $\{1, X_1, \ldots, X_{k-1}\}$. Noticing that the linearity of prediction operator, by (1.5.1), we get

$$S_k^o = \begin{cases} \sigma^{-1}X_k^o & (k = 1, \ldots, m), \\ \sigma^{-1}(X_k^o - \varphi_1 X_{k-1} - \cdots - \varphi_p X_{k-p}) & (k > m), \end{cases}$$

$$S_k - S_k^o = \sigma^{-1}(X_k - X_k^o) \quad (k \in \mathbb{Z}_+).$$

From this and (1.5.3), it follows that

$$X_{n+1}^o = \begin{cases} \sum_{j=1}^n \alpha_{nj}(X_{n+1-j} - X_{n+1-j}^o) & (1 \le n < m), \\ \varphi_1 X_n + \cdots + \varphi_p X_{n+1-p} + \sum_{j=1}^q \alpha_{nj}(X_{n+1-j} - X_{n+1-j}^o) & (n \ge m), \end{cases} \tag{1.5.4}$$

and the mean squared errors

$$V_{n+1} = E[(X_{n+1} - X_{n+1}^o)^2] = \sigma^2(S_{n+1} - S_{n+1}^o)^2 = \sigma^2\lambda_n, \tag{1.5.5}$$

where α_{nj} and λ_n are obtained from the innovation algorithm with h_{ij} which are stated in (1.5.2).

Especially, the predictions of the AR (p) and MA (q) processes are easier. By (1.5.4), the prediction of the AR (p) process is

$$X^o_{n+1} = \varphi_1 X_n + \cdots + \varphi_p X_{n+1-p} \quad (n \geq p)$$

and the prediction of the MA (q) process is

$$X^o_{n+1} = \sum_{j=1}^{\min\{p,q\}} \alpha_{nj}(X_{n+1-j} - X^o_{n+1-j}) \quad (n \geq 1),$$

where $\alpha_{nj} = \sigma^{-2}\gamma(n-j) = \sum_{l=0}^{q-|n-j|} \theta_l \theta_{l+|n-j|}$.

In general, the algorithm of the best linear prediction for an ARMA (p, q) process is as follows.

Step 1. Use the method given in Section 1.4.2 to compute the autocovariance function γ.

Step 2. Compute $\{h_{ij}\}_{i,j\in\mathbb{Z}}$ by (1.5.2).

Step 3. Use the recurrence formula of the innovation algorithm to find α_{nj} and λ_n by (1.2.5).

Step 4. The predictor X^o_{n+1} and the mean squared error are given by the formulas (1.5.4) and (1.5.5).

For example, consider the prediction of an ARMA (1, 1) process

$$X_t - \varphi X_{t-1} = Z_t + \theta Z_{t-1},$$
$$\{Z_t\} \sim \mathrm{WN}(0, \sigma),$$

where $|\varphi| < 1$. Now $m = p = q = 1$. Let

$$S_t = \sigma^{-1} X_t \qquad (t = 1),$$
$$S_t = \sigma^{-1}(X_t - \varphi X_{t-1}) \quad (t > 1).$$

By (1.5.2) and $h_{ij} = E[S_i S_j]$, we get

$$\begin{aligned}
h_{11} &= \sigma^{-2}\gamma(0) = 1 + \frac{(\varphi+\theta)^2}{1-\varphi^2} && \text{(by (1.4.9))},\\
h_{12} &= h_{21} = \sigma^{-2}(\gamma(1) - \varphi\gamma(0)) = \theta && \text{(by (1.4.10))},\\
h_{i+1,i} &= h_{i,i+1} = \theta_0\theta_1 + \theta_1\theta_2 = \theta_0\theta_1 = \theta && (i > 1),\\
h_{ii} &= \theta_0^2 + \theta_1^2 = 1 + \theta^2 && (i \geq 2),\\
h_{ij} &= \theta_0\theta_{|i-j|} + \theta_1\theta_{1+|i-j|} = 0 && \text{otherwise.}
\end{aligned}$$

By (1.5.4), the best linear predictor of X_{n+1} is

$$X^o_{n+1} = \varphi X_n + \alpha_{n1}(X_n - X^o_n) \quad (n \geq 1).$$

By the recurrence formula of the innovation algorithm, successively compute λ_0, α_{11}, λ_1, α_{22}, α_{21}, λ_2, α_{33}, α_{32}, α_{31}, ... as follows:

$$\lambda_0 = h_{11} = 1 + \frac{\varphi+\theta}{1-\varphi^2}, \qquad \alpha_{11} = \frac{h_{21}}{\lambda_0} = \frac{\theta}{\lambda_0},$$

$$\lambda_1 = h_{22} - \alpha_{11}^2\lambda_0 = 1 + \theta^2 - \frac{\theta^2}{\lambda_0}, \qquad \alpha_{22} = \frac{h_{31}}{\lambda_0} = 0, \qquad \alpha_{21} = \frac{h_{32} - \alpha_{11}\alpha_{22}\lambda_0}{\lambda_1} = \frac{\theta}{\lambda_1},$$

$$\lambda_2 = 1 + \theta^2 - \frac{\theta^2}{\lambda_1}, \qquad \alpha_{33} = \frac{h_{41}}{\lambda_0} = 0, \qquad \alpha_{32} = \frac{h_{42} - \alpha_{11}\alpha_{33}}{\lambda_1} = 0, \qquad \alpha_{31} = \frac{h_{43}}{\lambda_2} = \frac{\theta}{\lambda_2},$$

$$\vdots$$

$$\alpha_{n1} = \frac{\theta}{\lambda_{n-1}}.$$

So the best linear predictor of X_{n+1} satisfies

$$X_{n+1}^o = \varphi X_n + \frac{\theta}{\lambda_{n-1}}(X_n - X_n^o) \quad (n \ge 1),$$

where

$$\lambda_n = 1 + \theta^2 - \theta^2/\lambda_{n-1} \qquad (n \ge 1),$$
$$\lambda_0 = 1 + (\varphi+\theta)/(1-\varphi^2),$$

and the mean squared error satisfies

$$E[(X_{n+1} - X_{n+1}^o)^2] = \sigma^2\lambda_n.$$

Now we consider the h-step prediction of an ARMA (p, q) process $\{X_t\}$.

Denote by P_nX_{n+h} the best linear predictor of X_{n+h} in terms of $1, X_1, \dots, X_n$. For $n > m = \max\{p, q\}$, we find predictors $X_1^o, \dots, X_n^o$ and then find recursively predictors $P_nX_{n+1}, P_nX_{n+2}, \dots$ using the formula

$$P_nX_{n+h} = \sum_{l=1}^{p} \varphi_l P_nX_{n+h-l} + \sum_{l=h}^{q} \alpha_{n+h-1,l}(X_{n+h-l} - X_{n+h-l}^o) \quad (h \ge 1)$$

and the mean squared error is

$$E[(X_{n+h} - P_nX_{n+h})^2] = \sigma^2 \sum_{l=0}^{h-1} \left(\sum_{k=0}^{l} \chi_k \alpha_{n+h-k-1,l-k} \right)^2 \lambda_{n+h-j-1},$$

where

$$\chi_k = \sum_{v=1}^{\min\{p,k\}} \varphi_v \chi_{k-v} \quad (\chi_0 = 1).$$

If $\{X_t\}$ is a causal ARMA (p, q) process with Gaussian white noise

$$Z_t \sim \mathrm{WN}(0, \sigma^2),$$
$$Z_t \sim N(0, \sigma^2),$$

then the prediction error is

$$E[(X_{n+h} - P_n X_{n+h})^2] = \sigma^2 \sum_{l=0}^{h-1} \psi_l^2,$$

where ψ_l $(l = 0, 1, \ldots)$ satisfies $\theta(z)/\varphi(z) = \sum_{l=0}^{\infty} \psi_l z^l$.

1.5.2 Modeling

We will estimate the coefficients $\varphi_1, \ldots, \varphi_p, \theta_1, \ldots, \theta_q$ and the white noise variance σ^2 such that the autocovariance γ_X of the corresponding ARMA process approximates to the autocovariance function γ of the observation time series.

(a) The fitted AR model

From a causal AR (p) process

$$X_t - \sum_{k=1}^{p} \varphi_k X_{t-k} = Z_t,$$

the Yule–Walker equation given in Section 1.4.2 is as follows:

$$\begin{aligned} \gamma(0) - \boldsymbol{\Phi}^{\mathrm{T}} \gamma_p &= \sigma^2, \\ \Gamma_p \boldsymbol{\Phi} &= \gamma_p, \end{aligned}$$

where

$$\begin{aligned} \Gamma_p &= (\gamma(i-j))_{i,j=1,\ldots,p}, \\ \gamma_p &= (\gamma(1), \ldots, \gamma(p))^{\mathrm{T}}, \\ \boldsymbol{\Phi} &= (\varphi_1, \ldots, \varphi_p)^{\mathrm{T}}. \end{aligned}$$

In applications, we only know finitely many observations $x_0, \ldots, x_{N-1}$. Since $\bar{x} = \frac{1}{n} \sum_{k=0}^{N-1} x_k \approx 0$, the $\gamma(h)$ is estimated by

$$\hat{\gamma}(h) = \frac{1}{N} \sum_{0}^{N-1} x_n x_{n+h}.$$

In the Yule–Walker equation, replacing $\gamma(\nu)$ by the sample covariance $\hat{\gamma}(\nu)$, $\boldsymbol{\Phi} = (\varphi_1, \ldots, \varphi_p)^{\mathrm{T}}$ by its estimator $\hat{\boldsymbol{\Phi}} = (\hat{\varphi}_1, \ldots, \hat{\varphi}_p)^{\mathrm{T}}$, and σ by its estimator $\hat{\sigma}$, we get

$$\begin{aligned} \hat{\sigma}^2 &= \hat{\gamma}(0) - \hat{\boldsymbol{\Phi}}^{\mathrm{T}} \hat{\gamma}_p, \\ \hat{\Gamma}_p \hat{\boldsymbol{\Phi}} &= \hat{\gamma}_p. \end{aligned} \tag{1.5.6}$$

This is the *Yule–Walker equation of estimators*, where

$$\begin{aligned} \hat{\Gamma}_p &= (\hat{\gamma}(i-j))_{i,j=1,\ldots,p}, \\ \hat{\gamma}_p &= (\hat{\gamma}(1), \ldots, \hat{\gamma}(p))^{\mathrm{T}}. \end{aligned} \tag{1.5.7}$$

If $\hat{\gamma}(0) > 0$, then $\hat{\Gamma}_p$ is nonsingular. From this and (1.5.6),

$$\hat{\boldsymbol{\Phi}} = (\hat{\varphi}_1, \ldots, \hat{\varphi}_p)^{\mathrm{T}} = \hat{\Gamma}_p^{-1}\hat{\gamma}_p.$$

Clearly, $1 - \hat{\varphi}_1 z - \cdots - \hat{\varphi}_p z^p \neq 0$ ($|z| \leq 1$). So the fitted model

$$\begin{gathered} X_t - \hat{\varphi}_1 X_{t-1} - \cdots - \hat{\varphi}_p X_{t-p} = Z_t, \\ \{Z_t\} \sim \mathrm{WN}(0, \hat{\sigma}^2) \end{gathered} \tag{1.5.8}$$

is a causal AR (p) process. Therefore, its autocovariance function $\gamma_F(h)$ satisfies the Yule–Walker equation

$$\gamma_F(h) - \hat{\varphi}_1\gamma_F(h-1) - \cdots - \hat{\varphi}_p\gamma_F(h-p) = \begin{cases} 0 & (h = 1, \ldots, p), \\ \hat{\sigma}^2 & (h = 0). \end{cases} \tag{1.5.9}$$

Comparing this and (1.5.6),

$$\gamma_F(h) = \hat{\gamma}(h) \quad (h = 0, \ldots, p).$$

From this, it is seen that the autocovariances of the fitted model (1.5.8) at lags 0, 1, $\ldots, p$ coincide with the sample autocovariances.

For a causal AR (p) process $\{X_t\}$

$$\begin{gathered} X_t - \sum_1^p \varphi_k X_{t-k} = Z_t, \\ \{Z_t\} \sim WN(0, \sigma^2), \end{gathered}$$

the large-sample distribution of Yule–Walker estimator is

$$\hat{\boldsymbol{\Phi}} = (\hat{\varphi}_1, \ldots, \hat{\varphi}_p)^{\mathrm{T}} = \hat{\Gamma}_p^{-1}\hat{\gamma}_p$$

and the distribution is $\hat{\boldsymbol{\Phi}} \approx N(\boldsymbol{\Phi}, n^{-1}\sigma^2\Gamma_p^{-1})$, where $\boldsymbol{\Phi} = (\varphi_1, \ldots, \varphi_p)^{\mathrm{T}}$.

How do we select the order p? The partial autocorrelation function (PACF) is given in Section 1.4.3. The sample PACF is defined as

$$\begin{aligned} \hat{\beta}(0) &= 1, \\ \hat{\beta}(m) &= \hat{\Phi}_{mm} \quad (m \in \mathbb{Z}_+), \end{aligned}$$

where $\hat{\Phi}_{mm}$ is the last component of $\hat{\boldsymbol{\Phi}}_m = \hat{\Gamma}_m^{-1}\hat{\gamma}_m$. If the sample PACF $\hat{\beta}(m)$ ($0 \leq m \leq p$) is significantly different from zero and $\hat{\beta}(m) \approx 0$ ($m > p$). Precisely speaking, if

$$\begin{aligned} \hat{\beta}(m) &> 1.96n^{-\frac{1}{2}} \quad (0 \leq m \leq p), \\ \hat{\beta}(m) &< 1.96n^{-\frac{1}{2}} \quad (m > p), \end{aligned}$$

then a p-order AR model can fit the observation data.

(b) The fitted MA model

Given the observed data $\{x_1, \dots, x_n\}$, denote by $\hat{\gamma}(h)$ its sample autocovariance function. The sample covariance matrix is $(\hat{\gamma}(i-j))_{i,j=1,\dots,n}$. We can fit moving average models

$$X_t = Z_t + \hat{\alpha}_{m1}Z_{t-1} + \cdots + \hat{\alpha}_{mm}Z_{t-m} \quad (m \in \mathbb{Z}_+),$$
$$\{Z_t\} \sim WN(0, \hat{V}_m) \quad (m \in \mathbb{Z}_+),$$

and $\hat{\alpha}_{m1}, \dots, \hat{\alpha}_{mm}, \hat{V}_m$ are obtained by using the innovation algorithm with the autocovariance function matrix $(\gamma(i-j))_{i,j=1,\dots,n}$ replaced by the sample autocovariance function matrix $(\hat{\gamma}(i-j))_{i,j=1,\dots,n}$. Since we know that for a zero-mean stationary process, if its autocovariance function (ACVF) $\gamma(h) = 0$ for $h > q$, then this process can be represented as a moving average process of order q or less. Therefore, if its sample ACVF $\hat{\gamma}(h)$ satisfies the condition that $\hat{\gamma}(h)$ is significantly different from zero for $0 \le h \le q$ and $\hat{\gamma}(h) \approx 0$ for $h > q$, precisely speaking, if

$$\hat{\gamma}(h) > 1.96n^{-\frac{1}{2}} \quad (0 \le h \le q),$$
$$\hat{\gamma}(h) < 1.96n^{-\frac{1}{2}} \quad (h > q),$$

then a q-order MA model can fit the observation data.

(c) The fitted ARMA model

We use m-order Yule–Walker estimates

$$X_t - \hat{\varphi}_{m1}X_{t-1} - \cdots - \hat{\varphi}_{mm}X_{t-m} = Z_t,$$
$$\{Z_t\} \sim \mathrm{WN}(0, \hat{\sigma}_m)$$

to fit observed data $\{x_1, \dots, x_n\}$, where $m > \max\{p, q\}$. Then the *Hannam–Rissanen algorithm* is divided into three steps.

Step 1. Let $\hat{Z}_t = X_t - \hat{\varphi}_{m1}X_{t-1} - \cdots - \hat{\varphi}_{mm}X_{t-m}$ $(t = m+1, \dots, n)$.

Step 2. Choose $\beta = (\varphi_1, \dots, \varphi_p, \theta_1, \dots, \theta_q)$ such that the following sum of squares

$$S(\beta) = \sum_{t=m+1+q}^{n} (X_t - \varphi_1 X_{t-1} - \cdots - \varphi_p X_{t-p} - \theta_1 \hat{Z}_{t-1} - \cdots - \theta_q \hat{Z}_{t-q})^2$$

attains the minimal value. Let $\hat{\beta} = (H^{\mathrm{T}}H)^{-1}H^{\mathrm{T}}X_n$, where $X_n = (X_{m+1+q}, \dots, X_n)^{\mathrm{T}}$ and $H = (H_1|H_2)$, and

$$H_1 = \begin{pmatrix} X_{m+q} & X_{m+q-1} & \cdots & X_{m+q+1-p} \\ X_{m+q+1} & X_{m+q} & \cdots & X_{m+q+2-p} \\ \vdots & \vdots & \cdots & \vdots \\ X_{n-1} & X_{n-2} & \cdots & X_{n-p} \end{pmatrix},$$

$$H_2 = \begin{pmatrix} \hat{Z}_{m+q} & \hat{Z}_{m+q-1} & \cdots & \hat{Z}_{m+1} \\ \hat{Z}_{m+q+1} & \hat{Z}_{m+q} & \cdots & \hat{Z}_{m+2} \\ \vdots & \vdots & \cdots & \vdots \\ \hat{Z}_{n-1} & \hat{Z}_{n-2} & \cdots & \hat{Z}_{n-p} \end{pmatrix}.$$

It can be shown that $\hat{\beta} = (\hat{\varphi}_1, \ldots, \hat{\varphi}_p, \hat{\theta}_1, \ldots, \hat{\theta}_q)$ such that $S(\hat{\beta})$ attains the minimal value.

Step 3. For $t \le \max\{p, q\}$, let $\tilde{Z}_t = 0$, $V_t = 0$, and $W_t = 0$. For $\max\{p, q\} < t \le n$, let

$$\tilde{Z}_t = X_t - \sum_1^p \hat{\varphi}_j X_{t-j} - \sum_1^q \hat{\theta}_j \tilde{Z}_{t-j},$$

$$V_t = \sum_1^p \hat{\varphi}_j V_{t-j} + \tilde{Z}_t,$$

$$W_t = -\sum_1^p \hat{\theta}_j W_{t-j} + \tilde{Z}_t.$$

Take $\beta = \hat{\beta}^+$ such that

$$S^+(\beta) = \sum_{t=\max\{p,q\}+1}^{n} \left(\tilde{Z}_t - \sum_1^p \beta_j V_{t-j} - \sum_1^q \beta_{k+p} W_{t-k} \right)^2$$

attains the minimal value. The $\widetilde{\beta} = \hat{\beta}^+ + \hat{\beta}$ is a good estimator. Let $\widetilde{\beta} = (\tilde{\varphi}_1, \ldots, \tilde{\varphi}_p, \widetilde{\theta}_1, \ldots, \widetilde{\theta}_q)$. Then the ARMA (p, q) process

$$X_t - \tilde{\varphi}_1 X_{t-1} - \cdots - \tilde{\varphi}_p X_{t-p} = Z_t + \widetilde{\theta}_1 Z_{t-1} + \cdots + \widetilde{\theta}_q Z_{t-q}.$$

can fit the observed data.

(d) Maximum likelihood estimation

Suppose that $\{X_t\}$ is a zero-mean Gaussian time series and the covariance matrix Γ_n of $X_1, \ldots, X_n$ is nonsingular. The *likelihood* of X_n is

$$L(\Gamma_n) = (2\pi)^{-\frac{n}{2}} (\det \Gamma_n)^{-\frac{1}{2}} \exp\left(-\tfrac{1}{2} X_n^{\mathrm{T}} \Gamma_n^{-1} X_n\right).$$

It can be rewritten in the form

$$L(\Gamma_n) = \left((2\pi)^n V_0 \cdots V_{n-1}\right)^{-\frac{1}{2}} \exp\left(-\frac{1}{2} \sum_1^n \frac{(X_j - X_j^o)^2}{V_{j-1}} \right), \tag{1.5.10}$$

where X_j^o is the best linear prediction in terms of $X_{j-1}, \ldots, X_1$ and V_j is the corresponding mean squared error. If $\{X_t\}$ is not Gaussian, we always say (1.5.10) is the Gaussian likelihood of $X_1, \ldots, X_n$.

Suppose that data $\{X_t\}$ are from an ARMA (p, q) process. The one-step predictors X^o_{n+1} and the mean squared errors V_{n+1} are stated in (1.5.4) and (1.5.5). Therefore, X^o_j are the functions of $\varphi_1, \ldots, \varphi_p, \theta_1, \ldots, \theta_q$ and the Gaussian likelihood for an ARMA process is

$$L(\varphi, \theta, \sigma^2) = ((2\pi\sigma^2)^n \lambda_0 \cdots \lambda_{n-1})^{-\frac{1}{2}} \exp\left(-\frac{1}{2\sigma^2} \sum_1^n \frac{(X_j - X^o_j)^2}{\lambda_{j-1}} \right),$$

where $\varphi = (\varphi_1, \ldots, \varphi_p)$ and $\theta = (\theta_1, \ldots, \theta_q)$. Note that

$$\frac{\partial}{\partial \sigma^2} \log(\varphi, \theta, \sigma^2) = -\frac{n}{2\sigma^2} + \frac{1}{2\sigma^4} \sum_1^n \frac{(X_j - X^o_j)^2}{\lambda_{j-1}}.$$

Then $L(\varphi, \theta, \sigma^2)$ attains the minimal value when $\sigma^2 = \hat{\sigma}^2 = \frac{1}{n} \sum_{j=1}^n \frac{(X_j - X^o_j)^2}{\lambda_{j-1}}$. Denote

$$H(\varphi, \theta) = \sum_1^n \frac{(X_j - X^o_j)^2}{\lambda_{j-1}}.$$

Take $\varphi = \hat{\varphi}$ and $\theta = \hat{\theta}$ such that

$$l(\varphi, \theta) = \log \frac{H(\varphi, \theta)}{n} + \frac{1}{n} \sum_1^n \log \lambda_{j-1}$$

attains the minimal value and the estimate of σ^2 is

$$\hat{\sigma}^2 = \frac{1}{n - p - q} H(\hat{\varphi}, \hat{\theta}).$$

In this way, for an ARMA (p, q) process, we can give the estimators of parameters $\hat{\varphi}_1, \ldots, \hat{\varphi}_p, \hat{\theta}_1, \ldots, \hat{\theta}_q$ and $\hat{\sigma}^2$. The fitted ARMA (p, q) model is

$$X_t - \hat{\varphi}_1 X_{t-1} - \cdots - \hat{\varphi}_p X_{t-p} = Z_t + \hat{\theta}_1 Z_{t-1} + \cdots + \hat{\theta}_q Z_{t-q},$$
$$\{Z_t\} \sim \mathrm{WN}(0, \hat{\sigma}^2).$$

The selection of the orders p and q for general ARMA models is done by using the AICC statistic

$$\mathrm{AICC} = -2L\left(\varphi_p, \theta_q, \frac{H(\varphi_p, \theta_q)}{n} \right) + \frac{2(p + q + 1)n}{n - p - q - 2}.$$

Choose $p, q, \varphi_p, \theta_q$ such that AICC attains the minimal value.

1.6 Multivariate ARMA processes

Consider the k-dimensional time series $\boldsymbol{Y}_t = (Y_{t1}, \ldots, Y_{tk})^{\mathrm{T}}$. The purpose of multivariate time series analysis is to understand the relationship among the component time series. If $\boldsymbol{Y}_t$ has a constant mean vector $\mu = (\mu_1, \ldots, \mu_k)^{\mathrm{T}}$ for all t and its cross-covariance matrix $R(\boldsymbol{Y}_t, \boldsymbol{Y}_{t+l}^{\mathrm{T}}) = \Gamma(l)$ only depends on l, then the k-dimensional time series $\boldsymbol{Y}_t$ is called a *stationary vector process*, where $R(\boldsymbol{Y}_t, \boldsymbol{Y}_{t+l}^{\mathrm{T}}) = (E[Y_{ti}Y_{t+l,j}])_{i,j=1,\ldots,k}$ and each $E[Y_{ti}Y_{t+l,j}]$ is independent of t. Denote

$$\gamma_{ij}(l) = E[Y_{ti}Y_{t+l,j}],$$
$$\Gamma(l) = (\gamma_{ij}(l))_{i,j=1,\ldots,k}.$$

If $\{\boldsymbol{Y}_t\}$ is a k-dimensional zero-mean stationary vector process and

$$\boldsymbol{Y}_t - \sum_1^p \Phi_j \boldsymbol{Y}_{t-j} = \boldsymbol{Z}_t + \sum_1^q \Theta_j \boldsymbol{Z}_{t-j}, \tag{1.6.1}$$

then $\boldsymbol{Y}_t$ is called a *vector ARMA process*, where Φ_j, Θ_j are both $k \times k$ matrixes independent of t and $\boldsymbol{Z}_t$ is a k-dimensional white noise satisfying

$$\begin{aligned} E[\boldsymbol{Z}_t] &= 0, \\ E[\boldsymbol{Z}_t\boldsymbol{Z}_t^{\mathrm{T}}] &= \Sigma, \\ E[\boldsymbol{Z}_t\boldsymbol{Z}_{t+l}^{\mathrm{T}}] &= 0 \quad (l \neq 0). \end{aligned} \tag{1.6.2}$$

Here Σ is a $k \times k$ constant matrix. Denote $\{Z_t\} \sim \mathrm{WN}(0, \Sigma)$. Especially, if $\Phi_j = 0$ ($j = 1, \ldots, p$), then $\boldsymbol{Y}_t$ is called a *vector MA process*. If $\Theta_j = 0$ ($j = 1, \ldots, q$), then $\boldsymbol{Y}_t$ is called a *vector AR process*.

Introduce two matrix-valued polynomials

$$\Phi(z) = I - \Phi_1 z - \cdots - \Phi_p z^p,$$
$$\Theta(z) = I + \Theta_1 z + \cdots + \Theta_q z^q,$$

where I is the $k \times k$ identity matrix, z is a complex variable, and each component of matrices $\Phi(z)$, $\Theta(z)$ is a polynomial of degree p and q with matrix coefficients, respectively. Denote

$$\Phi_j = (\varphi_{mn}^{(j)})_{k\times k} \quad (j = 0, 1, \ldots, p), \quad \Phi_0 = I,$$
$$\Theta_j = (\theta_{mn}^{(j)})_{k\times k} \quad (j = 0, 1, \ldots, q), \quad \Theta_0 = I.$$

Then

$$\Phi(z) = \left(\sum_{j=0}^p \varphi_{mn}^{(j)} z^j \right)_{k\times k} \qquad \Theta(z) = \left(\sum_{j=0}^q \theta_{mn}^{(j)} z^j \right)_{k\times k}$$

Let B be backward shift operator. Then

$$B\boldsymbol{Y}_t = \boldsymbol{Y}_{t-1},$$
$$B^j\boldsymbol{Y}_t = \boldsymbol{Y}_{t-j},$$

and so

$$\Phi(B)\boldsymbol{Y}_t = (I - \Phi_1 B - \cdots - \Phi_p B^p)\boldsymbol{Y}_t = \boldsymbol{Y}_t - \Phi_1 \boldsymbol{Y}_{t-1} - \cdots - \Phi_p \boldsymbol{Y}_{t-p},$$
$$\Theta(B)\boldsymbol{Z}_t = (I + \Theta_1 B + \cdots + \Theta_p B^p)\boldsymbol{Z}_t = \boldsymbol{Z}_t + \Theta_1 \boldsymbol{Z}_{t-1} + \cdots + \Theta_q \boldsymbol{Z}_{t-p}.$$

Equation (1.6.1) can be rewritten in the form $\Phi(B)\boldsymbol{Y}_t = \Theta(B)\boldsymbol{Z}_t$, where $\{Z_t\} \sim \mathrm{WN}(0, \Sigma)$.

1.6.1 Vector MA processes

Let $\boldsymbol{Y}_t$ be an MA (q) process

$$\boldsymbol{Y}_t = \boldsymbol{Z}_t + \sum_1^q \Theta_j \boldsymbol{Z}_{t-j} = \Theta(B)\boldsymbol{Z}_t.$$

If the determinant of the matrix $\Theta(z)$ satisfies $\det \Theta(z) \neq 0$ $(|z| \leq 1)$, then the inverse matrix $\Theta^{-1}(z)$ exists on $|z| \leq 1$ and

$$\Theta^{-1}(z) = (\det \Theta(z))^{-1}\Theta^*(z) \quad (|z| \leq 1),$$

where $\Theta^*(z)$ is the associated matrix of $\Theta(z)$. Both each component $h_{ij}(z)$ of $\Theta^*(z)$ and $\det \Theta(z)$ are polynomials of z and $\det \Theta(z) \neq 0$ $(|z| \leq 1)$. Therefore, $h_{ij}(z)/\det \Theta(z)$ can be expanded into a power series and $\Theta^{-1}(z)$ can be expanded into the power series with matrix coefficients $\Theta^{-1}(z) = \sum_{j=0}^{\infty} \pi_j z^j$ $(|z| \leq 1)$. It is easy to show that $\pi_0 = I$. Replacing z by B,

$$\Theta^{-1}(B) = I + \sum_1^{\infty} \pi_j B^j$$

since $\|B\| = 1$. From this and (1.6.3), it follows that

$$\boldsymbol{Z}_t = \Theta^{-1}(B)\boldsymbol{Y}_t = \boldsymbol{Y}_t + \sum_{j=1}^{\infty} \pi_j \boldsymbol{Y}_{t-j}.$$

For a vector MA process, the cross-covariance matrix $\Gamma(l)$ is equal to

$$\Gamma(l) = \mathrm{Cov}(\boldsymbol{Y}_t, \boldsymbol{Y}_{t+l}) = E[\boldsymbol{Y}_t \boldsymbol{Y}_{t+l}^{\mathrm{T}}] = E\left[\left(\boldsymbol{Z}_t - \sum_1^q \Theta_j \boldsymbol{Z}_{t-j}\right)\left(\boldsymbol{Z}_{t+l} - \sum_1^q \Theta_j \boldsymbol{Z}_{t+l-j}\right)^{\mathrm{T}}\right]$$

$$= E\left[\left(\sum_0^q \Theta_j \boldsymbol{Z}_{t-j}\right)\left(\sum_0^q \Theta_j \boldsymbol{Z}_{t+l-j}\right)^{\mathrm{T}}\right] = \sum_{k=0}^q \sum_{j=0}^q \Theta_j E[\boldsymbol{Z}_{t-j}\boldsymbol{Z}_{t+l-k}^{\mathrm{T}}]\Theta_k^{\mathrm{T}},$$

where $\Theta_0 = I$. Again, by (1.6.2) and $\Theta_j = 0$ $(k > q)$,

$$E[\boldsymbol{Z}_{t-j}\boldsymbol{Z}_{t+l-k}^{\mathrm{T}}] = \delta_{j,k-l}\Sigma,$$

and so

$$\Gamma(l) = \sum_0^{q-l} \Theta_k \Sigma \Theta_{k+l}^{\mathrm{T}} \quad (l = 0, 1, \ldots, q).$$

Conversely, if the cross-covariance matrix is known, the MA coefficients Θ_k and Σ can be found by this formula.

For example, consider a bivariate MA (1) model $\boldsymbol{Y}_t = \boldsymbol{Z}_t + \Theta_1 \boldsymbol{Z}_{t-1}$, where

$$\boldsymbol{Y}_t = (Y_{1t}, Y_{2t})^{\mathrm{T}}, \quad \Theta_1 = \begin{pmatrix} \alpha_{11} & \alpha_{12} \\ \alpha_{21} & \alpha_{22} \end{pmatrix}.$$

Let $\boldsymbol{Z}_t = (Z_{1t}, Z_{2t})^{\mathrm{T}}$ and $\boldsymbol{Y}_t = \Theta(B)\boldsymbol{Z}_t$, and

$$\Theta(B) = I + \Theta_1 B = \begin{pmatrix} 1 + \alpha_{11}B & \alpha_{12}B \\ \alpha_{21}B & 1 + \alpha_{22}B \end{pmatrix}.$$

Then

$$\det(\Theta(B)) = (1 + \alpha_{11}B)(1 + \alpha_{22}B) - \alpha_{12}\alpha_{21}B^2. \tag{1.6.3}$$

The inverse matrix

$$\Theta^{-1}(B) = \frac{1}{\det(\Theta(B))} \begin{pmatrix} 1 + \alpha_{22}B & -\alpha_{12}B \\ -\alpha_{21}B & 1 + \alpha_{11}B \end{pmatrix}.$$

This implies $\Theta^{-1}(B)\boldsymbol{Y}_t = \boldsymbol{Z}_t$, and so

$$\begin{pmatrix} 1 + \alpha_{22}B & -\alpha_{12}B \\ -\alpha_{21}B & 1 + \alpha_{11}B \end{pmatrix} \boldsymbol{Y}_t = \boldsymbol{Z}_t \det(\Theta(B))$$

or

$$\begin{aligned} (1 + \alpha_{22}B)Y_{1t} &= \alpha_{12}BY_{2t} + \det(\Theta(B))Z_{1t}, \\ (1 + \alpha_{11}B)Y_{2t} &= \alpha_{21}BY_{1t} + \det(\Theta(B))Z_{2t}. \end{aligned}$$

Its solution is

$$\begin{aligned} Y_{1t} &= \alpha_{12}B(1 + \alpha_{22}B)^{-1}Y_{2t} + \det(\Theta(B))(1 + \alpha_{22}B)^{-1}Z_{1t}, \\ Y_{2t} &= \alpha_{21}B(1 + \alpha_{11}B)^{-1}Y_{1t} + \det(\Theta(B))(1 + \alpha_{11}B)^{-1}Z_{2t}. \end{aligned}$$

Such structure is called a *joint transfer function structure*. For convenience, assume that $\alpha_{12} = 0$. Then by (1.6.3) these equations can be rewritten in the form

$$\begin{aligned} Y_{1t} &= \det(\Theta(B))(1 + \alpha_{22}B)^{-1}Z_{1t} = (1 + \alpha_{11}B)Z_{1t}, \\ Y_{2t} &= a_{21}B(1 + \alpha_{11}B)^{-1}Y_{1t} + (1 + \alpha_{22}B)Z_{2t}. \end{aligned} \tag{1.6.4}$$

Even if $\alpha_{12} \neq 0$, the equations of a form similar to (1.6.4) could still be arrived at. From this it is seen that future values of the process Y_{2t} depend on the past of both Y_{1t} and Y_{2t}, whereas future values of Y_{1t} only depend on its own past and not on the past of Y_{2t}. In applications, the equations (1.6.4) often represent the model structure of most interest.

1.6.2 Vector AR processes

Let $\boldsymbol{Y}_t$ be an AR (p) process

$$\boldsymbol{Y}_t - \sum_1^p \Phi_j \boldsymbol{Y}_{t-j} = \boldsymbol{Z}_t \quad \text{or} \quad \Phi(B)\boldsymbol{Y}_t = \boldsymbol{Z}_t, \tag{1.6.5}$$

where $\Phi(z) = I - \Phi_1 z - \cdots - \Phi_p z^p$. If the determinant of $\Phi(z)$ satisfies $\det(\Phi(z)) \neq 0$ ($|z| \leq 1$), then the inverse matrix $\Phi^{-1}(z)$ exists and can be expanded into a power series with matrix coefficients

$$\Phi^{-1}(z) = \sum_0^\infty \Psi_j z^j \quad (|z| \leq 1).$$

Since $\boldsymbol{Y}_t$ is stationary and $\|B\| = 1$,

$$\Phi^{-1}(B) = \sum_{j=0}^\infty \Psi_j B^j.$$

From this and (1.6.5),

$$\boldsymbol{Y}_t = \Phi^{-1}(B)\boldsymbol{Z}_t = \sum_0^\infty \Psi_j B^j(\boldsymbol{Z}_t) = \sum_0^\infty \Psi_j \boldsymbol{Z}_{t-j}, \quad \Psi_0 = I.$$

Especially, for AR (1) model

$$\boldsymbol{Y}_t = \Phi \boldsymbol{Y}_{t-1} + \boldsymbol{Z}_t, \tag{1.6.6}$$

the condition $\det(\Phi(z)) = \det(I - \Phi z) \neq 0$ ($|z| \leq 1$) is equivalent to $\det(z^{-1}I - \Phi) \neq 0$ ($|z| \leq 1$), i.e., the absolute values of all eigenvalues of the matrix Φ are greater than or equal to 1. Repeatedly using (1.6.6), we get

$$\boldsymbol{Y}_t = \sum_0^n \Phi^j \boldsymbol{Z}_{t-j} + \Phi^{n+1} \boldsymbol{Y}_{t-n-1}.$$

From this, it is seen that when the initial value $\boldsymbol{Y}_{n-1}$ is known, the value of an AR (1) process $\boldsymbol{Y}_t$ can be deduced from the above formula.

From $\boldsymbol{Y}_{t-l} = \boldsymbol{Z}_{t-l} + \sum_{j=0}^\infty \Phi_j \boldsymbol{Z}_{t-l-j}$ and $E[\boldsymbol{Y}_{t-l}\boldsymbol{Z}_t^{\mathrm{T}}] = 0$ $(l > 0)$, it implies the *Yule–Walker equations*

$$\Gamma(0) = \sum_1^p \Gamma(-j)\Phi_j^{\mathrm{T}} + \Sigma,$$

$$\Gamma(l) = E[\boldsymbol{Y}_{t-l}\boldsymbol{Y}_t^{\mathrm{T}}] = \sum_1^p \Gamma(l-j)\Phi_j^{\mathrm{T}} \quad (l = 1, \ldots, p).$$

The AR (p) process can always be expressed in the form of a kp-dimensional vector AR (1) model

$$\boldsymbol{W}_t = (\boldsymbol{Y}_t^{\mathrm{T}}, \ldots, \boldsymbol{Y}_{t-p+1}^{\mathrm{T}})^{\mathrm{T}},$$

where $\boldsymbol{W}_t = \Phi \boldsymbol{W}_{t-1} + \boldsymbol{Z}_t$ with $\boldsymbol{Z}_t = (\boldsymbol{Z}_t^{\mathrm{T}}, \boldsymbol{0}^{\mathrm{T}}, \ldots, \boldsymbol{0}^{\mathrm{T}})^{\mathrm{T}}$ and

$$\Phi = \begin{pmatrix} \Phi_1 & \Phi_2 & \cdots & \cdots & \cdots & \cdots & \Phi_p \\ I & 0 & \cdots & \cdots & \cdots & \cdots & 0 \\ 0 & I & 0 & \cdots & \cdots & \cdots & \vdots \\ \vdots & \ddots & \ddots & \ddots & \cdots & \cdots & \vdots \\ 0 & \cdots & 0 & I & 0 & \cdots & 0 \end{pmatrix}$$

is a $kp \times kp$ matrix associated with $\Phi(B)$.

1.6.3 Multivariate ARMA (p, q) processes

Let $\{\boldsymbol{Y}_t\}_{t\in\mathbb{Z}}$ be a k-variate ARMA (p, q) model stated in (1.6.1) and (1.6.2). If $\det \Phi(z) \neq 0$ $(|z| \leq 1)$, then the matrix $\Phi(z)$ is nonsingular and its inverse matrix $\Phi^{-1}(z)$ exists. Multiplying both sides by $\Phi^{-1}(B)$,

$$\boldsymbol{Y}_t = \Phi^{-1}(B)\Theta(B)\boldsymbol{Z}_t = \Psi(B)\boldsymbol{Z}_t, \tag{1.6.7}$$

where $\Psi(B) = \Phi^{-1}(B)\Theta(B)$. Expand it into a power series in z with matrix coefficients

$$\Psi(z) = \sum_0^\infty \Psi_j z^j \quad (|z| \leq 1),$$

where Ψ_j is a $k \times k$ matrix. The corresponding operator is

$$\Psi(B) = \sum_{j=0}^\infty \Psi_j B^j.$$

By (1.6.7),

$$\begin{aligned}\sum_0^\infty \Theta_j z^j &= \left(I - \sum_1^p \Phi_j z^j\right)\left(I + \sum_1^q \Psi_j z^j\right)\\ &= I + (\Psi_1 - \Phi_1)z + \cdots + (\Psi_j - \Phi_1\Psi_{j-1} - \cdots - \Phi_p\Psi_{j-p})z^j + \cdots.\end{aligned}$$

Equating coefficient matrixes of various powers z^j, we get

$$\begin{aligned}\Psi_0 &= 1,\\ \Theta_j &= \Psi_j - \Phi_1\Psi_{j-1} - \cdots - \Phi_p\Psi_{j-p} \quad (j = 1, \ldots, p).\end{aligned}$$

From this and (1.6.7),

$$\boldsymbol{Y}_t = \Psi(B)\boldsymbol{Z}_t = \sum_0^\infty \Psi_j B^j \boldsymbol{Z}_t = \sum_0^\infty \Psi_j \boldsymbol{Z}_{t-j}.$$

1.7 State-space models

1.7.1 State-space representation

A state-space model for an m-dimensional time series $\{\boldsymbol{Y}_t\}$ consists of two equations. The first equation

$$\boldsymbol{Y}_t = K_t\boldsymbol{X}_t + \boldsymbol{W}_t \quad (t \in \mathbb{Z}_+). \tag{1.7.1}$$

is called the *observation equation* and the second equation

$$\boldsymbol{X}_{t+1} = L_t\boldsymbol{X}_t + \boldsymbol{V}_t \quad (t \in \mathbb{Z}_+), \tag{1.7.2}$$

is called the *state equation*, where K_t is an $m \times n$ matrix and L_t is an $n \times n$ matrix, and

$$\begin{aligned} \{\boldsymbol{W}_t\} &\sim \mathrm{WN}(0, R_t), \\ \{\boldsymbol{V}_t\} &\sim \mathrm{WN}(0, Q_t). \end{aligned} \tag{1.7.3}$$

Here R_t is an $m \times m$ matrix and Q_t is an $n \times n$ matrix.

Assume that

$$E[\boldsymbol{W}_t \boldsymbol{V}_s^{\mathrm{T}}] = 0, \quad E[\boldsymbol{X}_1 \boldsymbol{W}_s^{\mathrm{T}}] = 0, \quad E[\boldsymbol{X}_1 \boldsymbol{V}_s^{\mathrm{T}}] = 0 \quad (s, t \in \mathbb{Z}_+).$$

By (1.7.1) and (1.7.2),

$$\begin{aligned} \boldsymbol{X}_t &= (L_{t-1}, \dots, L_1)\boldsymbol{X}_1 + (L_{t-1}, \dots, L_2)\boldsymbol{V}_1 + \dots + L_{t-1}\boldsymbol{V}_{t-2} + \boldsymbol{V}_{t-1}, \\ \boldsymbol{Y}_t &= (K_t, \dots, K_1)\boldsymbol{X}_1 + (K_t, \dots, K_2)\boldsymbol{W}_1 + \dots + K_t\boldsymbol{W}_{t-1} + \boldsymbol{W}_t. \end{aligned}$$

This implies that for $1 \le s \le t$,

$$\begin{aligned} E[\boldsymbol{V}_t \boldsymbol{X}_s^{\mathrm{T}}] &= 0, \quad E[\boldsymbol{V}_t \boldsymbol{Y}_s^{\mathrm{T}}] = 0, \\ E[\boldsymbol{W}_t \boldsymbol{X}_s^{\mathrm{T}}] &= 0, \quad E[\boldsymbol{W}_t \boldsymbol{Y}_s^{\mathrm{T}}] = 0. \end{aligned}$$

The state-space model is the extension of the ARMA model.

For example, if $\{Y_t\}$ is a causal and invertible ARMA $(1, 1)$ model

$$\begin{aligned} Y_t - \varphi Y_{t-1} &= Z_t + \theta Z_{t-1}, \\ \{Z_t\} &\sim \mathrm{WN}(0, \sigma^2). \end{aligned}$$

The corresponding observation equation is $Y_t = (\theta, 1)\boldsymbol{X}_t$, where $\boldsymbol{X}_t = (X_{t-1}, X_t)^{\mathrm{T}}$, and the state equation is

$$\boldsymbol{X}_{t+1} = \begin{pmatrix} 0 & 1 \\ 0 & \varphi \end{pmatrix} \boldsymbol{X}_t + \boldsymbol{V}_t,$$

where $\boldsymbol{V}_t = (0, Z_{t+1})^{\mathrm{T}}$ and

$$\boldsymbol{X}_1 = \left(\sum_{k=0}^{\infty} \varphi^k Z_{-k}, \sum_{k=0}^{\infty} \varphi^k Z_{1-k} \right)^{\mathrm{T}}.$$

If $\{Y_t\}$ is a causal ARMA (p, q) model,

$$Y_t - \varphi_1 Y_{t-1} - \dots - \varphi_p Y_{t-p} = Z_t + \theta_1 Z_{t-1} + \dots + \theta_q Z_{t-q}.$$

Denote $r = \max\{p, q+1\}$, $\varphi_j = 0\ (j > p)$, $\theta_j = 0\ (j > q)$, and $\theta_0 = 1$. Then the observation equation is

$$\begin{aligned} Y_t &= (\theta_{r-1}, \dots, \theta_0)\boldsymbol{X}_t, \\ \boldsymbol{X}_t &= (U_{t-r+1}, \dots, U_t)^{\mathrm{T}}, \end{aligned}$$

where $\{U_t\}$ satisfies the causal AR (p) equation

$$U_t - \varphi_1 U_{t-1} - \dots - \varphi_p U_{t-p} = Z_t$$

and the state equation is

$$\boldsymbol{X}_{t+1} = \begin{pmatrix} 0 & 1 & 0 & \cdots & 0 \\ 0 & 0 & 1 & \cdots & 0 \\ \vdots & \vdots & \vdots & \ddots & 0 \\ 0 & 0 & 0 & \cdots & 1 \\ \varphi_r & \varphi_{r-1} & \varphi_{r-2} & \cdots & \varphi_1 \end{pmatrix} \boldsymbol{X}_t + \begin{pmatrix} 0 \\ 0 \\ \vdots \\ 0 \\ 1 \end{pmatrix} Z_{t+1} \quad (t \in \mathbb{Z}).$$

We extend ARMA models to ARIMA models which are nonstationary time series and are highly efficient in short-time forecasting. For $d = 0, 1, \dots,$ let

$$\tilde{Y}_n = (1 - B)^d Y_n = \sum_{k=0}^{d} (-1)^k \frac{d!}{k!(d-k)!} Y_{n-k}.$$

If $\{\tilde{Y}_n\}_{n\in\mathbb{Z}}$ is a causal ARMA (p, q) process, we say $\{Y_n\}_{n\in\mathbb{Z}}$ is an *ARIMA* (p, d, q) *process.*

Let $\{Y_t\}$ be an ARIMA model

$$(1 - \varphi B)(1 - B)Y_t = (1 + \theta B)Z_t,$$
$$\{Z_t\} \sim \mathrm{WN}(0, \sigma^2).$$

Then the observation equation is

$$Y_t = (\theta, 1, 1)(X_{t-1}, X_t, Y_{t-1})^{\mathrm{T}},$$

where

$$\begin{pmatrix} X_0 \\ X_1 \\ Y_0 \end{pmatrix} = \begin{pmatrix} \sum_0^\infty \varphi^j Z_{-j} \\ \sum_0^\infty \varphi^j Z_{1-j} \\ Y_0 \end{pmatrix},$$

$$\begin{pmatrix} X_t \\ X_{t+1} \\ Y_t \end{pmatrix} = \begin{pmatrix} 0 & 1 & 0 \\ 0 & \varphi & 0 \\ \theta & 1 & 1 \end{pmatrix} \begin{pmatrix} X_{t-1} \\ X_t \\ Y_{t-1} \end{pmatrix} + \begin{pmatrix} 0 \\ Z_{t+1} \\ 0 \end{pmatrix} \quad (t \in \mathbb{Z}_+).$$

For a general ARIMA process, a state-space representation can be found.

Consider a randomly varying trend with noise. Let Y_1 be a random variable and

$$\{V_t\} \sim WN(0, \sigma^2),$$
$$E[V_t Y_1] = 0 \quad (t \in \mathbb{Z}_+).$$

Define a process $\{Y_t\}$ as

$$Y_{t+1} = Y_t + \alpha + V_t = Y_1 + \alpha t + V_1 + \cdots + V_t,$$

where α is a constant. Let $\boldsymbol{X}_t = (Y_t, \alpha)^{\mathrm{T}}$ and $\boldsymbol{X}_t$ satisfy the state equation

$$\boldsymbol{X}_{t+1} = \begin{pmatrix} 1 & 1 \\ 0 & 1 \end{pmatrix} \boldsymbol{X}_t + \boldsymbol{V}_t \quad (t \in \mathbb{Z}),$$

where $\boldsymbol{V}_t = (V_t, 0)^{\mathrm{T}}$. The process $\{Y_t\}$ can be determined by the observation equation $Y_t = (1, 0)\boldsymbol{X}_t$.

1.7.2 Kalman prediction

Let $\{\boldsymbol{Y}_t\}_{t\in\mathbb{Z}}$ be a state-space model whose observation equation and state equation are stated in (1.7.1) and (1.7.2), respectively. We want to find the best linear prediction of the state $\boldsymbol{X}_t$ by the observation $\boldsymbol{Y}_1, \dots, \boldsymbol{Y}_{t-1}$ and a random vector $\boldsymbol{Y}_0$ satisfying $\boldsymbol{Y}_0 \perp \boldsymbol{V}_t$ and $\boldsymbol{Y}_0 \perp \boldsymbol{W}_t$ for all t. In many cases, choose $\boldsymbol{Y}_0 = (1, 1, \dots, 1)^{\mathrm{T}}$.

Let $\boldsymbol{X}_t = (X_{t1}, \dots, X_{tn})^{\mathrm{T}}$ be an n-dimensional random vector. The *best one-step linear predictor* $\boldsymbol{X}_t^o = (X_{t1}^o, \dots, X_{tn}^o)^{\mathrm{T}}$ of $\boldsymbol{X}_t$ in terms of t vectors

$$\boldsymbol{Y}_0 = (Y_{01}, \dots, Y_{0n}), \dots, \boldsymbol{Y}_{t-1} = (Y_{t-1,1}, \dots, Y_{t-1,n})$$

is defined as each X_{tk}^o is the best linear predictor of X_{tk} in terms of all the components

$$Y_{01}, \dots, Y_{0n}, Y_{11}, \dots, Y_{1n}, \dots, Y_{t-1,1}, \dots, Y_{t-1,n}$$

of t vectors $\boldsymbol{Y}_0, \dots, \boldsymbol{Y}_{t-1}$. By the orthogonality principle, we get

$$X_{tk}^o = \sum_{l=1}^{n} \alpha_{kl}^{(0)} Y_{0l} + \sum_{l=1}^{n} \alpha_{kl}^{(1)} Y_{1l} + \cdots + \sum_{l=1}^{n} \alpha_{kl}^{(t-1)} Y_{t-1,l} \quad (k = 1, \dots, n)$$

and $E[(X_{tk} - X_{tk}^o) Y_{\mu\nu}] = 0$.

Kalman prediction

For the state-space models (1.7.1) and (1.7.2), denote by $\boldsymbol{X}_t^o$ the best one-step linear predictor of $\boldsymbol{X}_t$ in terms of t observation vectors $\boldsymbol{Y}_0, \dots, \boldsymbol{Y}_{t-1}$ and denote by Ω_t the covariance matrix $E[(\boldsymbol{X}_t - \boldsymbol{X}_t^o)(\boldsymbol{X}_t - \boldsymbol{X}_t^o)^{\mathrm{T}}]$ of the error $\boldsymbol{X}_t - \boldsymbol{X}_t^o$. Then the linear predictor $\boldsymbol{X}_t$ is determined uniquely by initial conditions $\boldsymbol{X}_1^o$ and $\Omega_1 = E[(\boldsymbol{X}_1 - \boldsymbol{X}_1^o)(\boldsymbol{X}_1 - \boldsymbol{X}_1^o)^{\mathrm{T}}]$, and

$$\begin{aligned} \boldsymbol{X}_{t+1}^o &= L_t \boldsymbol{X}_t^o + \Theta_t \Delta_t^{-1} (\boldsymbol{Y}_t - K_t \boldsymbol{X}_t^o) \quad (t \in \mathbb{Z}_+), \\ \Omega_{t+1} &= L_t \Omega_t L_t^{\mathrm{T}} + Q_t - \Theta_t \Delta_t^{-1} \Theta_t^{\mathrm{T}} \quad (t \in \mathbb{Z}_+), \end{aligned}$$

where $\Theta_t = L_t \Omega_t K_t^{\mathrm{T}}$ and $\Delta_t = K_t \Omega_t K_t^{\mathrm{T}} + R_t$, and Δ_t^{-1} is the inverse of Δ_t. Here matrixes L_t, K_t, R_t, and Q_t are stated in (1.7.1), (1.7.2), and (1.7.3).

1.7.3 Kalman filtering and Kalman fixed point smoothing

For a state-space model, we find the estimator of the state vector $\boldsymbol{X}_t$ in terms of $\boldsymbol{Y}_0, \dots, \boldsymbol{Y}_t$. This estimator is called the *Kalman filter*, denoted by $\widetilde{\boldsymbol{X}}_t$.

Kalman filtering

Let $\boldsymbol{X}_t^o$, Ω_t, and Δ_t^{-1} be stated as in the Kalman prediction. Then

(a) the filtered estimate $\widetilde{\boldsymbol{X}}_t$ is

$$\boldsymbol{X}_t^o + \Omega_t K_t^{\mathrm{T}} \Delta_t^{-1} (\boldsymbol{Y}_t - L_t \boldsymbol{X}_t^o);$$

(b) the error covariance matrix is

$$E[(\boldsymbol{X}_t - \widetilde{\boldsymbol{X}}_t)(\boldsymbol{X}_t - \widetilde{\boldsymbol{X}}_t)^{\mathrm{T}}] = \Omega_t - \Omega_t K_t^{\mathrm{T}} \Delta_t^{-1} \Omega_t^{\mathrm{T}}.$$

For a state-space model, we find the estimator of the state vector $\boldsymbol{X}_t$ in terms of $\boldsymbol{Y}_0, \ldots, \boldsymbol{Y}_t, \ldots, \boldsymbol{Y}_s$. This estimator is called the *Kalman smoothing*, denoted by $\boldsymbol{X}_t^s$.

Kalman smoothing

Let $\boldsymbol{X}_t^o$ and Δ_t^{-1} be stated as in the Kalman prediction. Then

(a) the smoothed estimate $\boldsymbol{X}_t^s$ is determined by initial conditions $\boldsymbol{X}_t^o$ and $\Omega_{tt} = \Omega_t$, and

$$\boldsymbol{X}_t^s = \boldsymbol{X}_t^{s-1} + \Omega_{ts} G_s^{\mathrm{T}} \Delta_s^{-1} (\boldsymbol{Y}_s - G_s \boldsymbol{X}_s^o) \quad (s = t, t+1, \ldots),$$

where $\Omega_{t,s+1} = \Omega_{ts}(L_s - \Theta_s \Delta_s^{-1} K_s)^{\mathrm{T}}$;

(b) the error covariance matrix $E[(\boldsymbol{X}_t - \boldsymbol{X}_t^s)(\boldsymbol{X}_t - \boldsymbol{X}_t^s)^{\mathrm{T}}] = \Omega_t^s$ is determined by the recursive formula

$$\Omega_t^s = \Omega_t^{s-1} - \Omega_{ts} K_s^{\mathrm{T}} \Delta_s^{-1} K_s \Omega_{ts}^{\mathrm{T}}.$$

Further reading

[1] Bao C, Hao H, Li ZX. Integrated ARMA model method for damage detection of subsea pipeline system. *Engineering Structures*. 2013(48):176–192.

[2] Boularouk Y, Djeddour K. New approximation for ARMA parameters estimate. *Mathematics and Computers in Simulation*. 2015(118):116–122.

[3] David M, Ramahatana F, Trombe PJ, Lauret P. Probabilistic forecasting of the solar irradiance with recursive ARMA and GARCH models. *Solar Energy*. 2016(133):55–72.

[4] Flores JJ, Graff M, Rodriguez H. Evolutive design of ARMA and ANN models for time series forecasting. *Renewable Energy*. 2012(44):225–230.

[5] Galiana-Merino JJ, Pla C, Fernandez-Cortes A, Cuezva S, Ortiz J, Benavente D. Environmental wavelet tool: Continuous and discrete wavelet analysis and filtering for environmental time series. *Computer Physics Communications*. 2014(185):2758–2770.

[6] Ip RHL, Li WK, Leung KMY. Seemingly unrelated intervention time series models for effectiveness evaluation of large scale environmental remediation. *Marine Pollution Bulletin*. 2013(74): 56–65.

[7] Liu J, Deng Z. Information fusion Kalman predictor for two-sensor multichannel ARMA signal system with time-delayed measurements. *Procedia Engineering*. 2012(29):623–629.

[8] Liu Y, Wu J, Liu Y, Hu BX, Hao Y, Huo X, Fan Y, Yeh TJ, Wang ZL. Analyzing effects of climate change on streamflow in a glacier mountain catchment using an ARMA model. *Quaternary International*. 2015(358):137–145.

[9] Macciotta NPP, Vicario D, Pulina G, Cappio-Borlino A. Test day and lactation yield predictions in Italian Simmental cows by ARMA methods. *Journal of Dairy Science*. 2002(85):3107–3114.
[10] Kadri F, Harrou F, Chaabane S, Sun Y, Tahon C. Seasonal ARMA-based SPC charts for anomaly detection: Application to emergency department systems. *Neurocomputing*. 2016(173): 2102–2114.
[11] Kapetanios G. A note on an iterative least-squares estimation method for ARMA and VARMA models. *Economics Letters*. 2003(79):305–312.
[12] Krasnov H, Katra I, Friger M. Increase in dust storm related PM10 concentrations: A time series analysis of 2001–2015. *Environmental Pollution*. 2016(213):36–42.
[13] Piston N, Schob C, Armas C, Prieto I, Pugnaire FI. Contribution of co-occurring shrub species to community richness and phylogenetic diversity along an environmental gradient. *Perspectives in Plant Ecology, Evolution and Systematics*. 2016(19):30–39.
[14] Ran C, Deng Z. Self-tuning distributed measurement fusion Kalman estimator for the multichannel ARMA signal. *Signal Processing*. 2011(91):2028–2041.
[15] Soni K, Parmar KS, Kapoor S, Kumar N. Statistical variability comparison in MODIS and AERONET derived aerosol optical depth over Indo-Gangetic Plains using time series modeling. *Science of the Total Environment*. 2016(553):258–265.
[16] Takemura A. Exponential decay rate of partial autocorrelation coefficients of ARMA and short-memory processes. *Statistics & Probability Letters*. 2016(110):207–210.
[17] Tulbure MG, Broich M, Stehman SV, Kommareddy A. Surface water extent dynamics from three decades of seasonally continuous Landsat time series at subcontinental scale in a semi-arid region. *Remote Sensing of Environment*. 2016(178):142–157.
[18] Valipour M, Banihabib ME, Behbahani SMR. Comparison of the ARMA, ARIMA, and the autoregressive artificial neural network models in forecasting the monthly inflow of Dez Dam Reservoir. *Journal of Hydrology*. 2013(476):433–441.
[19] Voyant C, Muselli M, Paoli C, Nivet ML. Numerical weather prediction (NWP) and hybrid ARMA/ANN model to predict global radiation. *Energy*. 2012(39):341–355.
[20] Xu W, Gu R, Liu Y, Dai Y. Forecasting energy consumption using a new GM-ARMA model based on HP filter: The case of Guangdong Province of China. *Economic Modelling*. 2015(45):127–135.
[21] Zhang P, Qi W, Deng Z. Multi-channel ARMA signal covariance intersection fusion Kalman predictor. *Procedia Engineering*. 2012(29):609–615.

2 Chaos and dynamical systems

Complex dynamical systems exist widely in nature. In 1961, Edward Lorenz ran a numerical computer model to make a weather prediction. When he entered the initial condition 0.506 instead of 0.506127, the result was a completely different weather scenario. Later on, when Lorenz presented this result at the 139th meeting of the American Association for the Advancement of Science in 1972, Philip Merilees concocted "*Does the flap of a butterfly's wings in Brazil set off a tornado in Texas*" as the title of Lorenz's talk. This is the so-called *butterfly effect*. In general, for any complex dynamical system in environmental science, its trajectory often sensitively depends on initial conditions. If two initial conditions have a small difference and their trajectories after some time are exponential separation, such a system is called a *chaotic dynamical system*. If a set of initial conditions, after some time, is attracted to some subset which is invariant under dynamical evolution, such an invariant subset is called an *attractor* of the system. Attractors of chaotic dynamical systems often exhibit an unusual kind of self-similarity and have fractal dimension. The sensitivity dependence on initial conditions and the fractal dimension of attractors are two of the most important characteristics of dynamical systems. In environmental science, the equations underlying a dynamical system are often unknown. Using the delay-coordinate embedding technique, the dynamical system can be reconstructed from the trajectories which originate from a dynamical system.

2.1 Dynamical systems

In theory, a *dynamical system* is defined as a first-order differential equation acting on a phase space $\mathbb{R}^m$

$$\frac{\mathrm{d}}{\mathrm{d}t}\boldsymbol{x}(t) = \boldsymbol{F}(t, \boldsymbol{x}(t)) \quad (t \in \mathbb{R}), \tag{2.1.1}$$

where $\boldsymbol{x} = (x^{(1)}, \dots, x^{(m)})$ and $\boldsymbol{F} = (F_1, \dots, F_m)$. In the discrete case, define a *dynamical system* as an m-dimensional map

$$\boldsymbol{x}_{n+1} = \boldsymbol{F}(\boldsymbol{x}_n) \quad (n \in \mathbb{Z}), \tag{2.1.2}$$

where $\boldsymbol{x}_k = (x_k^{(1)}, \dots, x_k^{(m)})$. If $\boldsymbol{F}$ is not depend explicitly on t, i.e.,

$$\frac{\mathrm{d}}{\mathrm{d}t}\boldsymbol{x}(t) = \boldsymbol{F}(\boldsymbol{x}(t)) \quad (t \in \mathbb{R}), \tag{2.1.3}$$

then the system is called *autonomous*. If $\boldsymbol{F}$ satisfies the Lipschitz condition

$$\|\boldsymbol{F}(\boldsymbol{x}) - \boldsymbol{F}(\boldsymbol{x}')\| \le K\,\|\boldsymbol{x} - \boldsymbol{x}'\|^{\alpha} \quad (0 < \alpha \le 1),$$

DOI 10.1515/9783110424904-003

where $\boldsymbol{x}, \boldsymbol{x}' \in \mathbb{R}^m$, $K > 0$ is a constant, and $\|\cdot\|$ is the norm of $\mathbb{R}^m$, then the initial value problem

$$\begin{cases} \frac{\mathrm{d}}{\mathrm{d}t}\boldsymbol{x}(t) = \boldsymbol{F}(\boldsymbol{x}(t)) \quad (t \in \mathbb{R}), \\ \boldsymbol{x}(0) = \boldsymbol{a} \end{cases}$$

has a unique solution. Let $\boldsymbol{F} = (F_1, \ldots, F_m)$. Suppose that divergence of $\boldsymbol{F}$ in (2.1.3) is less than zero,

$$\operatorname{div} \boldsymbol{F} = \nabla \cdot \boldsymbol{F} = \frac{\partial F_1}{\partial x_1} + \frac{\partial F_2}{\partial x_2} + \cdots + \frac{\partial F_m}{\partial x_m} < 0.$$

Then the system is called a *dissipative system*. In this case, a set of initial conditions is contracted under the dynamical evolution. This set is attracted to some invariant subset which is called an *attractor* of the system.

For the discrete dynamical system (2.1.2), consider the Jacobian matrix of $\boldsymbol{F}$,

$$J_F = \left(\frac{\partial F_i}{\partial x^{(j)}}\right)_{m\times m} = \begin{pmatrix} \frac{\partial F_1}{\partial x^{(1)}} & \cdots & \frac{\partial F_1}{\partial x^{(m)}} \\ \vdots & \ddots & \vdots \\ \frac{\partial F_m}{\partial x^{(1)}} & \cdots & \frac{\partial F_m}{\partial x^{(m)}} \end{pmatrix},$$

if the absolute value of the determinant of J_F satisfies

$$|\det J_F| < 1,$$

then a set of initial conditions is contracted to an invariant subset which is called an attractor.

2.2 Henon and logistic maps

Simple dynamical systems can exhibit a completely unpredictable behavior. Here we give some examples, including the famous Henon and logistic maps.

2.2.1 Circular motion

For a dynamical system

$$\frac{\mathrm{d}\boldsymbol{x}_1}{\mathrm{d}t} = -\omega \boldsymbol{x}_2, \quad \frac{\mathrm{d}\boldsymbol{x}_2}{\mathrm{d}t} = \omega \boldsymbol{x}_1, \tag{2.2.1}$$

where ω is a constant, we have $\frac{\mathrm{d}^2\boldsymbol{x}_1}{\mathrm{d}t^2} = -\omega\frac{\mathrm{d}\boldsymbol{x}_2}{\mathrm{d}t} = -\omega^2\boldsymbol{x}_1$ or $\frac{\mathrm{d}^2\boldsymbol{x}_1}{\mathrm{d}t^2} + \omega^2\boldsymbol{x}_1 = 0$. The solution of this second-order differential equation

$$\begin{aligned} \boldsymbol{x}_1 &= a\cos\omega(t - t_0), \\ \boldsymbol{x}_2 &= a\sin\omega(t - t_0) \end{aligned}$$

is a trajectory of the dynamical system (2.2.1), where $a, t_0 \in \mathbb{R}$. Rewrite (2.2.1) into the vector form

$$\frac{\mathrm{d}\boldsymbol{x}}{\mathrm{d}t} = \boldsymbol{F}(\boldsymbol{x}) = A\boldsymbol{x},$$

where $\boldsymbol{x} = (x_1, x_2)^{\mathrm{T}}$ and $\boldsymbol{F} = (F_1, F_2) = (-\omega x_2, \omega x_1)$ and $A = \left(\begin{smallmatrix} 0 & -\omega \\ \omega & 0 \end{smallmatrix}\right)$. The divergence of $\boldsymbol{F}$ is

$$\nabla \cdot \boldsymbol{F} = \frac{\partial F_1}{\partial x_1} + \frac{\partial F_2}{\partial x_2} = \frac{\partial(-\omega x_2)}{\partial x_1} + \frac{\partial(\omega x_1)}{\partial x_2} = 0.$$

So the system has bounded solutions and is not dissipative.

2.2.2 Henon map

For the Henon map $\boldsymbol{x}_{n+1} = \boldsymbol{F}(\boldsymbol{x}_n)$ $(n \in \mathbb{Z})$, where $\boldsymbol{x}_n = (x_n^{(1)}, x_n^{(2)})$, $\boldsymbol{F} = (F_1, F_2)^{\mathrm{T}}$, and

$$F_1(\boldsymbol{x}_n) = a - (x_n^{(1)})^2 + b x_n^{(2)},$$
$$F_2(\boldsymbol{x}_n) = x_n^{(1)},$$

and a, b are constants, its Jacobian matrix is

$$J_{\boldsymbol{F}} = \begin{pmatrix} \frac{\partial F_1}{\partial x^{(1)}} & \frac{\partial F_1}{\partial x^{(2)}} \\ \frac{\partial F_2}{\partial x^{(1)}} & \frac{\partial F_2}{\partial x^{(2)}} \end{pmatrix} = \begin{pmatrix} -2x^{(1)} & b \\ 1 & 0 \end{pmatrix}$$

and the determinant $|\det J_{\boldsymbol{F}}| = |b|$. If $|b| < 1$, then the system is dissipative.

2.2.3 Fixed point

Let $\boldsymbol{x}_{n+1} = \boldsymbol{F}(\boldsymbol{x}_n)$, where $\boldsymbol{F}\colon \mathbb{R}^m \to \mathbb{R}^m$ is a map satisfying $\|\boldsymbol{F}(s) - \boldsymbol{F}(t)\| \le q\|s - t\|$ $(0 \le q < 1)$. Here $\|\cdot\|$ is the norm of the space $\mathbb{R}^m$. For any $\boldsymbol{x}_1 \in \mathbb{R}^m$,

$$\begin{cases} \boldsymbol{x}_2 = \boldsymbol{F}(\boldsymbol{x}_1), \\ \vdots \\ \boldsymbol{x}_{n+1} = \boldsymbol{F}(\boldsymbol{x}_n), \end{cases}$$

and so

$$\|\boldsymbol{x}_{n+1} - \boldsymbol{x}_n\| = \|\boldsymbol{F}(\boldsymbol{x}_n) - \boldsymbol{F}(\boldsymbol{x}_{n+1})\| \le q\,\|\boldsymbol{x}_n - \boldsymbol{x}_{n-1}\| \le \cdots \le q^{n-1}\|\boldsymbol{x}_2 - \boldsymbol{x}_1\|.$$

This implies that $\{\boldsymbol{x}_n\}$ is convergent. In fact,

$$\begin{aligned} \|\boldsymbol{x}_{n+p} - \boldsymbol{x}_n\| &\le \|\boldsymbol{x}_{n+p} - \boldsymbol{x}_{n+p-1}\| + \cdots + \|\boldsymbol{x}_{n+1} - \boldsymbol{x}_n\| \\ &\le (q^{n+p-2} + \cdots + q^{n-1})\|\boldsymbol{x}_2 - \boldsymbol{x}_1\| \\ &\le \frac{q^n}{1-q}\|\boldsymbol{x}_2 - \boldsymbol{x}_1\| \to 0 \quad (n \to \infty). \end{aligned}$$

Thus, there is a point $\boldsymbol{x}^* \in \mathbb{R}^m$ such that $\boldsymbol{x}_n \to \boldsymbol{x}^*$ $(n \to \infty)$.

Finally, we prove the uniqueness of $\boldsymbol{x}^*$, i.e., $\boldsymbol{x}^*$ is independent of the choice of the initial value.

Since $\boldsymbol{F}$ is continuous, clearly, from $\boldsymbol{x}_{n+1} = \boldsymbol{F}(\boldsymbol{x}_n)$ and $\boldsymbol{x}_n \to \boldsymbol{x}^*$, it follows that $\boldsymbol{x}^* = \boldsymbol{F}(\boldsymbol{x}^*)$. If there is an $\bar{\boldsymbol{x}} \neq \boldsymbol{x}^*$ such that $\bar{\boldsymbol{x}} = \boldsymbol{F}(\bar{\boldsymbol{x}})$, then

$$0 < \|\boldsymbol{x}^* - \bar{\boldsymbol{x}}\| = \|\boldsymbol{F}(\boldsymbol{x}^*) - \boldsymbol{F}(\bar{\boldsymbol{x}})\| \le q\,\|\boldsymbol{x}^* - \bar{\boldsymbol{x}}\| < \|\boldsymbol{x}^* - \bar{\boldsymbol{x}}\|.$$

This is a contradiction. So $\boldsymbol{x}^* = \bar{x}$.

From this, it is seen that sometimes the solution of a dynamical equation tends to a point which is independent of the choice of the initial value. This point is just an attractor of this system.

2.2.4 Linear mapping

Consider the dynamical system

$$\boldsymbol{x}_{n+1} = \alpha \boldsymbol{x}_n + \beta \quad (n \in \mathbb{Z}), \tag{2.2.2}$$

where $\boldsymbol{x}_n, \alpha, \beta \in \mathbb{R}$ and α, β are constants. Let $\boldsymbol{x}_n = \gamma \boldsymbol{y}_n + \delta$. Then (2.2.2) becomes

$$\gamma \boldsymbol{y}_{n+1} + \delta = \alpha \gamma \boldsymbol{y}_n + \alpha\delta + \beta.$$

Choose δ such that $\delta = \alpha\delta + \beta$, i.e., $\delta = \frac{\beta}{1-\alpha}$. So

$$\boldsymbol{y}_{n+1} = \alpha \boldsymbol{y}_n \quad (n \in \mathbb{Z}), \tag{2.2.3}$$

i.e., $\boldsymbol{y}_{n+1} = \boldsymbol{F}(\boldsymbol{y}_n)$, where $\boldsymbol{F}(\boldsymbol{y}) = \alpha \boldsymbol{y}$.

By (2.2.3), $\boldsymbol{y}_n = \boldsymbol{y}_0 \alpha^n$. If $|\alpha| < 1$, then $\{\boldsymbol{y}_n\}$ is exponentially decaying wherever $\boldsymbol{y}_0$ is any number. Clearly, $\boldsymbol{y}_n \to 0$ $(n \to \infty)$. Thus, the trajectory is attracted to the fixed point $\boldsymbol{y}_0 = 0$, i.e., $\boldsymbol{y}_0$ is an attractor. If $|\alpha| > 1$, then $\{\boldsymbol{y}_n\}$ is exponentially increasing. If $\alpha = -1$, then it is an oscillatory system between two values.

For a dynamical system $\boldsymbol{x}_{n+1} = \boldsymbol{F}(\boldsymbol{x}_n)$ $(n \in \mathbb{Z})$, its fixed points are the solutions of $\boldsymbol{x} = \boldsymbol{F}(\boldsymbol{x})$. Start from a point $\boldsymbol{x}_0$ in a small neighborhood of a fixed point $\boldsymbol{x}^*$, define a small number ε_n such that $\boldsymbol{x}_n = \boldsymbol{x}^* + \varepsilon_n$. So

$$\boldsymbol{x}^* + \varepsilon_{n+1} = \boldsymbol{x}_{n+1} = \boldsymbol{F}(\boldsymbol{x}^* + \varepsilon_n) \approx \boldsymbol{F}(\boldsymbol{x}^*) + \boldsymbol{F}'(\boldsymbol{x}^*)\varepsilon_n.$$

Note that $\boldsymbol{x}^* = \boldsymbol{F}(\boldsymbol{x}^*)$. Then

$$\varepsilon_{n+1} = \boldsymbol{F}'(\boldsymbol{x}^*)\varepsilon_n. \tag{2.2.4}$$

Furthermore, when $|\boldsymbol{F}'(\boldsymbol{x}^*)| < 1$, $\boldsymbol{x}_n \to \boldsymbol{x}^*$ $(n \to \infty)$. This means that $\boldsymbol{x}^*$ is an attractor.

2.2.5 Logistic map

The logistic map is $x_{n+1} = F(x_n) = Ax_n(1 - x_n)$, where A is a constant.

Consider a fixed point $x_1^* = 0$. Then

$$F'(x_1^*) = A(1 - 2x_1^*) = A,$$

and so $|F'(x_1^*)| < 1$ if and only if $|A| < 1$. This means that $x_1^* = 0$ is an attractor when $-1 < A < 1$. For another fixed point $x_2^* = 1 - \frac{1}{A}$,

$$F'(x_2^*) = A(1 - 2(1 - A^{-1})) = -A + 2,$$

and so $|F'(x_2^*)| < 1$ if and only if $1 < A < 3$. This means that $x_2^* = 1 - A^{-1}$ is an attractor when $1 < A < 3$.

Now consider the case $A = 3$. When $A = 3$, $F'(x_2^*) = -1$. By (2.2.4),

$$\begin{aligned} \varepsilon_{n+1} &= -\varepsilon_n, \\ \varepsilon_{n+2} &= \varepsilon_n \quad (n \in \mathbb{Z}), \end{aligned}$$

i.e., $x_{n+2} = x_n$. So $\{x_n\}$ is an oscillatory trajectory between two values under a small perturbation.

To obtain such an oscillatory trajectory, consider the map $T^2(x) := F(F(x))$ and find its fixed point through the system of equations

$$\begin{aligned} y_2 &= F(y_1) = Ay_1(1 - y_1), \\ y_1 &= F(y_2) = Ay_2(1 - y_2). \end{aligned} \tag{2.2.5}$$

Its solution is

$$\begin{aligned} y_1^* &= \frac{A + 1 + \sqrt{(A + 1)(A - 3)}}{2A}, \\ y_2^* &= \frac{A + 1 - \sqrt{(A + 1)(A - 3)}}{2A}. \end{aligned} \tag{2.2.6}$$

Clearly, if $A \geq 3$, the solution exists. Then y_1^* and y_2^* are fixed points of the map F^2. From this and (2.2.5), it follows that

$$(F^2)'(y_1^*) = F'(F(y_1^*))F'(y_1^*) = F'(y_2^*)F'(y_1^*) = A^2(1 - 2y_1^*)(1 - 2y_2^*).$$

By (2.2.6),

$$\begin{aligned} y_1^* + y_2^* &= 1 + A^{-1}, \\ y_1^* y_2^* &= \frac{1}{4A^2}((A + 1)^2 - (A + 1)(A - 3)) = \frac{A + 1}{A^2}. \end{aligned}$$

So $(F^2)'(y_1^*) = 1 - (A + 1)(A - 3)$.

Similarly, $(F^2)'(y_2^*) = (F^2)'(y_1^*) = 1 - (A + 1)(A - 3)$.

From

$$\begin{aligned} (F^2)'(y_1^*) &= (F^2)'(y_2^*) = 1 \quad &&\text{when } A = 3, \\ (F^2)'(y_1^*) &= (F^2)'(y_2^*) = -1 \quad &&\text{when } A = 1 + \sqrt{6}, \end{aligned}$$

similar to the case $A = 3$, it is easily shown that when $A = 1 + \sqrt{6}$, the trajectories are oscillatory among four values.

Therefore, when A increases from 3, the trajectories are oscillatory among two values, then four values, eight values, ... When A approximates the value 3.56995, periods of oscillations tend to infinity. When the parameter $A > 3.56995$, the system begins to exhibit chaotic behavior and have the following characteristics:
- sensitively depend on initial conditions;
- nonperiodicity;
- the attractors have fractal structures (i.e. strange attractors).

The interval $[3.56995, 4]$ for the parameter A is often called the *chaotic domain.*

2.3 Lyapunov exponents

For a dynamical system, if two initial conditions have a small difference δx and their difference after time t becomes $\delta x e^{\lambda t}$, then λ is called the Lyapunov exponent. Numerically, one can calculate the maximal Lyapunov exponent $\lambda_{\max}$ as follows. Choose two very close initial points and let their distance be $d_0 \ll 1$. After time t_i, their difference will become d_i $(i \in \mathbb{Z}_+)$. Then the maximal Lyapunov exponent can be computed by

$$\lambda_{\max} = \lim_{N\to\infty} \frac{\sum_1^N \log \frac{d_i}{d_0}}{\sum_1^N t_i}.$$

Negative Lyapunov exponents are characteristic of dissipative or nonconservative systems. Such systems exhibit asymptotic stability; the more negative the exponent, the greater the stability. A positive Lyapunov exponent is characteristic of chaotic systems. Nearby points, no matter how close, will diverge to any arbitrary separation. In a dynamical system there exist many different Lyapunov exponents, but the sum of all Lyapunov exponents cannot be positive in a physically meaningful system.

Let x and y be two close trajectories for a dynamical system with dimension m

$$\mathbf{x}_{n+1} = F(\mathbf{x}_n), \quad \mathbf{y}_{n+1} = F(\mathbf{y}_n).$$

Then

$$\mathbf{y}_{n+1} - \mathbf{x}_{n+1} = F(\mathbf{y}_n) - F(\mathbf{x}_n) = J_n(\mathbf{x}_n)(\mathbf{y}_n - \mathbf{x}_n) + O(\|\mathbf{y}_n - \mathbf{x}_n\|^2),$$

where $J_n(\mathbf{x}_n)$ is the $m \times m$ Jacobian matrix of F at $\mathbf{x}_n$. Let $\delta_n = \mathbf{y}_n - \mathbf{x}_n$. It is rewritten into $\delta_{n+1} = J_n \delta_n$ $(n \in \mathbb{Z}_+)$, where $J_n = J_n(\mathbf{x}_n)$. This implies that

$$\delta_{N+1} = J_N \delta_N = J_N J_{N-1} \delta_{N-1} = \cdots = U_N \delta_1.$$

where $U_N = \prod_{n=1}^N J_n$.

In order to compute Lyapunov exponents, we need to find δ_1 such that $\|\delta_{n+1}\|$ attains the maximal value under the condition $\|\delta_1\| = 1$ (i.e., $\delta_1^{\mathrm{T}}\delta_1 = 1$) by using the Lagrange multipliers method. Introduce a function

$$H(u) = (U_N u)^{\mathrm{T}}(U_N u) - \lambda(u^{\mathrm{T}}u - 1).$$

Note that

$$(U_N u)^{\mathrm{T}}(U_N u) = u^{\mathrm{T}}U_N^{\mathrm{T}}U_N u = u^{\mathrm{T}}G_N u,$$

where $G_N = U_N^{\mathrm{T}}U_N$ is an $m \times m$ real symmetric positive definite matrix and u is an m-dimensional vector and $u^{\mathrm{T}}G_N u$ is a quadratic form of u. So

$$H(u) = u^{\mathrm{T}}G_N u - \lambda(u^{\mathrm{T}}u - 1).$$

Let $G_N = (\alpha_{ij})_{m\times m}$ and $u = (u_1, \dots, u_m)$. Then

$$H(u) = \sum_{i=1}^{m}\sum_{j=1}^{m}\alpha_{ij}u_i u_j - \lambda\left(\sum_{1}^{m}u_i^2 - 1\right).$$

This implies that

$$\frac{\partial H}{\partial u_l} = 2\sum_{i=1}^{n}\alpha_{il}u_i - 2\lambda u_l \quad (l = 1, \dots, m).$$

Since $\frac{\partial H}{\partial u_l} = 0\,(l = 1, \dots, m)$ and $\|u_l\| = 1$, when λ is the eigenvalue of the matrix G_N and u is the corresponding unit eigenvector, $H(u)$ attains the minimal value. Denote its eigenvalues by $\lambda_1^{(N)} \ge \lambda_2^{(N)} \cdots \ge \lambda_m^{(N)}$. The corresponding unit eigenvectors $e_1^{(N)}, \dots, e_m^{(N)}$ satisfy $G_N e_i^{(N)} = \lambda_i^{(N)} e_i^{(N)}$. Then the Lyapunov exponents are

$$\lambda_i = \lim_{N\to\infty}\frac{1}{2N}\log|\lambda_i^{(N)}| \quad (i = 1, \dots, m).$$

In the one-dimensional case, the Jacobian matrix $F'(x_n)$ on x_n is a real number. The corresponding Lyapunov exponent is

$$\lambda = \lim_{N\to\infty}\frac{1}{N}\sum_{1}^{N}\log|F'(x_n)|.$$

2.4 Fractal dimension

From an intuitive notion of dimension, a set of finite points is zero-dimensional, a straight line is one-dimensional, a plane is two-dimensional, and a cube is three-dimensional. One often determines the dimension by degrees of freedom. For example, the equation with a parameter

$$x = x(t), \quad y = y(t) \qquad (0 \le t \le 1),$$

where $x(t)$ and $y(t)$ are continuous functions, is considered to be a one-dimensional continuous curve since it only has one degree of freedom. The equation

$$x = x(u, v), \quad y = y(u, v) \qquad (u, v \in [0, 1]^2),$$

where $x(u, v)$ and $y(u, v)$ are continuous functions, is considered to be a two-dimensional continuous surface.

However, the approach in which one uses degrees of freedom to determine dimension is not precise. In 1890, Peano constructed a continuous curve

$$x = x_p(t), \quad y = y_p(t) \qquad (0 \le t \le 1),$$

such that its trajectory fills the whole unit square $[0, 1]^2$, i.e., its dimension is two.

Peano's curve is constructed as follows.

First, the segment $S = [0, 1]$ is divided equally into four small segments

$$S_1^{(1)} = \left[0, \tfrac{1}{4}\right], \quad S_2^{(1)} = \left[\tfrac{1}{4}, \tfrac{1}{2}\right],$$
$$S_3^{(1)} = \left[\tfrac{1}{2}, \tfrac{3}{4}\right], \quad S_4^{(1)} = \left[\tfrac{3}{4}, 1\right].$$

Correspondingly, the unit square $Q = [0, 1]^2$ is equally divided into four small squares

$$Q_1^{(1)} = \left[0, \tfrac{1}{2}\right] \times \left[0, \tfrac{1}{2}\right], \quad Q_2^{(1)} = \left[\tfrac{1}{2}, 1\right] \times \left[0, \tfrac{1}{2}\right],$$
$$Q_3^{(1)} = \left[0, \tfrac{1}{2}\right] \times \left[\tfrac{1}{2}, 1\right], \quad Q_4^{(1)} = \left[\tfrac{1}{2}, 1\right] \times \left[\tfrac{1}{2}, 1\right]$$

and each small segment $S_k^{(1)}$ $(k = 1, 2, 3, 4)$ is divided equally into four small segments

$$S_{4(k-1)+1}^{(2)}, \quad S_{4(k-1)+2}^{(2)}, \quad S_{4(k-1)+3}^{(2)}, \quad S_{4k}^{(2)} \qquad (k = 1, 2, 3, 4).$$

So the segment S is divided into 2^4 segments $S_1^{(2)}, S_2^{(2)}, \dots, S_{16}^{(2)}$. Correspondingly, each $Q_k^{(1)}$ $(k = 1, 2, 3, 4)$ is divided equally into four small squares. So the square Q is divided equally into 16 small squares.

Continuing this procedure to the n-th steps. Then the segment S is equally divided into 4^n small segments $S_1^{(n)}, S_2^{(n)}, \dots, S_{4^n}^{(n)}$ and the square Q is equally divided into 4^n small squares $Q_1^{(n)}, \dots, Q_{4^n}^{(n)}$ $(n = 1, 2, \dots)$.

Now we construct a continuous map from $[0, 1]$ to $[0, 1]^2$ one-to-one as follows.

For a point $a \in S$, there is a sequence of segments $S_{k_1}^{(1)}, \dots, S_{k_n}^{(n)}$ such that

$$S_{k_1}^{(1)} \supset S_{k_2}^{(2)} \supset \cdots \supset S_{k_n}^{(n)} \supset \cdots .$$

The sequence $\{S_{k_n}^{(n)}\}_{n \in \mathbb{Z}_+}$ converges to the point a. The corresponding small square $\{Q_{k_n}^{(n)}\}_{n \in \mathbb{Z}_+}$ converges to a unique point b as $n \to \infty$. Let a corresponds to b, i.e., the map φ from $[0, 1]$ into $[0, 1]^2$ is given by $\varphi(a) = b$.

Conversely, for any $b' \in [0, 1]^2$, there is a sequence $\{Q_{k'_n}^{(n)}\}_{n \in \mathbb{Z}_+}$ such that

$$Q_{k'_1}^{(1)} \supset Q_{k'_2}^{(2)} \supset \cdots \supset Q_{k'_n}^{(n)} \supset \cdots$$

and $\{Q_{k_n'}^{(n)}\}_{n\in\mathbb{Z}_+}$ converges to b'. The corresponding small segment $\{S_{k_n'}^{(n)}\}_{n\in\mathbb{Z}_+}$ converges to a unique point a', and so $\varphi(a') = b'$. This implies that $\varphi(a) = b$ is a one-to-one map from $[0, 1]$ to $[0, 1]^2$ and $\varphi([0, 1]) = [0, 1]^2$.

Finally, we prove that φ is a continuous map. Let $a_n \to a$ ($a_n \in S$), $b_n = \varphi(a_n)$, and $b = \varphi(a)$. We need only to prove that $b_n \to b$.

By the above construct, there are $k_1 < k_2 < \cdots$ such that

$$\begin{aligned} S_{k_1}^{(1)} \supset S_{k_2}^{(2)} \supset \cdots \supset S_{k_n}^{(n)} \supset \cdots, \\ Q_{k_1}^{(1)} \supset Q_{k_2}^{(2)} \supset \cdots \supset Q_{k_n}^{(n)} \supset \cdots, \end{aligned}$$

and $a \in S_{k_n}^{(n)}$, $b \in Q_{k_n}^{(n)}$ ($n \in \mathbb{Z}_+$). For $\varepsilon > 0$, choose $m \in \mathbb{Z}_+$ such that

$$\sqrt{2}\,(4^{-m}) < \varepsilon.$$

By $a_n \to a$, there is a $p \in \mathbb{Z}_+$ such that $a_n \in S_{k_m}^{(m)}$, $b_n \in Q_{k_m}^{(m)}$ ($n > p$). Note that $b \in Q_{k_m}^{(m)}$ and $Q_{k_m}^{(m)}$ is a square with the diameter $\sqrt{2}\,(4^{-m})$. Then

$$\|b_n - b\| \le \sqrt{2}\,(4^{-m}) < \varepsilon.$$

So $b_n \to b$. This implies that $\varphi(a) = b$ is a one-to-one continuous map from $[0, 1]$ to $[0, 1]^2$. Let

$$\begin{aligned} \varphi(a) &= (\varphi_1(a), \varphi_2(a)), \\ b &= (b_1, b_2). \end{aligned}$$

Then the map $b = \varphi(a)$ can be rewritten into

$$b_1 = \varphi_1(a), \quad b_2 = \varphi_2(a) \qquad (0 \le a \le 1),$$

where both φ_1 and φ_2 are continuous. So it is a continuous curve. But it is two-dimensional.

From this, it is seen that the notion of dimension needs to be described more precisely. There are several ways to define dimension.

2.4.1 Box-counting dimension

Cover a bounded data set A by boxes with diameter ε. Denote the minimal number of boxes by $N(\varepsilon)$. Then the box-counting dimension D_1 is defined by

$$D_1 = \lim_{\varepsilon\to 0} \frac{\log N(\varepsilon)}{\log \frac{1}{\varepsilon}}.$$

By this definition, the segment $[0, 1]$ is one-dimensional, the square $[0, 1]^2$ is two-dimensional, the cube $[0, 1]^3$ is three-dimensional. This definition coincides with the intuitive notion of dimension.

2.4.2 Information dimension

In the definition of box-counting dimension, it is not considered how many points of the set lie in each box. The information dimension overcomes this shortcoming and is defined as

$$D_2 = \lim_{\varepsilon \to 0} \frac{H(\varepsilon)}{\log \frac{1}{\varepsilon}},$$

where

$$H(\varepsilon) = -\sum_{1}^{N(\varepsilon)} P_i \log P_i$$

and P_i is the relative frequency that the point (in the set) occurs in the i-th box of the covering.

2.4.3 Correlation dimension

In the higher dimensional case, the computations of box-counting dimension and the information dimension are too complicated. For convenience, the correlation dimension is introduced. It is defined as

$$D_3 = \lim_{\varepsilon \to 0} \frac{\log C(\varepsilon)}{\log \varepsilon},$$

where $C(\varepsilon) = \frac{2}{N(N-1)} \sum_{i=1}^{N} \sum_{j=i+1}^{N} \Theta(\varepsilon - \|x_i - x_j\|)$ and Θ is the Heaviside step function

$$\Theta(x) = 0 \quad (x \le 0),$$
$$\Theta(x) = 1 \quad (x > 0).$$

2.4.4 Self-similarly dimension

A segment may be divided into 2 subsegments with similarity ratio $\frac{1}{2}$. A square may be divided into 4 subsquares with similarity ratio $\frac{1}{2}$. A cube may be divided into 8 subcubes with similarity ratio $\frac{1}{2}$. The numbers 2, 4, and 8 can be rewritten as 2^1, 2^2, and 2^3. Here the exponents 1, 2, and 3 coincide with their intuitive dimensions.

In the real world, many diagrams can be divided into b similar small diagrams with similarity ratio $\frac{1}{a}$, where a is a positive integer, and each small diagram can be divided into b similar smaller diagrams with similarity ratio $\frac{1}{a}$. If this procedure can continue forever, then we call such a diagram *self-similar* and its dimension

$$D_4 = \frac{\log b}{\log a}.$$

For example, the following Canton set is a self-similar set.

First, divide $[0,1]$ into three parts $[0,\frac{1}{3}]$, $(\frac{1}{3},\frac{2}{3})$, $[\frac{2}{3},1]$. Remove the middle part $G_1=(\frac{1}{3},\frac{2}{3})$. The residual part is

$$E_1=[0,1]\setminus\left(\tfrac{1}{3},\tfrac{2}{3}\right)=\left[0,\tfrac{1}{3}\right]\bigcup\left[\tfrac{2}{3},1\right].$$

Secondly, divide $[0,\frac{1}{3}]$ and $[\frac{2}{3},1]$ into three parts, respectively,

$$\begin{array}{lll}\left[0,\tfrac{1}{9}\right], & \left(\tfrac{1}{9},\tfrac{2}{9}\right), & \left[\tfrac{2}{9},\tfrac{1}{3}\right],\\ \left[\tfrac{2}{3},\tfrac{7}{9}\right], & \left(\tfrac{7}{9},\tfrac{8}{9}\right), & \left[\tfrac{8}{9},1\right].\end{array}$$

Remove $G_2=(\frac{1}{9},\frac{2}{9})\bigcup(\frac{7}{9},\frac{8}{9})$. The residual part is

$$E_2=[0,1]\setminus(G_1\cup G_2)=\left[0,\tfrac{1}{9}\right]\bigcup\left[\tfrac{2}{9},\tfrac{1}{3}\right]\bigcup\left[\tfrac{2}{3},\tfrac{7}{9}\right]\bigcup\left[\tfrac{8}{9},1\right].$$

Continuing this procedure, we obtain a set $G_\infty=\bigcup_{n=1}^\infty G_n$. Clearly, G_∞ is an open set with measure 1. Let $C=[0,1]\setminus G_\infty$. Then C is a closed set with measure 0. The set C is called the *Canton set.*

The Canton set can be divided into 2 similar small sets C_1 and C_2 with similarity ratio $\frac{1}{3}$. Both C_1 and C_2 can be divided into 2 similar smaller sets with similarity ratio $\frac{1}{3}$. This procedure can continue forever. Therefore, the Canton set C is a self-similar set with dimension $D_4=\frac{\log 2}{\log 3}$.

For the Canton set, its dimension is a fractal. Fractal dimensional diagrams are ubiquitous in nature. For chaotic dynamical systems, its attractor is often fractal dimensional. A fractal dimensional diagram has a complicated geometric structure. Self-similarity is the most important characteristic of fractal diagrams.

2.5 Prediction

Consider the linear prediction. For a given time series S_n $(n=1,\ldots,N)$ of measurements, we want to predict the following measurement S_{N+1}. The predictor S_{n+1}^o of the measurement S_{n+1} can be expressed by a linear combination of $S_{n-m+1},\ldots,S_n$, i.e.,

$$S_{n+1}^o=\sum_{i-1}^m\alpha_iS_{n-m+i}\quad(n=m,\ldots,N),\tag{2.5.1}$$

where α_i's are m undetermined coefficients which are independent of n. Choose α_i's such that

$$g(\alpha_1,\ldots,\alpha_m)=\sum_{n=m}^{N-1}(S_{n+1}^o-S_{n+1})^2\tag{2.5.2}$$

attains the minimal value. By the extremum principle, it is necessary to find the derivatives $\frac{\partial g}{\partial\alpha_i}$. By (2.5.1) and (2.5.2),

$$\frac{1}{2}\frac{\partial g}{\partial\alpha_i}=\sum_{n=m}^{N-1}(S_{n+1}^o-S_{n+1})\frac{\partial S_{n+1}^o}{\partial\alpha_i}=\sum_{n=m}^{N-1}\left(\sum_{j=1}^m\alpha_jS_{n-m+j}-S_{n+1}\right)S_{n-m+i}.$$

So $\frac{\partial g}{\partial \alpha_i} = 0\,(i = 1, \ldots, m)$ are equivalent to the system of linear equations $\sum_{j=1}^{m} c_{ij}\alpha_j = d_i$ $(i = 1, \ldots, m)$, where

$$c_{ij} = \sum_{n=m}^{N-1} S_{n-m+j} S_{n-m+i},$$

$$d_i = \sum_{n=m}^{N-1} S_{n-m+i} S_{n+1}.$$

The solutions are denoted by $\alpha_1^o, \ldots, \alpha_m^o$. By (2.5.1), we get the prediction of S_{N+1}

$$S_{N+1}^o = \sum_{i=1}^{m} \alpha_i^o S_{N-m+i}.$$

For the nonlinear prediction, we consider the vector

$$\boldsymbol{S}_n = (S_{n-(m-1)}, S_{n-(m-2)}, \ldots, S_{n-1}, S_n),$$

where m is the embedding dimension. In order to predict S_{N+1}, a good method is to find a known embedding vector $\boldsymbol{R}_N$ closest to $\boldsymbol{S}_N$. Then we use R_{N+1} to predict S_{N+1}.

More generally, we choose a small $\varepsilon > 0$ and find L known embedding vectors $\boldsymbol{R}_N^k$ $(k = 1, \ldots, L)$ such that $\|\boldsymbol{R}_N^k - \boldsymbol{S}_N\| \le \varepsilon$. The prediction of S_{N+1} is

$$\hat{S}_{N+1} = \frac{1}{L} \sum_{k=1}^{L} R_{N+1}^k.$$

2.6 Delay embedding vectors

For a dynamical system, instead of measuring the actual states $\boldsymbol{x}_n$, we observe a scalar time series depending on the states $S_n = S(\boldsymbol{x}_n)$. This scalar is a projection of variables of a system on the real axis. Due to the reduction of dimensionality and nonlinearity of the projection process, it is difficulty to reconstruct the state space of the original system by this scalar time series. When the attractor dimension is much smaller than the dimension of the state space, it is enough to construct a new space such that the attractor in the new space is equivalent to the original one.

Consider a dynamical system $x_{n+1} = F(x_n)$ $(n = 1, \ldots, N)$ with the attractor A. To reconstruct the attractor A, we construct the m-dimensional vector by the scalar time series $\{S_n\}$

$$\boldsymbol{S}_n = (S_{n-(m-1)}, S_{n-(m-2)}, \ldots, S_{n-1}, S_n)$$

and τ is the decay time and m is the embedding dimension.

An embedding of a compact smooth manifold M into the space $\mathbb{R}^m$ is defined as a map which is a one-to-one continuously differentiable map with the nonsingular Jacobian matrix. The following theorem shows how to choose m such that the attractor A can embed into the space $\mathbb{R}^m$.

If $\mathbf{x}_n \in A$, then $S_n \in \mathbb{R}$, and so $\mathbf{S}_n \in \tilde{A} \subset \mathbb{R}^m$, where $\tilde{A}$ is the image of A in the embedding process. This implies $d_{\tilde{A}} = d_A$, where d_B is the dimension of B.

Since the embedding map is one-to-one continuously differentiable, the self-intersection must not occur in $\tilde{A}$. Otherwise, there are two directions at the self-intersection, which will destroy the one-to-one correspondence. In an m-dimensional space, the intersection of a d_1-dimensional subspace and a d_2-dimensional subspace in general is a d_I-dimensional subspace, where $d_I = d_1 + d_2 - m$. When these two subspaces do not intersect, then $d_I < 0$. Now $\tilde{A}$ does not intersect with itself, so $d_I = 2d_A - m < 0$, i.e., $m > 2d_A$. Therefore, $m > 2d_A$ is a necessary condition that A is embedded in $\mathbb{R}^m$.

On the other hand, we have the following theorem.

Delay embedding theorem. *Let d_A be the box-counting dimension of the attractor A. If $m > 2d_A$, then except for some special cases, the map $\mathbf{S}_n = (S_{n-(m-1)}, \ldots, S_n)$ from $\mathbf{x}_n \in A$ into $\mathbb{R}^m$ is an embedding map.*

The whole embedding procedure can be visualized in the following diagram:

$$\begin{array}{ccc} \mathbf{x}_n \in A & \stackrel{\mathbf{F}}{\longrightarrow} & \mathbf{x}_{n+1} \in A \\ \downarrow & & \downarrow \\ S_n \in \mathbb{R} & & S_{n+1} \in \mathbb{R} \\ \downarrow & & \downarrow \\ \mathbf{S}_n \in \tilde{A} \subset \mathbb{R}^m & \stackrel{G}{\longrightarrow} & \mathbf{S}_{n+1} \in \tilde{A} \in \mathbb{R}^m \end{array}$$

Since the attractor is an invariant subset, from $\mathbf{S}_n \in \tilde{A}$, it follows that $\mathbf{S}_{n+1} \in \tilde{A}$. The new dynamical system $\mathbf{S}_{n+1} = G(\mathbf{S}_n)$ is uniquely determined by $\mathbf{x}_{n+1} = \mathbf{F}(\mathbf{x}_n)$.

A large embedding dimension requires knowledge of long time series. To reduce the cost of calculation, one often chooses $d_A < m < 2d_A$ if the self-intersection is neglectable.

Finally, we choose the suitable delay time τ such that τ is large enough and $\mathbf{S}_n$ and $\mathbf{S}_{n+1}$ are rather independent, and at the same time τ is not so large that they are completely independent in a statistical sense. Let p_i be the probability that $\mathbf{S}_n$ is in the i-th bin of histogram, and let p_{ij} be the probability that $\mathbf{S}_n$ is in the bin i and $\mathbf{S}_{n+\tau}$ is in the j-th bin. The mutual information for time delay τ is

$$I(\tau) = \sum_{i,j} p_{ij}(\tau) \log p_{ij}(\tau) - 2 \sum_i p_i \log p_i .$$

The first minimum of the mutual information is a good candidate for time delay.

2.7 Singular spectrum analysis

Singular spectrum analysis is a technique to identify recurrent patterns and adaptively enhance signal-to-noise ratio in a dynamical system. Consider the dynamical system

$\boldsymbol{x}_{n+1} = F(\boldsymbol{x}_n)$. Let $S_n = S(\boldsymbol{x}_n)$ $(n = 1, 2, 3, \dots, N)$ be an observed scalar time series from this dynamical system. Define

$$\boldsymbol{Y}_1 = (S_{N-2(m-1)}, \dots, S_{N-m+1})^{\mathrm{T}} \quad \dots \quad \boldsymbol{Y}_m = (S_{N-(m-1)}, \dots, S_N)^{\mathrm{T}}.$$

where $m \ll N$. Let $\boldsymbol{Y} = (\boldsymbol{Y}_1, \dots, \boldsymbol{Y}_m)$. Then the covariance matrix $\Sigma_{\boldsymbol{YY}}$ of $\boldsymbol{Y}$ is

$$\Sigma_{\boldsymbol{YY}} = (\mathrm{Cov}(\boldsymbol{Y}_i, \boldsymbol{Y}_j))_{m\times m}\,.$$

Since the matrix $\Sigma_{\boldsymbol{YY}}$ is a real symmetric matrix, its eigenvalues are $\lambda_1 \ge \lambda_2 \ge \dots \ge \lambda_m$ and the corresponding eigenvectors $\boldsymbol{e}_1, \boldsymbol{e}_2, \dots, \boldsymbol{e}_m$ form an orthonormal basis. In general, the main recurrent patterns are just the first several eigenvectors. Let

$$\alpha_k = (\boldsymbol{Y}, \boldsymbol{e}_k) \quad (k = 1, \dots, m).$$

Then

$$\boldsymbol{Y} = \sum_{k=1}^{m} \alpha_k \boldsymbol{e}_k. \tag{2.7.1}$$

Since $\boldsymbol{e}_k$ is a normal eigenvector of $\Sigma_{\boldsymbol{YY}}$,

$$\begin{aligned} \mathrm{Var}(\alpha_k) &= \boldsymbol{e}_k^{\mathrm{T}}(\lambda_k \boldsymbol{e}_k) = \lambda_k(\boldsymbol{e}_k^{\mathrm{T}}\boldsymbol{e}_k) = \lambda_k \quad (k = 1, \dots, m), \\ \mathrm{Cov}(\alpha_k, \alpha_l) &= \boldsymbol{e}_k^{\mathrm{T}}\Sigma_{\boldsymbol{YY}}\boldsymbol{e}_l = \boldsymbol{e}_k^{\mathrm{T}}(\lambda_l \boldsymbol{e}_l) = 0 \qquad (k \ne l). \end{aligned} \tag{2.7.2}$$

By (2.7.1) and (2.7.2), it follows that the total variance of $\boldsymbol{Y}$ is

$$\mathrm{Var}(\boldsymbol{Y}) = E[\|\boldsymbol{Y}\|_2^2] = E\left[\sum_1^m \alpha_k^2\right] = \sum_1^m E[\alpha_k^2] = \sum_1^m \lambda_k$$

and the ratio $\lambda_L / \sum_{k=1}^m \lambda_k$ is defined as the contribution of $\boldsymbol{e}_L$ to Y.

2.8 Recurrence networks

Recurrence networks are a new approach to analyzing the recurrence properties of complex dynamical systems. For N observed trajectories S^n $(n = 1, \dots, N)$ from a dynamical system, their time delay embedding vectors

$$\boldsymbol{S}_n^{\mathrm{T}} = (S_{T-(m-1)\tau}^n, \dots, S_{T-\tau}^n, S_T^n),$$

where m is the embedding dimension and τ is the delay. Define a recurrence network associated with these delay time vectors: each S^n $(n = 1, \dots, N)$ is considered as a vertex. A simple approach is to introduce an edge between two vertices S^i and S^j when the correlation coefficient between $\boldsymbol{S}_i^{\mathrm{T}}$ and $\boldsymbol{S}_j^{\mathrm{T}}$ is larger than a given threshold. Another approach is to introduce an edge between two vertices based on the distance. For $i \ne j$, if the distance between S^i and S^j is less than a given threshold ε, where ε

is reasonably small (in particular, much smaller that the attractor diameter), then the vertices S^i and S^j are considered to be connected by an edge. In detail,

For a given threshold $\varepsilon > 0$, let

$$R_{ij} = H(\varepsilon - \text{distance}(S^i, S^j)) \quad (i, j = 1, \ldots, m),$$

where H is the Heaviside function

$$H(\alpha) = \begin{cases} 1 & (\alpha \geq 0), \\ 0 & (\alpha < 0) \end{cases}$$

and $\|\cdot\|$ is the norm of the space $\mathbb{R}^m$. If the pair i, j $(i \neq j)$ such that $R_{ij} = 1$, it means that S^i and S^j are close. In this case, we introduce an edge between S^i and S^j. If $R_{ij}(\varepsilon) = 0$, it means that S^i and S^j are not close. In this case, we do not introduce an edge between S^i and S^j. The matrix $(R_{ij}(\varepsilon))_{m\times m}$ is called a *recurrence matrix*.

Let

$$A_{ij}(\varepsilon) = R_{ij}(\varepsilon) - \delta_{ij} \quad (i, j = 1, \ldots, m), \tag{2.8.1}$$

where δ_{ij} is the Kronecker delta, i.e., $\delta_{ij} = 0$ $(i \neq j)$ and $\delta_{ij} = 1$ $(i = j)$. The matrix $(A_{ij})_{m\times m}$ is called an *adjacency matrix*. When $i \neq j$, $A_{ij}(\varepsilon) = R_{ij}(\varepsilon)$.

The topological characteristics of recurrence networks can capture the fundamental properties of dynamical systems.

2.8.1 Local recurrence rate

First, we measure the importance of a vertex in a complex network, the local recurrence rate of a vertex $\mathbf{S}_v$ $(v = 1, \ldots, N)$ is

$$k_v = \sum_{i=1}^{N} A_{vi} \quad (A_{vv} = 0)$$

which is the number of vertices $i \neq v$ connected directly with v. Therefore, the *local connectivity* is defined as

$$\rho_v = \frac{1}{N-1} \sum_{i=1}^{N} A_{vi}.$$

Local connectivity gives the relationship between local edge density and local correlation dimension.

2.8.2 Global recurrence rate (edge density)

We discuss the mean of all local recurrence rates

$$\bar{k} = \frac{1}{N} \sum_{l=1}^{N} k_l = \frac{1}{N} \sum_{i=1}^{N} \sum_{l=1}^{N} A_{li}.$$

Note that the correlation coefficients are symmetric, $R_{ij} = R_{ji}$ and $R_{ii} = 1$. By (2.8.1), $A_{li} = A_{il}$ and $A_{ll} = 0$. This implies that

$$\bar{k} = \frac{2}{N} \sum_{i=1}^{N} \sum_{l=1}^{i-1} A_{il}.$$

Define the global recurrence rate $\bar{\rho}$ as $\bar{\rho} = \frac{1}{N} \sum_{l=1}^{N} \rho_l$. Furthermore,

$$\bar{k} = \frac{1}{N} \sum_{1}^{N} (N-1)\rho_l = (N-1)\bar{\rho}.$$

and so

$$\bar{\rho} = \frac{2}{N(N-1)} \sum_{i<l} A_{il} = \frac{2}{N(N-1)} \sum_{i<l} \left(H(\varepsilon - \text{distance}(S^i, S^j)) - \delta_{il} \right).$$

Let L be the total number of edges in the recurrence network. Then

$$L = \sum_{i<l} A_{il} = \sum_{i=1}^{N} \sum_{l=1}^{i-1} A_{il}.$$

This implies that $\bar{k} = \frac{2L}{N}$.

2.8.3 Clustering coefficients

Let N_v^{Δ} be the total number of triangles including vertex v, and k_v be the number of vertices S^v $(i \neq v)$ which are connected directly with S^v. Then

$$N_v^{\Delta} \leq \frac{1}{2} k_v (k_v - 1).$$

Define the *local clustering coefficient* as

$$c_v = \frac{2}{k_v(k_v-1)} N_v^{\Delta} \quad (k_v \neq 0, 1).$$

Define the *global clustering coefficient* as the average of all local clustering coefficients $C = \frac{1}{N} \sum_{v=1}^{N} c_v$. For a sufficiently large N, the global clustering coefficient is determined by the scale ε of resolution. The C is a characteristic parameter which reflects the topology of a network. Let $k_v^* = \frac{1}{k_v} \sum_{i=1}^{N} A_{vi} k_i$, which expresses the mean density of state in the next neighborhood.

Let $\Delta k_v = k_v - k_v^*$, where $\Delta k_v > 0$ indicates the local maxima of phase space density and $\Delta k_v < 0$ corresponds to the local density minima.

2.8.4 Path and distance

A path from S^i to S^j is an ordered sequence of the vertices $\{S^{k_1}, \dots, S^{k_\tau}\}$, where $k_1 = i$ and $k_\tau = j$, and there is an edge between S^{k_l} and $S^{k_{l+1}}$ $(l = 0, \dots, \tau - 1)$. The number τ of edges is called the *length* of the path.

The *distance* between any two vertices S^i and S^j of the network is defined as the length of the shortest path between them, denoted by l_{ij}. The *average distance* $\tilde{L}$ is defined as

$$\tilde{L} = \frac{2}{N(N-1)} \sum_{i<j} l_{ij}.$$

Further reading

[1] Antoci A, Gori L, Sodini M. Nonlinear dynamics and global indeterminacy in an overlapping generations model with environmental resources. *Communications in Nonlinear Science and Numerical Simulation*. 2016(38):59–71.

[2] Biagioni S, Krashevska V, Achnopha Y, Saad A, Sabiham S, Behling H. 8000 years of vegetation dynamics and environmental changes of a unique inland peat ecosystem of the Jambi Province in Central Sumatra, Indonesia. *Palaeogeography, Palaeoclimatology, Palaeoecology*. 2015(440):813–829.

[3] Chen Q, van Dam T, Sneeuw N, Collilieux X, Weigelt M, Rebischung P. Singular spectrum analysis for modeling seasonal signals from GPS time series. *Journal of Geodynamics*. 2013(72): 25–35.

[4] Ding Z, Yi G, Tam VWY, Huang T. A system dynamics-based environmental performance simulation of construction waste reduction management in China. *Waste Management*. 2016(51): 130–141.

[5] Donner RV, Zou Y, Donges JF, Marwan N, Kurths J. Recurrence networks – A novel paradigm for nonlinear time series analysis. *New Journal of Physics*. 2010(12): 033025.

[6] Ghil M, Allen MR, Dettinger MD, Ide K, Kondrashov D, Mann ME, Robertson AW, Saunders A, Tian Y, Varadi F, Yiou P. Advanced spectral methods for climatic time series, *Review of Geophysics*. 2002(40):1003–1043.

[7] Hassani H, Xu Z, Zhigljavsky A. Singular spectrum analysis based on the perturbation theory. *Nonlinear Analysis: Real World Applications*. 2011(12):2752–2766.

[8] Khan MAR, Poskitt DS. Forecasting stochastic processes using singular spectrum analysis: Aspects of the theory and application. *International Journal of Forecasting*. 2016: in press.

[9] Malta T, Santos PT, Santos AMP, Rufino M, Silva A. Long-term variations in Ibero-Atlantic sardine (*Sardina pilchardus*) population dynamics: Relation to environmental conditions and exploitation history. *Fisheries Research*. 2016(179):47–56.

[10] Moroz IM, Cropp R, Norbury J. Chaos in plankton models: Foraging strategy and seasonal forcing. *Ecological Modelling*. 2016(332):103–111.

[11] Roques L, Chekroun MD. Probing chaos and biodiversity in a simple competition model. *Ecological Complexity*. 2011(8):98–104.

[12] Palus M, Novotna D. Detecting modes with nontrivial dynamics embedded in colored noise: enhanced Monte Carlo SSA and the case of climate oscillations. *Physics Letters A*. 1998(248): 191–202.

[13] Rose L, Bhaskaran PK. The role of environmental forcing on tidal dynamics along complex near-shore waters off Bangladesh. *Ocean Engineering*. 2016(116):68–81.
[14] Sterk AE, Vitolo R, Broer HW, Simo C, Dijkstra HA. New nonlinear mechanisms of midlatitude atmospheric low-frequency variability. *Physica D: Nonlinear Phenomena*. 2010(239):702–718.
[15] Sugihara G, May RM. Nonlinear forecasting as a way of distinguishing chaos from measurement error in time series. *Nature*. 1990(344):734–741.
[16] Takens F. Detecting strange attractors in turbulence, in: Rand DA, Young LS (eds), *Dynamical Systems and Turbulence*, Lecture Notes in Mathematics, vol. 898, pp. 366–381, Springer-Verlag, 1981.
[17] Varsakelis C, Anagnostidis P. On the susceptibility of numerical methods to computational chaos and superstability. *Communications in Nonlinear Science and Numerical Simulation*. 2016(33):118–132.
[18] Vitanov NK, Hoffmann NP, Wernitz B. Nonlinear time series analysis of vibration data from a friction brake: SSA, PCA, and MFDFA. *Chaos, Solitons & Fractals*. 2014(69):90–99.
[19] Vitanov NK, Sakai K, Dimitrova ZI. SSA, PCA, TDPSC, ACFA: Useful combination of methods for analysis of short and nonstationary time series. *Chaos, Solitons & Fractals*. 2008(37):187–202.
[20] Wang S, Huang GH, Baetz BW, Huang W. A polynomial chaos ensemble hydrologic prediction system for efficient parameter inference and robust uncertainty assessment, Original Research Article. *Journal of Hydrology*. 2015(530):716–733.
[21] Zhang F, Zhang G. Dynamics of a low-order atmospheric circulation chaotic model. *Optik – International Journal for Light and Electron Optics*. 2016(127):4105–4108.
[22] Zhao Q, Silverman E, Fleming K, Boomer GS. Forecasting waterfowl population dynamics under climate change – Does the spatial variation of density dependence and environmental effects matter? *Biological Conservation*. 2016(194):80–88.

3 Approximation

Environmental data provide huge amounts of information, but it is complex to process due to the size, variety, and dynamic nature of the data. Approximation of environmental data is one efficient method for extracting meaningful information, so it is widely applied in environmental modeling and prediction as well as decision-making and environmental impact assessment. The fundamental problem of approximation theory is to approximate to complicated data with one or several functions that can be computed easily. These approximation tools include trigonometric and algebraic polynomials, spline functions, rational functions, and wavelets. For high-dimensional data, its approximation rate often becomes slow when the dimensionality increases. This is the so-called *dimension curse* problem. In order to solve it, hyperbolic cross approximations have received much attention in recent years.

3.1 Trigonometric approximation

We state classical results in trigonometric approximation, including Fourier expansion, the decay rate of Fourier coefficients, uniform approximation and square approximation. In particular, we give the Fourier sine (cosine) expansion with a polynomial term such that we can reconstruct the original function with few Fourier sine (cosine) coefficients. Furthermore, we also give some results on random approximation.

3.1.1 Fourier series

Let f be a 2π-periodic function on $\mathbb{R}$ and integrable on $[-\pi, \pi]$, denoted by $f \in L_{2\pi}$. The simplest approximation tool is trigonometric polynomials of degree n,

$$T_n(x) = \frac{a_0}{2} + \sum_1^n (a_k \cos(kx) + b_k \sin(kx)) \quad (k \in \mathbb{Z}_+),$$

where a_k and b_k are real constants. The series

$$\frac{a_0}{2} + \sum_1^\infty (a_n \cos(nx) + b_n \sin(nx))$$

is called a *trigonometric series*. If the trigonometric series converges to $f(x)$ uniformly, i.e.,

$$f(x) = \frac{a_0}{2} + \sum_1^\infty (a_n \cos(nx) + b_n \sin(nx)) \tag{3.1.1}$$

DOI 10.1515/9783110424904-004

uniformly, then $f(x)$ must be a 2π-periodic continuous function on $\mathbb{R}$, denote by $f \in C_{2\pi}$. Applying termwise integration gives $a_0 = \frac{1}{\pi} f(x)\,dx$ and

$$a_k = \frac{1}{\pi}\int_{-\pi}^{\pi} f(x)\cos(kx)\,dx \quad (k \in \mathbb{Z}_+),$$
$$b_k = \frac{1}{\pi}\int_{-\pi}^{\pi} f(x)\sin(kx)\,dx \quad (k \in \mathbb{Z}_+). \tag{3.1.2}$$

Let $f \in L_{2\pi}$. Then numbers a_0, a_k, and b_k determined by (3.1.2) are called *Fourier coefficients* of f. The corresponding trigonometric series (3.1.1) is called the *Fourier series* of f.

If $f(x)$ is a 2π-periodic even function, then $b_n = 0$ $(n \in \mathbb{Z}_+)$. Its Fourier series is $\frac{a_0}{2} + \sum_{n=1}^{\infty} a_n\cos(nx)$, where $a_n = \frac{2}{\pi}\int_0^{\pi} f(x)\cos(nx)\,dx$ $(n = 0, 1, \dots)$ are called the *Fourier cosine coefficients.*

If $f(x)$ is a 2π-periodic odd function, then $a_0 = 0$, $a_n = 0$ $(n \in \mathbb{Z}_+)$. Its Fourier series is $\sum_{n=1}^{\infty} b_n\sin(nx)$, where $b_n = \frac{2}{\pi}\int_0^{\pi} f(x)\sin(nx)\,dx$ $(n \in \mathbb{Z}_+)$ are called the *Fourier sine coefficients.*

If f is a $2l$-periodic function, a simple transform shows that its Fourier series is

$$\frac{a_0}{2} + \sum_1^{\infty}\left(a_n\cos\frac{\pi n x}{l} + b_n\sin\frac{\pi n x}{l}\right),$$

where $a_0 = \frac{1}{l}\int_{-l}^{l} f(x)\,dx$ and

$$a_n = \frac{1}{l}\int_{-l}^{l} f(x)\cos\frac{\pi n x}{l}\,dx,$$
$$b_n = \frac{1}{l}\int_{-l}^{l} f(x)\sin\frac{\pi n x}{l}\,dx \qquad (n \in \mathbb{Z}_+).$$

For example, let $f(x) \in L_{2\pi}$. If

$$f(x) = \begin{cases} 0, & -\pi \le x < 0, \\ x, & 0 \le x < \pi, \end{cases}$$

then its Fourier series is

$$\frac{\pi}{4} + \sum_1^{\infty}\frac{(-1)^k - 1}{k^2\pi}\cos(kx) + \frac{(-1)^{k+1}}{k}\sin(kx).$$

If $f(x) = |x|$ $(-\pi \le x < \pi)$, then $f(x)$ is an even function and its Fourier series is

$$f(x) = \frac{\pi}{2} - \frac{4}{\pi}\sum_1^{\infty}\frac{\cos(2n-1)x}{(2n-1)^2}.$$

If $f(x) = \frac{\pi - x}{2}$ $(0 \le x < 2\pi)$, then $f(x)$ is an odd function and its Fourier series is

$$f(x) = \sum_1^{\infty} \frac{\sin(nx)}{n}.$$

The Riemann–Lebesgue lemma states that if $f \in L_{2\pi}$, then Fourier coefficients $a_k \to 0$, $b_k \to 0$ $(k \to \infty)$.

The decay rate of Fourier coefficients depends on the smoothness of functions, while the smoothness of functions are descripted by the differentiability of functions and Lipschitz indices. If $f \in C(\mathbb{R})$ and

$$|f(y) - f(x)| \le M|y - x|^{\alpha} \quad ((x, y) \in \mathbb{R}),$$

then we say $f \in \mathrm{lip}_M\, \alpha$ $(0 < \alpha \le 1)$.

Let $f \in C_{2\pi}$ and $f^{(m)} \in \mathrm{lip}_M\, \alpha$. Using integration by parts, the Fourier coefficients are estimated as

$$|a_k| \le \frac{(2\pi)^{\alpha} M}{k^{m+\alpha}}, \quad |b_k| \le \frac{(2\pi)^{\alpha} M}{k^{m+\alpha}} \qquad (k \in \mathbb{Z}_+). \tag{3.1.3}$$

Denote partial sums of the Fourier series of f by $s_n(f; x)$

$$s_n(f; x) = \frac{a_0}{2} + \sum_1^n (a_k \cos(kx) + b_k \sin(kx)).$$

A rough estimate of the partial sum approximation is

$$|f(x) - s_n(f; x)| \le 2M \sum_{n+1}^{\infty} \frac{1}{k^{m+\alpha}} \le 2M \int_n^{\infty} \frac{1}{t^{m+\alpha}}\, dt = \frac{2M}{(m + \alpha - 1)n^{m+\alpha-1}}$$

and the precise result is

$$f(x) - s_n(f; x) = O\left(\frac{\log n}{n^{m+\alpha}}\right) \quad (x \in \mathbb{R})$$

and the bound of the term "O" has a precise estimate.

It is well known that there exists a continuous function with period 2π such that partial sums of its Fourier series do not converge to it, but the arithmetic mean of partial sums of its Fourier series must converge to it uniformly, i.e., the following proposition holds.

Let $f \in C_{2\pi}$ and $s_n(f; x)$ be the partial sums of its Fourier series. Define the Fejér sums as

$$\sigma_n(f; x) = \frac{1}{n} \sum_0^{n-1} s_{\nu}(f; x).$$

Then $\sigma_n(f; x)$ converges to $f(x)$ on $\mathbb{R}$ uniformly. This implies the Weierstrass theorem as follows. For any $f \in C_{2\pi}$ and $\epsilon > 0$, there exists a trigonometric polynomial $T(x)$

such that $|f(x) - T(x)| < \epsilon$ $(x \in \mathbb{R})$. Weierstrass theorem shows that any 2π-periodic continuous function can be uniformly approximated arbitrarily by trigonometric polynomials.

The Fourier series is also expressed in a complex form. Let $f \in L_{2\pi}$. Its complex Fourier coefficients are

$$c_n = \frac{1}{2\pi} \int_{-\pi}^{\pi} f(x)\, \mathrm{e}^{-\mathrm{i}nx}\, \mathrm{d}x \quad (n \in \mathbb{Z}).$$

Its Fourier series is $\sum_{n\in\mathbb{Z}} c_n \mathrm{e}^{\mathrm{i}nx}$. Comparing this with (3.1.2), we get that $c_0 = a_0$ and

$$c_k = \tfrac{1}{2}(a_k - \mathrm{i} b_k) \quad (k > 0),$$
$$c_k = \tfrac{1}{2}(a_k + \mathrm{i} b_k) \quad (k < 0).$$

So the partial sums $s_n(f; x) = \sum_{k=-n}^{n} c_k \mathrm{e}^{\mathrm{i}kx}$.

3.1.2 Uniform approximation

Denote by H_n^{T} the set of trigonometric polynomials of degree n. Let $f \in C_{2\pi}$. Consider the derivation $\Delta(T) = \max|T(x) - f(x)|$, where $T \in H_n^{\mathrm{T}}$ is any trigonometric polynomial. The greatest lower bound

$$E_n(f) = \inf_{T\in H_n^{\mathrm{T}}} \{\Delta T\}$$

is called the *best approximation* of f in H_n^{T}. If $T(x)$ is such that $\max|T(x) - f(x)| = E_n(f)$, then $T(x)$ is called the *best approximation trigonometric polynomial* in H_n^{T}.

From Weierstrass theorem, it follows that the best approximations $E_n(f)$ are monotone decreasing and tend to zero as $n \to \infty$. The best approximation polynomial exists and is unique. It is characterized as follows.

For $f \in C_{2\pi}$, $E_n(f)$ and $T_n(x)$ are its best approximation and best approximation polynomial in H_n, respectively, if and only if there exist $2n + 2$ points $0 \le x_1 < x_2 < \cdots < x_{2n+2} < 2\pi$ such that

$$|T_n(x_k) - f(x_k)| = E_n(f) \qquad (k = 1, \ldots, 2n + 2),$$
$$T_n(x_{k+1}) - f(x_{k+1}) = -(T_n(x_k) - f(x_k)) \quad (k = 1, \ldots, 2n + 2).$$

So the error curve $T_n(x) - f(x)$ is undamply oscillatory.

Jackson inequality describes the relationship between the smoothness of functions and approximation rate by trigonometric polynomials as follows.

Jackson inequality

If $f \in C_{2\pi}$ and $f^{(p)} \in \operatorname{lip} \alpha$ $(0 < \alpha \le 1)$, then

$$E_n(f) = O\left(\frac{1}{n^{p+\alpha}}\right).$$

Finding the best approximation polynomial is difficult. But the Vallee–Poussin sum $V_n(f;x) = \frac{1}{n}\sum_{j=n}^{2n-1} s_j(f;x)$, where $s_n(f;x)$ is the partial sum of Fourier series of the function f, satisfies

$$\max|V_n(f;x) - f(x)| = O\left(\frac{1}{n^{p+\alpha}}\right).$$

3.1.3 Square approximation

If f is a 2π-periodic function and $\int_{-\pi}^{\pi} f^2(x)\,dx < \infty$, then we say $f \in L_{2\pi}^2$. If $f \in L_{2\pi}^2$, from $|f| \le \frac{1}{2}(1+f^2)$, it follows that $f \in L_{2\pi}$. We use a trigonometric polynomial $T(x)$ to approximate $f \in L_{2\pi}^2$ in the mean square sense. Define the *square approximation error* as $\left(\int_{-\pi}^{\pi}|f(x) - T(x)|^2\,dx\right)^{\frac{1}{2}}$.

Expand $f \in L_{2\pi}^2$ into Fourier series

$$f(x) = \frac{a_0}{2} + \sum_1^{\infty}(a_n\cos(nx) + b_n\sin(nx)).$$

Then, for square approximation, its partial sum

$$s_n(f;x) = \frac{a_0}{2} + \sum_1^{n}(a_k\cos(kx) + b_k\sin(kx))$$

attains the best approximation of f in H_n^{T}, i.e., for any trigonometric polynomial $T_n(x)$ of degree $\le n$,

$$\int_{-\pi}^{\pi}(f(x) - s_n(f;x))^2\,dx \le \int_{-\pi}^{\pi}(f(x) - T_n(x))^2\,dx.$$

In applications, the Parseval identity

$$\frac{1}{\pi}\int_{-\pi}^{\pi}|f(x)|^2\,dx = \frac{a_0^2}{2} + \sum_{n=1}^{\infty}(a_n^2 + b_n^2)$$

is often used. Let f be an m-order continuously differentiable function. Consider the difference

$$f(x) - s_n(f;x) = \sum_{n+1}^{\infty}(a_k\cos(kx) + b_k\sin(kx)).$$

By the Parseval identity and (3.1.3), the mean square error is estimated by

$$\frac{1}{\pi}\int_{-\pi}^{\pi}(f(x) - s_n(f;x))^2\,dx = \sum_{n+1}^{\infty}(a_k^2 + b_k^2) = 2L^2\sum_{n+1}^{\infty}\frac{1}{k^{2m}} \le 2L^2\int_n^{\infty}\frac{1}{t^{2m}}\,dt = \frac{2L^2}{(2m-1)n^{2m-1}}.$$

3.1.4 Fourier sine expansion with polynomial terms

In applications, the target functions are often smooth functions on an interval. One periodically extends it to the whole real axis, the obtained function, in general, is a discontinuous function on the real axis and then expands it into a Fourier series, the Fourier coefficients decay very slow. Thus one cannot use Fourier coefficients to compress data. To solve this problem, a Fourier sine expansion with simple polynomial terms is given. Its Fourier coefficients decay fast. This gives a good approximation tool. The main idea is as follows.

Suppose that the function f is l-order continuous and differentiable on $[0,1]$, i.e., $f^{(l)} \in C[0,1]$.

Step 1. We construct a polynomial $\alpha_n(t)$ of degree $2n+1$ such that their derivatives of even orders satisfy

$$\alpha_n^{(2j)}(0) = 0, \quad \alpha_n^{(2j)}(1) = 0 \quad (0 \le j \le n-1),$$
$$\alpha_n^{(2n)}(0) = 0, \quad \alpha_n^{(2n)}(1) = 1.$$

Such $\alpha_n(t)$ can be represented as follows:

$$\begin{aligned} \alpha_0(t) &= t, \\ \alpha_1(t) &= \frac{1}{6}t(t^2-1), \\ \alpha_n(t) &= \frac{1}{(2n+1)!}t^{2n+1} + \sum_0^{n-1} b_k t^{2k+1} \quad (n = 2,3,\dots), \end{aligned} \tag{3.1.4}$$

where the coefficients $\{b_k\}_{k=0,\dots,n-1}$ satisfy

$$\sum_j^{n-1} \frac{(2k+1)!}{(2k-2j+1)!} b_k = -\frac{1}{(2n-2j+1)!} \quad (j = 0,\dots,n-1).$$

Step 2. We give a decomposition $f(t) = g(t) + h(t)$, where

$$g(t) = \sum_0^{[\frac{l-1}{2}]} \left(f^{(2i)}(0)\alpha_i(1-t) + f^{(2i)}(1)\alpha_i(t)\right) \tag{3.1.5}$$

is an algebraic polynomial of degree $\le l$ and $h(t)$ is the residual. Moreover, it is easily checked that $h^{(l)} \in C[0,1]$ and $h^{(2i)}(0) = h^{(2i)}(1) = 0$ $(i = 0,1,\dots,m)$, where $m = [\frac{l-1}{2}]$.

Step 3. For the residual h, we do an odd extension and then do periodic extension to get a 2-periodic odd function which is a continuously differentiable function of order l on $\mathbb{R}$. We expand it into Fourier sine series

$$h(t) = \sum_1^{\infty} b_n(h)\sin(\pi n t),$$

where $b_n(h) = 2\int_0^1 h(t)\sin(\pi n t)\,dt$ and $b_n(h) = O\left(\frac{1}{n^l}\right)$, and so the coefficients decay fast.

Step 4. We get the Fourier expansion with a polynomial term

$$f(t) = g(t) + \sum_{1}^{\infty} b_n(h)\sin(\pi n t).$$

Its partial sum

$$\varphi_N(t) = g(t) + \sum_{1}^{N} b_n(h)\sin(\pi n t)$$

is a good approximation tool, where $g(t)$ is a polynomial of degree $\leq l$ which is determined by the derivative values of f at $x = 0$ and $x = 1$ (see (3.1.5)), and $b_n(h) = O(\frac{1}{n^l})$.

The square approximation error is

$$\int_0^1 |\varphi_N(t) - f(t)|^2 \mathrm{d}t = \frac{1}{2}\sum_{N+1}^{\infty} b_n^2(h) = O\left(\frac{1}{N^{2l-1}}\right)$$

and so f can be reconstructed by the derivative values of f at ends of $[0, 1]$ and few Fourier coefficients.

3.1.5 Approximations for random processes

If the coefficients of a trigonometric (or algebraic) polynomial are random variables, this polynomial is called a *random trigonometric (or algebraic) polynomial.* A random process $\xi(t)$ on $[0, 1]$ can be well approximated by a combination of random algebraic polynomials and random trigonometric polynomials.

(a) Calculus of random processes

Let $\xi = \xi(t)$ $(t \in T)$ be a random process, $t_0 \in T$ and Y be a random variable. If $\lim_{t\to t_0} E[(\xi(t) - Y)^2] = 0$, then $\xi(t)$ *has the limit* Y at t_0. If $\lim_{t\to t_0} \xi(t) = \xi(t_0)$, then $\xi(t)$ *is continuous* at t_0. If $\lim_{t\to t_0} \frac{\xi(t)-\xi(t_0)}{t-t_0} = Y$, then $\xi(t)$ *has the derivative* Y at t_0. Write $\xi'(t_0) = Y$. Suppose that $\xi(t)$ and $\eta(t)$ are both differentiable random processes and $\zeta(t)$ is a differentiable deterministic function. Then the following operation rules hold:

$$E[\xi'(t)] = (E[\xi(t)])',$$
$$(\alpha\xi(t) + \beta\eta(t))' = \alpha\xi'(t) + \beta\eta'(t),$$
$$(\xi(t)\zeta(t))' = \xi'(t)\zeta(t) + \xi(t)\zeta'(t).$$

Let $\xi = \xi(t)$ $(a \leq t \leq b)$ be a random process. Given a partition $a = t_0 < t_1 < \cdots < t_n = b$, arbitrarily take $\theta_k \in [t_{k-1}, t_k]$ $(k = 1, \ldots, n)$. If the limit $\lim_{\delta\to 0}\sum_{k=1}^{n}\xi(\theta_k)(t_k - t_{k-1}) = I$

exists, where $\delta = \max_k |t_k - t_{k-1}|$, then the random process ξ is *integrable* on $[a, b]$. Write $\int_a^b \xi(t)\,dt = I$. If $\xi(t)$ is integrable on $[a, b]$, then

$$E\left[\int_a^b \xi(t)\,dt\right] = \int_a^b E[\xi(t)]\,dt.$$

If $\xi(t)$ is a continuously differentiable random process on $[a, b]$, then $\int_a^b \xi'(t)\,dt = \xi(b) - \xi(a)$.

(b) A good approximation tool for random processes

Let a random process $\xi(t)$ be l-order continuously differentiable on $[0, 1]$. Then the random polynomial

$$\eta(t) = \sum_0^{[\frac{l-1}{2}]} (\xi^{(2i)}(0)\alpha_i(1-t) + \xi^{(2i)}(1)\alpha_i(t))$$

is of degree $\le l$, where $\alpha_i(t)$ is stated in (3.1.4). So the random process can be decomposed into

$$\xi(t) = \eta(t) + \zeta(t),$$

called the *decomposition formula*, where the residual $\zeta(t)$ is a random process satisfying $\zeta^{(l)} \in C([0, 1])$ and $\zeta^{(2l)}(0) = \zeta^{(2l)}(1) = 0$ $(i = 1, \dots, m)$, where $m = [\frac{l-1}{2}]$. Expand $\zeta(t)$ into Fourier sine series

$$\zeta(t) = \sum_1^\infty c_n(\zeta) \sin(\pi n t),$$

where the Fourier sine coefficients $c_n(\zeta)$ are random variables.

When $l = 2m + 1$ is odd, noticing that $\zeta(t) = \xi(t) - \eta(t)$ and using integration by parts l times,

$$\begin{aligned} c_n(\zeta) &= 2\int_0^1 \zeta(t)\sin(\pi n t)\,dt = \frac{2}{(\pi n)^l}\int_0^1 \zeta^{(l)}(t)\cos(\pi n t)\,dt \\ &= \frac{2}{(\pi n)^l}\int_0^1 (\xi^{(l)}(t) - \eta^{(l)}(t))\cos(\pi n t)\,dt. \end{aligned}$$

Denote $\tau = \eta^{(2m+1)}(t)$. Since $\eta(t)$ is a random polynomial of degree $2m + 1$, τ is independent of t, and so

$$\int_0^1 \eta^{(2m+1)}(t)\cos(\pi n t)\,dt = \tau \int_0^1 \cos(\pi n t)\,dt = 0.$$

So

$$c_n(\zeta) = \frac{2}{(\pi n)^l} \int_0^1 \xi^{(l)}(t) \cos(\pi n t)\, dt. \tag{3.1.6}$$

So $|E[c_n(\zeta)]| \le \frac{2}{(\pi n)^l} E_l$, where $E_l = \max_{0\le t\le 1} |E_l[\xi^{(l)}(t)]|$.

Note that $\operatorname{Var}(c_n(\zeta)) \le E[c_n^2(\zeta)]$. From (3.1.6), it follows that

$$E[c_n^2(\zeta)] = \frac{4}{(\pi n)^{2l}} \int_0^1 \int_0^1 E[\xi^{(l)}(t)\xi^{(l)}(s)] \cos(\pi n t)\cos(\pi n s)\, dt\, ds.$$

By the Schwarz inequality from probability theory,

$$|E[\xi^{(l)}(t)\xi^{(l)}(s)]| \le (E[(\xi^{(l)}(t))^2] E[(\xi^{(l)}(s))^2])^{\frac{1}{2}} \le V_l,$$

and so $\operatorname{Var}(c_n(\zeta)) \le \frac{4V_l}{(\pi n)^{2l}}$, where $V_l = \max_{0\le t\le 1}(E[(\xi^{(l)}(t))^2])$.

When $l = 2m$ is even, the above conclusion is also valid. So, for $l \in \mathbb{Z}_+$, the following proposition holds.

Let a random process $\xi \in C^l([0,1])$. Then

$$\xi(t) = \sum_0^{[\frac{l-1}{2}]} \left(\xi^{(2i)}(0)\alpha_i(1-t) + \xi^{(2i)}(1)\alpha_i(t)\right) + \sum_1^{\infty} c_n(\zeta)\sin(\pi n t), \tag{3.1.7}$$

$$E[c_n(\zeta)] \le \frac{2E_l}{(\pi n)^l},$$

$$\operatorname{Var}(c_n(\zeta)) \le \frac{4V_l}{(\pi n)^{2l}},$$

where $c_n(\zeta) = \frac{2}{(\pi n)^l}\int_0^1 \xi^{(l)}(t)\cos(\pi n t)\, dt$, $E_l = \max_{0\le t\le 1}|E[\zeta^{(l)}(t)]|$, and $V_l = \max_{0\le t\le 1}$ $(E[(\xi^{(l)}(t))^2])$.

Consider the partial sum of the series in (3.1.7)

$$\varphi_N(\xi;t) = \sum_0^{[\frac{l-1}{2}]} \left(\xi^{(2i)}(0)\alpha_i(1-t) + \xi^{(2i)}(1)\alpha_i(t)\right) + \sum_1^{N} c_n(\zeta)\sin(\pi n t).$$

The mean square error is

$$\int_0^1 E[(\varphi_N(t) - \xi(t))^2]dt = \frac{1}{2}\sum_{k+1}^{\infty} E[|c_n(\zeta)|^2] \le \frac{V_l}{(2l-1)(N\pi)^{2l-1}},$$

where V_l is stated as above. The $\varphi_N(\xi;t)$ is a good approximation tool, it is a combination of a random polynomial of degree $\le l$ and a random trigonometric polynomial of degree N, where l is the order of derivative of f and N determines the size of error.

3.2 Multivariate approximation and dimensionality reduction

Suppose that f is a 2π-periodic function and is integrable on $[-\pi, \pi]^m$, denoted by $f \in L_{2\pi}(\mathbb{R}^m)$. Define its Fourier coefficients as

$$c_{\boldsymbol{n}}(f) = \frac{1}{(2\pi)^m} \int_{[-\pi,\pi]^m} f(\boldsymbol{t})\, \mathrm{e}^{-\mathrm{i}\boldsymbol{n}\cdot\boldsymbol{t}}\, \mathrm{d}\boldsymbol{t},$$

where $\boldsymbol{n} \cdot \boldsymbol{t} = n_1 t_1 + \cdots + n_m t_m$ and $\mathrm{d}\boldsymbol{t} = \mathrm{d}t_1 \cdots \mathrm{d}t_m$. The series $\sum_{\boldsymbol{n}\in\mathbb{Z}^m} c_{\boldsymbol{n}}(f)\, \mathrm{e}^{\mathrm{i}\boldsymbol{n}\cdot\boldsymbol{t}}$ is called the *Fourier series* of f. By the Riemann–Lebesgue lemma, $c_{\boldsymbol{n}}(f) \to 0$ as $\|\boldsymbol{n}\| = (\sum_{i=1}^m n_i^2)^{\frac{1}{2}} \to \infty$.

Denote

$$D^{(l_1,\dots,l_m)} = \frac{\partial^{l_1+\cdots+l_m}}{\partial t_1^{l_1} \cdots \partial t_m^{l_m}}.$$

If $f \in L_{2\pi}(\mathbb{R}^m)$ and $D^{(l_1,\dots,l_m)} f \in C(\mathbb{R}^m)$, then Fourier coefficients satisfy

$$c_{\boldsymbol{n}}(f) = O(n_1^{-l_1} \cdots n_m^{-l_m}).$$

If $\int_{[-\pi,\pi]^m} |f(\boldsymbol{t})|^2\, \mathrm{d}\boldsymbol{t} < \infty$, then the Parseval identity

$$(2\pi)^m \int_{[-\pi,\pi]^m} |f(\boldsymbol{t})|^2\, \mathrm{d}\boldsymbol{t} = \sum_{\boldsymbol{n}\in\mathbb{Z}^m} |c_{\boldsymbol{n}}(f)|^2$$

holds. For the partial sums

$$s_N(f; \boldsymbol{t}) = \sum_{n_1=-N}^{N} \cdots \sum_{n_m=-N}^{N} c_{\boldsymbol{n}}(f)\, \mathrm{e}^{\mathrm{i}\boldsymbol{n}\cdot\boldsymbol{t}},$$

the *square approximation errors* satisfy

$$e_N^2(f) = \int_{[-\pi,\pi]^m} |s_N(f; \boldsymbol{t}) - f(\boldsymbol{t})|^2\, \mathrm{d}\boldsymbol{t} = \frac{1}{(2\pi)^m} \left(\sum_{\boldsymbol{n}\in\mathbb{Z}^m} - \sum_{n_1=-N}^{N} \cdots \sum_{n_m=-N}^{N} \right) |c_{\boldsymbol{n}}(f)|^2.$$

If $f \in L_{2\pi}(\mathbb{R}^m)$ and $D^{(l,\dots,l)} f \in C(\mathbb{R}^m)$, then $c_{\boldsymbol{n}}(f) = O\left((n_1 \cdots n_m)^{-l}\right)$. This implies that

$$e_N^2(f) = O\left(\sum_{N}^{\infty} \frac{1}{k^{2l}} \right) = O\left(\frac{1}{N^{2l-1}} \right).$$

Note that the number N_c of Fourier coefficients in $s_N(f; \boldsymbol{t})$ satisfies $N_c \sim N^m$. Thus

$$e_N^2(f) = O\left(N_c^{-\frac{2l-1}{m}} \right). \tag{3.2.1}$$

3.2.1 Dimensionality reduction

From (3.2.1), it is seen that the approximation rate becomes very slow as the dimensionality increases. To explain how to solve this dimensionality curse problem, we take a sine series as an example.

Let f be a 2π-periodic function on $\mathbb{R}^m$ and $D^{(l,\dots,l)} \in C(\mathbb{R}^m)$. If f is an odd function, i.e.,

$$\begin{aligned} &f(t_1,\dots,t_{i-1},-t_i,t_{i+1},\dots,t_m) \\ &\quad = -f(t_1,\dots,t_{i-1},t_i,t_{i+1},\dots,t_m) \quad (\boldsymbol{t}=(t_1,\dots,t_m)\in\mathbb{R}^m) \end{aligned}$$

for any $1 \le i \le m$, then $f(\boldsymbol{t})$ can be expanded into a Fourier sine series $f(\boldsymbol{t}) = \sum_{\boldsymbol{n}\in\mathbb{Z}_+^m} c_{\boldsymbol{n}}(f)T_{\boldsymbol{n}}(\boldsymbol{t})$, where

$$\begin{aligned} T_{\boldsymbol{n}}(\boldsymbol{t}) &= \prod_1^m \sin(n_j t_j), \\ c_{\boldsymbol{n}}(f) &= \frac{2^m}{\pi^m}\int_{[0,\pi]^m} f(\boldsymbol{t})T_{\boldsymbol{n}}(\boldsymbol{t})\,\mathrm{d}\boldsymbol{t}. \end{aligned}$$

The square approximation errors of partial sums $s_N(f;\boldsymbol{t}) = \sum_{n_1,\dots,n_m=1}^N c_{\boldsymbol{n}}(f)T_{\boldsymbol{n}}(\boldsymbol{t})$ satisfy

$$e_N^2(f) = O\left(\frac{1}{N_c^{\frac{2l-1}{m}}}\right), \tag{3.2.2}$$

where N_c is the number of Fourier sine coefficients in the partial sums $s_N(f;\boldsymbol{t})$.

To solve the dimensionality curse problem, we consider the hyperbolic cross truncations of Fourier sine series. First, we rewrite Fourier sine series in the form

$$f(\boldsymbol{t}) = \sum_{\boldsymbol{p}\in\mathbb{Z}_+^m}\ \sum_{\boldsymbol{q}\in\{0,1\}^m} c_{2\boldsymbol{p}+\boldsymbol{q}}(f)T_{2\boldsymbol{p}+\boldsymbol{q}}(\boldsymbol{t}),$$

where $\{0,1\}^m$ is the set of vertices of the unit cube $[0,1]^m$. Define the hyperbolic cross truncations of Fourier sine series of f as

$$s_N^{(h)}(f;\boldsymbol{t}) = \sum_{\substack{1\le|\boldsymbol{p}|\le N \\ 1\le p_1,\dots,p_m\le N-1}}\ \sum_{\boldsymbol{q}\in\{0,1\}^m} c_{2\boldsymbol{p}+\boldsymbol{q}}(f)T_{2\boldsymbol{p}+\boldsymbol{q}}(\boldsymbol{t}),$$

where $\boldsymbol{p} = (p_1,\dots,p_m)$ and $|\boldsymbol{p}| = \prod_1^m p_k$. The square approximation errors of hyperbolic cross truncations are

$$(e_N^{(h)}(f))^2 = \int_{[0,\pi]^m} |s_N^{(h)}(f;\boldsymbol{t}) - f(\boldsymbol{t})|^2\,\mathrm{d}\boldsymbol{t} = O\left(\frac{\log^{m-1}N}{N^{2l-1}}\right).$$

The number N_c of Fourier sine coefficients in $s_N^{(h)}(f;\boldsymbol{t})$ are $N_c \sim N\log^{m-1}N$, so

$$(e_N^{(h)}(f))^2 = O\left(\frac{\log^{2l(m-1)}N_c}{N_c^{2l-1}}\right),$$

where m is the dimensionality. Comparing this with (3.2.2), since $\log N_c$ increases very slowly, the dimensionality curse problem is basically solved.

3.2.2 Bivariate Fourier cosine expansion with simple polynomial factors

A good approximation tool to compress data has been proposed recently. Suppose that $f(t_1, t_2) \in C([0, 1]^2)$. If we do the even extension from $[0, 1]^2$ to $[-1, 1]^2$ such that $f(t_1, t_2) \in C([-1, 1]^2)$ and

$$f(t_1, t_2) = f(-t_1, t_2) = f(t_1, -t_2) = f(-t_1, -t_2) \quad ((t_1, t_2) \in [-1, 1]^2)$$

and then do periodic extension from $[-1, 1]$ to $\mathbb{R}^2$, we get a 2-periodic even continuous function. So $f(t_1, t_2)$ can be expanded into the following Fourier cosine series:

$$f(t_1, t_2) = \sum_{n_1, n_2 \in \mathbb{Z}} c_{n_1,n_2}(f) \cos(\pi n_1 t_1) \cos(\pi n_2 t_2),$$

where

$$c_{n_1,n_2} = \lambda_{n_1,n_2} \int_{[0,1]^2} f(t_1, t_2) \cos(\pi n_1 t_1) \cos(\pi n_2 t_2)\, dt_1\, dt_2$$

and

$$\lambda_{n_1,n_2} = \begin{cases} 1 & (n_1 = n_2 = 0), \\ 4 & (n_1 \neq 0, n_2 \neq 0), \\ 2 & \text{otherwise}. \end{cases}$$

Suppose that f is a bivariate function on the unit square $[0, 1]^2$ and $D^{(3,3)}f$ is continuous on $[0, 1]^2$. If we expand f into a bivariate Fourier series, by discontinuity, the Fourier coefficients decay very slowly. To reconstruct f by the fewest Fourier cosine coefficients, a new approximation tool is constructed. The main idea is to give a decomposition of bivariate functions for solving the discontinuity problem of partial derivative functions after periodic extension and use hyperbolic cross truncation to solve a dimension curse problem.

For $D^{(3,3)}f \in C([0, 1]^2)$, the algorithm is as follows:

Step 1. Based on a fundamental polynomial $p(t) = \frac{1}{2}t^2 - \frac{1}{6}$, we construct a bivariate polynomial

$$\begin{aligned} q(t_1, t_2) = {} & D^{(1,1)}f(0, 0)p(1 - t_1)p(1 - t_2) - D^{(1,1)}f(0, 1)p(1 - t_1)p(t_2) \\ & - D^{(1,1)}f(1, 0)p(t_1)p(1 - t_2) + D^{(1,1)}f(1, 1)p(t_1)p(t_2). \end{aligned} \tag{3.2.3}$$

Denote $f_1(t_1, t_2) = f(t_1, t_2) - q(t_1, t_2)$. Clearly,

$$D^{(1,1)}f_1(0, 0) = D^{(1,1)}f_1(0, 1) = D^{(1,1)}f_1(1, 0) = D^{(1,1)}f_1(1, 1) = 0.$$

Step 2. Define

$$\begin{aligned} f_2(t_1, t_2) = {} & -D^{(1,0)}f_1(0, t_2)p(1 - t_1) + D^{(1,0)}f_1(1, t_2)p(t_1) \\ & - D^{(0,1)}f_1(t_1, 0)p(1 - t_2) + D^{(0,1)}f_1(t_1, 1)p(t_2). \end{aligned}$$

Expand the first factor of each term on the right-hand side of this formula into Fourier cosine series

$$f_2(t_1,t_2) = -\left(\sum_0^\infty c_n^{(1)}\cos(\pi n t_2)\right)p(1-t_1) + \left(\sum_0^\infty c_n^{(2)}\cos(\pi n t_2)\right)p(t_1)$$
$$-\left(\sum_0^\infty c_n^{(3)}\cos(\pi n t_1)\right)p(1-t_2) + \left(\sum_0^\infty c_n^{(4)}\cos(\pi n t_1)\right)p(t_2). \quad (3.2.4)$$

Clearly, $c_n^{(i)} = O(\frac{1}{n^3})$ $(i = 1,2,3,4)$.

Step 3. Let $f_3 = f_1 - f_2$. Then we get a decomposition formula $f = q + f_2 + f_3$, where $q = q(t_1,t_2)$ is a bivariate polynomial. Clearly, $D^{(1,1)}f_3(t_1,t_2) = 0$ on the boundary of $[0,1]^2$.

Step 4. Expand f_3 into a bivariate Fourier cosine series

$$f_3(t_1,t_2) = \sum_{n_1,n_2=0}^{\infty} c_{n_1,n_2}(f_3)\cos(\pi n_1 t_1)\cos(\pi n_2 t_2). \quad (3.2.5)$$

Clearly, $|c_{n_1,n_2}| = O(\frac{1}{n_1^3 n_2^3})$. Substituting (3.2.4) and (3.2.5) into the decomposition $f = q + f_2 + f_3$, we get a Fourier cosine expansion of f with a polynomial term and polynomial factors. The Fourier cosine coefficients decay fast. The hyperbolic cross truncation of this expansion is

$$s_N^{(h)}(f_3;t_1,t_2) = \sum_{n_1=0}^{N-1} c_{n_1,0}(f_3)\cos(\pi n_1 t_2)$$
$$+ \sum_{n_2=1}^{N-1}\sum_{n_1=0}^{\left[\frac{N-1}{n_2}\right]-1} c_{n_1,n_2}(f_3)\cos(\pi n_1 t_1)\cos(\pi n_2 t_2)$$

which contains $N + \sum_{n_1=1}^{N-1}\left[\frac{N-1}{n_1}\right] \sim N\log N$ Fourier cosine coefficients.

Step 5. We construct good approximation tools

$$R_N^{(h)}(f;t_1,t_2) = q(t_1,t_2) + s_N^{(b)}(f_2;t_1,t_2) + s_N^{(h)}(f_3;t_1,t_2), \quad (3.2.6)$$

where $q(t_1,t_2)$ and $s_N^{(h)}(f_3;t_1,t_2)$ are stated as above, while $s_N^{(b)}(f_2;t_1,t_2)$ is given by replacing ∞ by $N-1$ in (3.2.4)

$$s_N^{(b)}(f_2,t_1,t_3) = -\left(\sum_0^{N-1} c_n^{(1)}\cos(\pi n t_2)\right)p(1-t_1) + \left(\sum_0^{N-1} c_n^{(2)}\cos(\pi n t_2)\right)p(t_1)$$
$$-\left(\sum_0^{N-1} c_n^{(3)}\cos(\pi n t_1)\right)p(1-t_2) + \left(\sum_0^{N-1} c_n^{(4)}\cos(\pi n t_1)\right)p(t_2).$$

In (3.2.6), $q(t_1,t_2)$ is a bivariate polynomial which is determined by derivative values of f at (0, 0), (1, 0), (0, 1), and (1, 1), $s_N^{(b)}(f_2;t_1,t_2)$ is a combination of cosine polynomials and fixed polynomial factors, and $s_N^{(h)}(f_3;t_1,t_2)$ is a bivariate cosine polynomial.

Step 6. Estimate approximation error. We obtain the following estimate:

$$\iint_{\mathbb{R}^2} |R_N^{(h)}(f; t_1, t_2) - f(t_1, t_2)|^2 \, dt_1 \, dt_2 = \|R_N^{(h)}(f) - f\|^2 = O\left(\frac{\log N}{N^5}\right).$$

Denote by N_c the number of cosine coefficients in $R_N^{(h)}(f; t_1, t_2)$. So

$$N_c = 5N + \sum_1^{N-1} \left[\frac{N-1}{n}\right] \sim N \log N,$$

and so

$$\|R_N^{(h)}(f) - f\|^2 = O\left(\frac{\log^6 N_c}{N_c^5}\right).$$

If we directly expand f into Fourier cosine series

$$f(t_1, t_2) = \sum_{n_1, n_2 = 0}^{\infty} c_{n_1, n_2}(f) \cos(\pi n_1 t_1) \cos(\pi n_2 t_2)$$

and approximate to $f(t_1, t_2)$ by its partial sum

$$s_n(f; t_1, t_2) = \sum_{n_1, n_2 = 0}^{N-1} c_{n_1, n_2}(f) \cos(\pi n_1 t_1) \cos(\pi n_2 t_2),$$

then $\|s_N(f) - f\|^2 \sim N_c^{-\frac{3}{2}}$, where N_c is the number of cosine coefficients in $s_n(f; t_1, t_2)$.

Similarly, we can construct the Fourier sine expansion with a polynomial term and polynomial factors such that Fourier sine coefficients decay fast, and reconstruct f using fewest Fourier sine coefficients. Generally speaking, the Fourier cosine expansion is better. Therefore, we only introduce the Fourier cosine expansion here.

3.3 Polynomial approximation

Here we discuss the approximation problem of functions on a finite interval $[a, b]$ by algebraic polynomials.

3.3.1 The best uniform approximation and its numerical computation

A continuous function on a closed interval can be approximated by a polynomial, precisely speaking, if $f(x) \in C([a, b])$, then for any given $\epsilon > 0$, there exists a polynomial $P(x)$ such that $|P(x) - f(x)| < \epsilon$ $(x \in [a, b])$.

In fact, if $f \in C[-1, 1]$, then the sequence of Bernstein polynomials of f

$$(B_n f)(x) = \sum_0^n f\left(\frac{k}{n}\right) \frac{n!}{k!(n-k)!} x^k (1-x)^{n-k} \quad (n \in \mathbb{Z}_+)$$

converges uniformly to $f(x)$ on $[0, 1]$ as $n \to \infty$.

Let $f \in C([a,b])$. Denote by H_n the set of polynomials of degree $\le n$. Then the *best approximation error* of f in H_n is defined as

$$E_n(f) = \inf_{P \in H_n} \left\{ \max_{a \le x \le b} |P(x) - f(x)| \right\}.$$

If $\tilde{P} \in H_n$ and $E_n(f) = \max_{a \le x \le b} |\tilde{P}(x) - f(x)|$, then $\tilde{P}$ is called the *best approximation polynomial* of f in H_n.

The best approximation polynomial exists and is unique. It is characterized as follows.

Let $f \in C([a,b])$. Then $Q(x)$ is the best approximation polynomial of f in H_n if and only if there exist $n+2$ points $a \le x_1 < x_2 < \cdots < x_{n+2} \le b$ such that

$$|Q(x_k) - f(x_k)| = \max_{a \le x \le b} |Q(x) - f(x)| \qquad (k = 1, \dots, n+2),$$
$$Q(x_k) - f(x_k) = -(Q(x_{k+1}) - f(x_{k+1})) \quad (k = 1, \dots, n+1).$$

This characterization shows that if $Q(x)$ is the best approximation polynomial, then the deviation $Q(x) - f(x)$ is an undamped oscillation. It gives a numerical method to find out the best approximation polynomial. Suppose that $f \in C([a,b])$ and its derivative exists on $[a,b]$, and its n-th best approximation polynomial is $P_n(x) = \sum_{k=0}^{n} \alpha_k x^k$. Then the best approximation error E_n, the best approximation polynomial $P_n(x)$, and $n+2$ deviation points $y_1, \dots, y_{n+2}$ should satisfy the following $2n+4$ equations:

$$\begin{aligned} (f(y_k) - P_n(y_k))^2 &= (E_n)^2, \\ (y_k - a)(y_k - b)(f'(y_k) - P_n'(y_k)) &= 0 \quad (k = 1, 2, \dots, n+2). \end{aligned} \tag{3.3.1}$$

To solve this system of equations, Remez gave a numerical procedure as follows.

(a) Take initial values $a \le y_1 < y_2 < \cdots < y_{n+2} \le b$.
(b) Solve out unknown numbers $a_0, a_1, \dots, a_n$ and E_n from the system of linear equations

$$a_0 + a_1 y_k + \cdots + a_n y_k^n - (-1)^k E_n = f(y_k) \quad (k = 1, 2, \dots, n+2).$$

So the polynomial $P_n^0(x) = \sum_{k=0}^{n} a_k^0 x^k$ and E_n^0 are obtained.
(c) Find extreme points $x_1, x_2, \dots, x_{n+2}$ of $f(x) - P_n(x)$, where $a \le x_1 < x_2 < \cdots < x_{n+2} \le b$.
(d) Replace y_k by x_k in (b), solve out $a_0^1, a_1^1, \dots, a_n^1$ and E_n^1, and then obtain the polynomial $P_n^1(x)$. Again, take extreme points $x_1^1, x_2^1, \dots, x_{n+2}^1$ of $f(x) - P_n^1(x)$, where $a \le x_1^1 < x_2^1 < \cdots < x_{n+2}^1 \le b$.

Continuing this procedure, the obtained coefficients $a_0^\nu, a_1^\nu, \dots, a_n^\nu$ and E_n^ν satisfy

$$a_k^\nu \to \alpha_k,$$
$$E_n^\nu \to E_n \quad (\nu \to \infty).$$

So the best approximation polynomial $P_n(x) = \sum_{k=0}^{n} \alpha_k x^k$ and the best approximation error E_n are obtained.

Weierstrass theorem shows that for any $f \in C[a, b]$, the approximation error $E_n(f)$ in H_n monotonically tends to zero as $n \to \infty$. The decay rate of $E_n(f)$ depends on the smoothness of f.

Let $f^{(p)} \in \operatorname{lip} \alpha$ $(0 < \alpha \leq 1)$ on $[a, b]$. Then the best approximation of f in H_n satisfies $E_n(f) \leq \frac{M}{n^{p+\alpha}}$, where M is a constant independent of n. The following proposition indicates that the polynomial approximation becomes better at ends of the interval $[-1, 1]$.

Let $f \in C[-1, 1]$. Then there exists a polynomial sequence $\{P_n(x)\}$ of degree n such that for $-1 \leq x \leq 1$,

$$|f(x) - P_n(x)| \leq M(\Delta_n(x))^{p+\alpha} \quad (0 < \alpha \leq 1;\ p = 0, 1, \dots)$$

if and only if $f^{(p)} \in \operatorname{lip} \alpha$, where M is a constant independent of n and x, and

$$\Delta_n(x) = \max\left(\frac{\sqrt{1 - x^2}}{n}, \frac{1}{n^2}\right).$$

3.3.2 Orthogonal polynomials

Now we consider a polynomial $P_n(x)$ of degree n such that $\int_a^b (f(x) - P_n(x))^2 \, dx$ is as small as possible. Such an approximation is called the *square approximation* of polynomials. It is different from uniform approximation. If a sequence of polynomials converges uniformly to a function, then the function is continuous. Therefore, if a function has points of discontinuity, then the function cannot be approximated by polynomials uniformly. However, a piecewise continuous function can be approximated by polynomials in the square sense. More generally, we consider the weighted square approximation.

The notation $f \in L_\rho^2([a, b])$ means that the function f on $[a, b]$ satisfies

$$\int_a^b \rho(x) f^2(x) \, dx < \infty,$$

where $\rho(x) \geq 0$ and $\int_a^b \rho(x) dx = 1$. Define the *inner product* of $f, g \in L_\rho^2([a, b])$ as

$$(f, g)_\rho = \int_a^b \rho(x) f(x) g(x) \, dx.$$

If $(f, g)_\rho = 0$, then f and g are *orthogonal* with weight ρ. Define the *norm* of $f \in L_\rho^2([a, b])$ as $\|f\|_\rho = (f, f)_\rho^{1/2}$.

Let $\{P_n\}_{n=0,1,\dots}$ be a system of n-degree polynomials and ρ be a weight function on $[a,b]$. If, for $n, m = 0, 1, \dots,$

$$(P_n, P_m)_\rho = \int_a^b \rho(x) P_n(x) P_m(x)\,\mathrm{d}x = 0 \quad (n \neq m),$$

then $\{P_n\}$ is said to be a *system of orthogonal polynomials on* $[a,b]$ *with weight* ρ. Again, if $(P_n, P_n)_\rho = 1$ $(n = 0, 1, \dots)$, then $\{P_n\}$ is said to be a system of a *normal orthogonal polynomial system.*

For any weight function $\rho(x)$ on $[a,b]$, there is a normal orthogonal polynomial system $\{\omega_n(x)\}_{n=0,1,\dots}$ with weight $\rho(x)$, where $\omega_n(x)$ is just a polynomial of degree n. Any $f \in L^2_\rho([a,b])$ can be expanded into Fourier series with respect to the system $\{\omega_n(x)\}_{n=0,1,\dots}$ as

$$f(x) = \sum_0^\infty c_k \omega_k(x),$$

where $c_k = \int_a^b f(x)\omega_k(x)\rho(x)\,\mathrm{d}x$, and the Parseval identity holds

$$\int_a^b \rho(x) f^2(x)\,\mathrm{d}x = \sum_{k=0}^\infty c_k^2.$$

Let $s_n(x)$ be its partial sum: $s_n(x) = \sum_{k=0}^n c_k \omega_k(x)$. Then

$$\int_a^b \rho(x)(f(x) - s_n(x))^2\,\mathrm{d}x \to 0 \quad (n \to \infty).$$

For any polynomial $P_n(x)$ of degree n,

$$\int_a^b \rho(x)(f(x) - P_n(x))^2\,\mathrm{d}x \le \int_a^b \rho(x)(f(x) - s_n(x))^2\,\mathrm{d}x, \tag{3.3.2}$$

i.e., $s_n(x)$ is the best square approximation polynomial in all n-degree polynomials and the best square approximation error is $E_n = \left(\sum_{k=n+1}^\infty c_k^2\right)^{\frac{1}{2}}$.

We introduce several important orthogonal polynomials often used in applications.

(a) Chebyshev polynomials

$$T_n(x) = \cos(n \arccos x)$$

are orthogonal polynomials with weight $(1 - x^2)^{-\frac{1}{2}}$ on $[-1, 1]$. So $T_0(x) = 1$ and $T_1(x) = x$, and the recurrence formula holds

$$T_{n+1}(x) = 2xT_n(x) - T_{n-1}(x) \quad (n \ge 1).$$

Note that

$$\int_{-1}^{1} T_n(x)T_m(x)\frac{\mathrm{d}x}{\sqrt{1-x^2}} = \int_0^{\pi} T_n(\cos\theta)T_m(\cos\theta)\,\mathrm{d}\theta$$

$$= \int_0^{\pi} \cos(n\theta)\cos(m\theta)\,\mathrm{d}\theta = 0 \quad (n \neq m),$$

$$\int_{-1}^{1} T_n^2(x)\frac{\mathrm{d}x}{\sqrt{1-x^2}} = \int_0^{\pi} \cos^2(nx)\,\mathrm{d}x = \begin{cases} \frac{\pi}{2} & (n \neq 0), \\ \pi & (n = 0). \end{cases}$$

The system

$$\hat{T}_0(x) = \frac{1}{\sqrt{\pi}},$$

$$\hat{T}_n(x) = \sqrt{\frac{2}{\pi}}\cos(n \arccos x) \quad (n \in \mathbb{Z}_+)$$

is a normal orthogonal system with weight $\frac{1}{\sqrt{1-x^2}}$ on $[-1, 1]$.

If $\int_{-1}^{1} \frac{f^2(x)}{\sqrt{1-x^2}}\,\mathrm{d}x < \infty$, then f can be expanded into Fourier–Chebyshev series

$$f(x) = \sum_0^{\infty} c_n^{\mathrm{T}}\hat{T}_n(x), \quad \text{where} \quad c_n^{\mathrm{T}} = \int_{-1}^{1} f(x)\hat{T}_n(x)\frac{\mathrm{d}x}{\sqrt{1-x^2}}.$$

Let $x = \cos\theta$. Then $f(\cos\theta) = \alpha_0 + \sum_1^{\infty} \alpha_n \cos(n\theta)$, where

$$\alpha_0 = \frac{1}{\pi}\int_0^{\pi} f(\cos\theta)\,\mathrm{d}\theta,$$

$$\alpha_n = \frac{2}{\pi}\int_0^{\pi} f(\cos\theta)\cos(n\theta)\,\mathrm{d}\theta \quad (n \in \mathbb{Z}_+).$$

This is just the Fourier cosine expansion of $f(\cos\theta)$. Chebyshev polynomials have the following properties:

$$|T_n(x)| \le 1 \qquad (-1 \le x \le 1),$$

$$T_n\left(\cos\frac{k\pi}{n}\right) = (-1)^k \quad (0 \le k \le n),$$

and their zeros are $x_k^{(n)} = \cos\frac{(2k+1)\pi}{2n}$ $(k = 0, 1, \ldots, n)$. In all monic polynomials $P_n(x) = x^n + \sum_{k=1}^{n} a_k x^{n-k}$, the polynomial $\frac{1}{2^{n-1}}T_n(x)$ is such that $\max_{|x|\le 1}|P_n(x)|$ attains the minimal value $\frac{1}{2^{n-1}}$. If we choose the zeros of the Chebyshev polynomial as the interpolation nodes, then a good interpolation approximation formula is given in Chapter 4.

(b) Legendre polynomials

$$P_n(x) = \frac{1}{2^n n!} \frac{\mathrm{d}^n}{\mathrm{d}x^n}(x^2 - 1)^n$$

are orthonormal polynomials with the weight $\rho(x) = 1$ on $[-1, 1]$, and $P_0(x) = 1$, $P_1(x) = x$, $P_2(x) = \frac{1}{2}(3x^2 - 1)$, and the recurrence formula holds

$$P_{n+1}(x) = \frac{2n+1}{n+1} x P_n(x) - \frac{n}{n+1} P_{n-1}(x) \quad (n \geq 2).$$

The Legendre polynomial $P_n(x)$ has n real zeros $x_\nu = \cos\theta_\nu$, where $\frac{2\nu-1}{2n+1}\pi \leq \theta_\nu \leq \frac{2\nu}{2n+1}\pi$ $(\nu = 1, \ldots, n)$. If we choose zeros of a Legendre polynomial as nodes, a good numerical integral formula is given in Chapter 6. The system

$$\hat{P}_0(x) = \sqrt{\frac{1}{2}} P_0(x),$$
$$\hat{P}_n(x) = \sqrt{\frac{2n+1}{2}} P_n(x) \quad (n \in \mathbb{Z}_+)$$

is a normal orthogonal system on $[-1, 1]$ with the weight $\rho(x) = 1$. If $f \in L^2([-1, 1])$, then f can be expanded into Fourier–Legendre series

$$f(x) = \sum_0^\infty c_n^{(P)} \hat{P}_n(x), \quad \text{where} \quad c_n^{(P)} = \int_{-1}^{1} f(x)\hat{P}_n(x)\,\mathrm{d}x.$$

Its partial sums $s_n^{(P)}(f; x) = \sum_{k=0}^n c_k^{(P)} \hat{P}_k(x)$ satisfying

$$\int_{-1}^{1} (f(x) - s_n^{(P)}(f; x))^2\,\mathrm{d}x \to 0 \quad (n \to \infty).$$

From (3.3.2), it is seen that $s_n^{(P)}(f; x)$ is the best square approximation polynomial in all polynomials of degree n and the best square approximation error is

$$e_n(f) = \left(\sum_{k=n+1}^{\infty} (c_k^{(P)})^2 \right)^{\frac{1}{2}}.$$

(c) Jacobian polynomials

$$J_n^{(\alpha,\beta)}(x) = (1 - x)^{-\alpha}(1 + x)^{-\beta} \frac{(-1)^n}{2^n\, n!} \frac{\mathrm{d}^n}{\mathrm{d}x^n} \left((1 - x)^{n+\alpha}(1 + x)^{n+\beta}\right)$$

are orthogonal polynomials on $[-1, 1]$ with the weight $(1 - x)^\alpha(1 + x)^\beta$ $(\alpha, \beta > -1)$.

- When $\alpha = \beta = 0$, $J_n^{(\alpha,\beta)}(x) = P_n(x)$ (Legendre polynomial).
- When $\alpha = \beta = -\frac{1}{2}$, $J_n^{(\alpha,\beta)} = T_n(x)$ (Chebyshev polynomial).

(d) Laguerre polynomials and Hermite polynomials

Laguerre polynomials and Hermite polynomials are both orthogonal polynomials on infinite intervals. One uses them to solve the square approximation problems on infinite intervals. The Laguerre polynomials

$$L_n(x) = \mathrm{e}^x \frac{\mathrm{d}^n}{\mathrm{d}x^n}(x^n \mathrm{e}^{-x}) \quad (n = 0, 1, \dots)$$

are orthogonal with weight e^{-x} on $[0, \infty)$, and $L_0(x) = 1$, $L_1(x) = -x + 1$, and the recurrence formula holds

$$nL_n(x) = (-x + 2n - 1)L_{n-1}(x) - (n-1)L_{n-2}(x) \quad (n \geq 2).$$

The Hermite polynomials

$$H_n(x) = \mathrm{e}^{x^2} \frac{\mathrm{d}^n}{\mathrm{d}x^n} \mathrm{e}^{-x^2} \quad (n = 0, 1, \dots)$$

are orthogonal with weight e^{-x^2} on $(-\infty, \infty)$, and $H_0(x) = 1$, $H_1(x) = 2x$, and the recurrence formula holds

$$H_n(x) = 2xH_{n-1}(x) - 2(n-1)H_{n-2}(x) \quad (n \geq 2).$$

3.4 Spline approximation and rational approximation

In this section, we discuss approximation by piecewise polynomials and the rational approximation. For the approximation by piecewise polynomials, we consider not only the one-dimensional case but also the high-dimensional case. For functions defined in a domain, we use differentiability to describe the smoothness. Sobolev space is the most important and best known smoothness space.

For $1 \leq p < \infty$ and $r \in \mathbb{Z}_+$, define the *Sobolev space* $W^r(L^p(\Omega))$ as the set of all functions f on Ω which have all derivatives $D^\nu f(|\nu| \leq r)$ in $L^p(\Omega)$, i.e.,

$$\int_\Omega |D^\nu f|^p \,\mathrm{d}t < \infty \quad (|\nu| \leq r),$$

where $D^\nu f = \frac{\partial^{|\nu|} f}{\partial t_1^{\nu_1} \cdots \partial t_d^{\nu_d}}$. Here $\nu = (\nu_1, \dots, \nu_d)$ and $|\nu| = \nu_1 + \cdots + \nu_d$.

For $p = \infty$ and $r \in \mathbb{Z}_+$, define the Sobolev space $W^r(L^\infty(\Omega))$ as the set of all functions f on Ω whose all derivatives $D^\nu f$ $(|\nu| \leq r)$ are continuous on Ω and $\max_{t \in \Omega} |D^\nu f(t)| < \infty$ $(|\nu| \leq r)$, i.e., $W^r(L^\infty(\Omega)) = C^r(\Omega)$.

3.4.1 Approximation by piecewise polynomial based on fixed partitions

A partition Δ of Ω is a finite set, $\Delta = \{C\}$ of subdomains C (such as polyhedrons) which are pairwise disjoint and union to Ω. Define a polynomial of degree r on each subdomain C to obtain a piecewise polynomial relative to the partition Δ. Denote the set of these piecewise polynomials by $S^r(\Delta)$. The *approximation error* of f by elements of $S^r(\Delta)$ is defined as

$$e_\Delta(f)_p := \inf_{S\in S^r(\Delta)} \left(\int_\Omega |f(\mathbf{t}) - S(\mathbf{t})|^p \,\mathrm{d}t\right)^{\frac{1}{p}} \quad (p < \infty),$$

$$e_\Delta(f)_\infty := \inf_{s\in S^r(\Delta)} \left(\max_{\mathbf{t}\in\Delta} |f(\mathbf{t}) - S(\mathbf{t})|\right).$$

Let $\Omega = [0, 1]^d$ and let $S^r(\Delta)$ be the space of piecewise polynomials of order r relative to Δ, where $\Delta = \{C\}$ is a partition of Ω. Denote $\operatorname{diam}(\Delta) := \max_{C\in\Delta}(\operatorname{diam}(C))$. Then, for $f \in W^r(L_p(\Omega))$, the approximation error $e_\Delta(f)_p$ in the space $S^r(\Delta)$ satisfies

$$e_\Delta(f)_p \le L_{r,p}(\operatorname{diam}(\Delta))^r \sum_{|\nu|=r} \left(\int_\Omega |D^\nu f|^p \,\mathrm{d}t\right)^{\frac{1}{p}},$$

where $L_{r,p}$ is a constant independent of f and Δ.

In the one-dimensional case, suppose that f is an r-order continuously differentiable function on $[0, 1]$, i.e., $f \in C^r([0, 1])$. Let $T: 0 = t_0 < t_1 < \cdots < t_N = 1$ be a partition of $[0, 1]$ and $S^r(T)$ be the set of piecewise polynomials of order r relative to the partition T. Then the approximation error of f by elements of $S^r(T)$

$$e_T(f)_\infty := \inf_{S\in S^r(T)} \left(\max_{0\le t\le 1} |f(t) - S(t)|\right)$$

satisfies

$$\sigma_T(f)_\infty \le L_r \max_{0\le t\le 1} |f^{(r)}(t)| \delta_T^r,$$

where $\delta_T = \max_{0\le k<N} |t_{k+1} - t_k|$ and the constant L_r is independent of f and T.

Especially, suppose that T is a partition with equally spaced points, i.e., $t_k = \frac{k}{N}$ $(k = 0, 1, \ldots, N)$ and $\delta_T = \frac{1}{N}$. Then for an r-order continuously differentiable function on $[0, 1]$, the approximation error of f by piecewise polynomials of order r relative to the partition is as follows:

$$\sigma_T(f)_\infty \le L_r \max_{0\le t\le 1} |f^{(r)}(t)| \frac{1}{N^r}, \tag{3.4.1}$$

where the constant L_r is independent of f and N.

If we try to approximate to $f \in C^r([0, 1])$ by an algebraic polynomial $P(x)$ such that the approximation error satisfies (3.4.1), then the degree of $P(x)$ must be greater than or equal to N. For a large N, this is very difficult. From this, we see an advantage of piecewise polynomial approximation.

3.4.2 Approximation by piecewise polynomial via free partitions

In approximation by piecewise polynomials, if partitions Δ are not fixed, then we freely choose partitions depending on the target function f and control the number N of subdomains.

Let $r \in \mathbb{Z}_+$ and for each $N \in \mathbb{Z}_+$, let $\Sigma_{N,r}$ be the space of piecewise polynomials of order r with N pieces on $\Omega = [0, 1]$, that is, for each element $S \in \Sigma_{N,r}$, there exists a partition Λ of Ω consisting of N disjoint intervals $I \subset \Omega$ and r-degree polynomials P_I such that $S = \sum_{t\in\Lambda} P_I\chi_I$, where χ_I is the characteristic function of the interval I.

For $0 < p \le \infty$, define the *approximation error* as

$$\sigma_{N,r}(f)_p = \inf_{S\in\Sigma_{N,r}} \left(\int_0^1 |f(t) - S(t)|^p \, dt \right)^{\frac{1}{p}} \quad (0 \le p < \infty),$$

$$\sigma_{N,r}(f)_\infty = \inf_{S\in\Sigma_{N,r}} \left(\max_{0\le t\le 1} |f(t) - S(t)| \right).$$

The relationship between the approximation error and the smoothness of the target function is given by Petrushev. The advantage is seen there.

The following example also explains the advantage of free partitions. Let $f \in C([0, 1])$ be a piecewise monotone function and Σ_N be the set of piecewise constant functions on $[0, 1]$ with N pieces. Then there is a constant $M > 0$ such that the approximation error

$$\sigma_N(f) = \inf_{S\in\Sigma_N} \left(\max_{0\le t\le 1} |f(t) - S(t)| \right)$$

satisfies $\sigma_N(f) = O\left(\frac{1}{N}\right)$. However, if $f \in \operatorname{lip} \alpha$, we approximate f by piecewise polynomials based on a fixed equally spaced partition T with N pieces, the approximation error $e_T(f)$ cannot attain $O(\frac{1}{n^\alpha})$. In applications, if f has different smoothness on each subinterval, we may choose a partition depending on f such that there are more nodes in which f is not smooth and fewer nodes in which f is smooth. This gives a good approximation.

Consider piecewise polynomial approximation with not only free knots but also free degrees of the polynomial pieces. The theory for this type of approximation is far from complete.

Let Σ_N^* denote the set of all polynomials $S = \sum_{I\in\Delta} P_I\chi_I$, where Δ is a partition and for each $I \in \Delta$, there is a polynomial P_I of degree r_I with $\sum_{I\in\Delta} r_I \le N$. Denote

$$\sigma_N^*(f)_p = \inf_{S\in\Sigma_N^*} \left(\int_0^1 |f(t) - S_N(t)|^p \, dt \right)^{\frac{1}{p}}.$$

Clearly, $\sigma_{Nr}^*(f)_p \le \sigma_{N,r}(f)_p$. For example, when $f(x) = x^\beta$ $(x \in [0, 1], \beta > 0)$,

$$\sigma_{N,r}(f)_p \approx \frac{1}{N^r},$$

$$\sigma_N^*(f)_p \le Ce^{-(\sqrt{2}-1)\sqrt{N}}.$$

3.4.3 Rational approximation

Let $R_n(\mathbb{R}^d)$ denote the space of rational functions in d variables. Each $R \in R_n$ is a quotient $R = P/Q$, where P and Q are two d variate polynomials of total degree $\leq n$. Define the approximation error as

$$r_n(f)_p := \inf_{R \in R_n} \|f - R\|_{L_p(\Omega)}.$$

For many functions with less singularity, rational approximation is very efficient.

For example, $f(t) = |t|$ on $[0, 1]$ is approximated by a polynomial of degree $\leq n$, the approximation error is $E_n(f)_\infty \approx \frac{1}{n}$. However, Newmann showed that if it is approximated by the rational function of total degree $\leq n$, the approximation error $r_n(f)_\infty \leq 3\mathrm{e}^{-\sqrt{n}}$.

In fact, let

$$r_n(t) = t\,\frac{p(t) - p(-t)}{p(t) + p(-t)},$$

where $p(t) = \prod_{k=1}^{n-1}(t + \mathrm{e}^{-\frac{k}{\sqrt{n}}})$. A direct estimate will deduce the above result.

(a) Pade approximation

Suppose that $f(x)$ can be expanded into a power series on $[-1, 1]$

$$f(x) = c_0 + \sum_{k \in \mathbb{Z}_+} c_k x^k.$$

We approximate to f by a rational function $R_{mn}(x) = p_m(x)/q_n(x)$, where

$$p_m(x) = \sum_0^m a_k x^k, \quad q_n(x) = \sum_0^n b_j x^j. \tag{3.4.2}$$

The coefficients $\{a_k\}_{k=0,\dots,m}$ and $\{b_j\}_{j=0,\dots,n}$ satisfy the system of linear equations

$$\begin{aligned}
a_0 &= c_0 b_0,\\
a_1 &= c_1 b_0 + c_0 b_1,\\
&\ \vdots\\
a_m &= c_m b_0 + c_{m-1} b_1 + \cdots + c_{m-n} b_n,
\end{aligned}$$

$$\begin{aligned}
c_{m+1} b_0 + c_m b_1 + \cdots + c_{m-n+1} b_n &= 0,\\
c_{m+2} b_0 + c_{m+1} b_1 + \cdots + c_{m-n+2} b_n &= 0,\\
&\ \vdots\\
c_{m+n} b_0 + c_{m+n-1} b_1 + \cdots + c_m b_n &= 0.
\end{aligned}$$

This is a system of $m+n+1$ linear equations with $m+n+2$ unknown. So this system has a nonzero solution, say $a_0, a_1, \dots, a_m, b_0, b_1, \dots, b_n$, and all its solutions are $ka_0, ka_1, \dots, ka_m, kb_0, kb_1, \dots, kb_n$ for $k \in \mathbb{R}$. From this and (3.4.2), the unique rational function $R_{mn}(x) = p_m(x)/q_n(x)$ with order (m, n) approximating to f is determined. The rational function $R_{mn}(x)$ is called the *Pade approximant*. When $m = n$ or $m = n + 1$, the Pade approximation often gives an exact approximation.

(b) Mahley approximation

Replace the power series expansion in the Pade approximation by the Chebyshev series expansion to obtain the Mahley approximation.

Suppose that a function $f(x)$ can be expanded into a Chebyshev series uniformly and absolutely as follows:

$$f(x) = \sum_{0}^{\infty} c_k T_k(x),$$

where $T_k(x)$ are Chebyshev polynomials, and approximate to $f(x)$ by a rational function $R_{mn}(x) = p_m(x)/q_n(x)$ of degree (m, n), where

$$p_m(x) = \sum_{0}^{m} a_k T_k(x), \quad q_n(x) = \sum_{0}^{n} b_j T_j(x).$$

Then the coefficients $\{a_k\}_{k=0,\dots,m}$ and $\{b_j\}_{j=0,\dots,n}$ are the nonzero solution of the system of linear equations

$$a_0 = b_0 c_0 + \frac{1}{2} \sum_{1}^{n} b_\nu c_\nu,$$

$$a_k = b_0 c_k + \frac{1}{2} b_k c_0 + \frac{1}{2} \sum_{\nu=1}^{n} b_\nu (c_{\nu+k} + c_{|\nu-k|}) \quad (k = 1, 2, \dots, m+n).$$

Pade approximation is based on the power series expansion. It sometimes has a good effect only in a neighborhood at the center for the power series. Comparing these two kinds of approximations, the effect of Mahley approximation is better than that of Pade approximation. Especially when the convergence rate of power series expansion is slow, while the convergence rate of Chebyshev series expansion is quick, their difference is large.

3.5 Wavelet approximation

Wavelet theory provides a simple and powerful decomposition of the target function into a series. Multiresolution approximation and the N-term approximation of wavelet series are often used. Moreover, these algorithms can be generalized easily to the high-dimensional case.

3.5.1 Fourier transforms and wavelet transforms

The *Fourier transform* of $f \in L(\mathbb{R})$ is defined as

$$\hat{f}(\omega) = \int_{\mathbb{R}} f(t)\,\mathrm{e}^{-\mathrm{i}t\omega}\,\mathrm{d}t \quad (\omega \in \mathbb{R}).$$

If $\hat{f} \in L(\mathbb{R})$, the *inverse transform* is

$$f(t) = \frac{1}{2\pi}\int_{\mathbb{R}} \hat{f}(\omega)\,\mathrm{e}^{\mathrm{i}t\omega}\,\mathrm{d}\omega \quad (t \in \mathbb{R}).$$

For example, the Fourier transform of

$$f(t) = \chi_{[0,1]}(t) \quad \text{is} \quad \hat{f}(\omega) = \frac{2\sin\omega}{\omega},$$

where $\chi_{[0,1]}$ is the characteristic function of $[0,1]$; the Fourier transform of

$$f(t) = \mathrm{e}^{-t^2} \quad \text{is} \quad \hat{f}(\omega) = \sqrt{\pi}\,\mathrm{e}^{-\frac{\omega^2}{4}}.$$

The Fourier transform has the following properties:

- If $f \in L^1(\mathbb{R})$, then $\lim_{\omega\to\infty}\hat{f}(\omega) = 0$.
- If $f^{(m)}(t)$ is continuous on $\mathbb{R}$ and $f^{(k)} \in L^1(\mathbb{R})$ for each k, then $\hat{f}(\omega) = O\left(\frac{1}{|\omega|^m}\right)$.
- Operational rules

$$(f(t+\alpha))^\wedge = \mathrm{e}^{\mathrm{i}\alpha\omega}\hat{f}(\omega),$$
$$(\mathrm{e}^{\mathrm{i}bt}f(t)^\wedge) = \hat{f}(\omega - b),$$
$$(f(\lambda t))^\wedge = \frac{1}{|\lambda|}\hat{f}\left(\frac{\omega}{\lambda}\right).$$

- Convolution formula $(f * g)^\wedge(\omega) = \hat{f}(\omega)\hat{g}(\omega)$, where the convolution product of $f, g \in L^1(\mathbb{R})$ is defined as $(f * g)(t) = \int_{\mathbb{R}} f(t-x)g(x)\,\mathrm{d}x$.

The space $L^2(\mathbb{R})$ is the set of functions f satisfying $\int_{\mathbb{R}} |f(t)|^2\,\mathrm{d}t < \infty$.

In the space $L^2(\mathbb{R})$,

- The inner product is defined as $(f, g) = \int_{\mathbb{R}} f(t)\bar{g}(t)\,\mathrm{d}t$.
- The norm is defined as $\|f\|_{L^2} = \left(\int_{\mathbb{R}} |f(t)|^2\,\mathrm{d}t\right)^{\frac{1}{2}}$.
  If $\|f_n - f\|_{L^2} \to 0\ (n \to \infty)$, we say $\lim_{n\to\infty} f_n = f$.
- Fourier transform is defined as $\hat{f}(\omega) = \lim_{R\to\infty}\int_{-R}^{R} f(t)\,\mathrm{e}^{-\mathrm{i}\omega t}\,\mathrm{d}t$.
- Parseval identity $\|\hat{f}\|_{L^2} = \|f\|_{L^2}$.

Fourier transform of a signal can provide only global frequency information. To obtain the frequency content of the signal as it evolves with time, one introduces the wavelet transform.

A wavelet is a damped function $\psi \in L^2(\mathbb{R})$ with zero average. The *wavelet transform* of $f \in L^2(\mathbb{R})$ is defined by

$$(W_\psi f)(a, b) = \frac{1}{\sqrt{|a|}} \int_{\mathbb{R}} f(t)\bar{\psi}\left(\frac{t-b}{a}\right) \mathrm{d}t \quad (a \neq 0, b \in \mathbb{R}),$$

where $\bar{\psi}$ means the conjugate of ψ. Its inverse formula is

$$f(t) = \frac{1}{c_\psi} \iint_{\mathbb{R}^2} (W_\psi f)(a, b) \frac{1}{\sqrt{|a|}} \psi\left(\frac{t-b}{a}\right) \frac{\mathrm{d}a}{a^2} \mathrm{d}b \quad (t \in \mathbb{R}),$$

where $c_\psi = \int_{\mathbb{R}} \frac{|\hat{\psi}(\omega)|^2}{|\omega|} \mathrm{d}\omega < \infty$. In applications, Morlet wavelets $\psi^M(t)$ and Mexican hat wavelet $\psi^H(t)$ are often used. They have respectively the representations

$$\psi^M(t) = \pi^{-\frac{1}{4}} \mathrm{e}^{\mathrm{i}t\theta} \mathrm{e}^{-\frac{t^2}{2}},$$
$$\psi^H(t) = -\frac{1}{\Gamma(2.5)} (1 - t^2) \mathrm{e}^{-\frac{t^2}{2}},$$

where θ is a parameter and $\Gamma(t)$ is the gamma function.

3.5.2 Multiresolution analyses and wavelet bases

Let $\psi \in L^2(\mathbb{R})$. If the integral translations and dyadic dilations of a function ψ

$$\psi_{mn}(t) = 2^{\frac{m}{2}} \psi(2^m t - n) \quad (m, n \in \mathbb{Z})$$

form a normal orthogonal basis for $L^2(\mathbb{R})$, then $\{\psi_{mn}\}_{m,n\in\mathbb{Z}}$ is called a *wavelet basis* and ψ is called a *wavelet*. For example, let

$$\psi_h(t) = \begin{cases} -1, & 0 \le t < \frac{1}{2}, \\ 1, & \frac{1}{2} < t < 1, \\ 0, & \text{otherwise}. \end{cases} \tag{3.5.1}$$

Then the Haar system $\{2^{\frac{m}{2}} \psi_h(2^m t - n)\}$ $(m, n \in \mathbb{Z})$ is a normal orthogonal basis for $L^2(\mathbb{R})$. This is the simplest wavelet basis. The construction of a wavelet basis is based on multiresolution analysis (MRA).

An MRA consists of a sequence $\{V_m\}_{m\in\mathbb{Z}}$ of closed subspaces in $L^2(\mathbb{R})$ satisfying

(a) $V_m \subset V_{m+1}$, $\overline{\bigcup_{m\in\mathbb{Z}} V_m} = L^2(\mathbb{R})$, $\bigcap_{m\in\mathbb{Z}} V_m = \{0\}$;
(b) $f \in V_m$ if and only if $f(2\cdot) \in V_{m+1}$ $(m \in \mathbb{Z})$;
(c) there exists a $\varphi \in V_0$ such that $\{\varphi(t-n)\}_{n\in\mathbb{Z}}$ is a normal orthogonal basis of V_0, where φ is called a *scaling function*.

For example, for any $m \in \mathbb{Z}$, let

$$V_m^{(h)} = \left\{f : f \in L^2(\mathbb{R}) \text{ and is a constant in each interval } \left(\frac{n}{2^m}, \frac{n+1}{2^m}\right), \ n \in \mathbb{Z}\right\}.$$

Then $\{V_n^{(h)}\}_{m\in\mathbb{Z}}$ is an MRA with scaling function $\varphi_h(t) = \chi_{[0,1]}(t)$. Another example is

$$V_m^{(s)} = \{f; f \in L^2(\mathbb{R}),\ \hat{f}(\omega) = 0\ (|\omega| > 2^m\pi)\} \quad (m \in \mathbb{Z}),$$

where $\hat{f}$ is the Fourier transform of f. Then $\{V_m^{(s)}\}$ is an MRA with scaling function $\varphi_s(t) = \frac{\sin(\pi t)}{\pi t}$.

One uses an MRA to construct a wavelet basis as follows.

Step 1. Find the *transfer function* H satisfying the *bi-scale equation* $\hat{\varphi}(2\cdot) = H\hat{\varphi}$, and H is a 2π-periodic function and $\int_{-\pi}^{\pi} |H(\omega)|^2\, d\omega < \infty$.

Step 2. Expand $H(\omega)$ into Fourier series $H(\omega) = \sum_{n\in\mathbb{Z}} c_n e^{in\omega}$, where c_n are *bi-scale coefficients*.

Step 3. Find the wavelet using the *wavelet formula*

$$\psi(t) = -2 \sum_{n\in\mathbb{Z}} (-1)^n \bar{c}_{1-n} \varphi(2t - n)$$

or $\hat{\psi}(\omega) = e^{-i\frac{\omega}{2}} \bar{H}(\frac{\omega}{2} + \pi)\hat{\varphi}(\frac{\omega}{2})$.

Step 4. The integral translations and dyadic dilations of ψ derive a wavelet basis $\{\psi_{m,n}(t)\}_{m,n\in\mathbb{Z}}$, where

$$\psi_{mn}(t) = 2^{\frac{m}{2}} \psi(2^m t - n).$$

For example, from the scaling function $\varphi_h(t) = \chi_{[0,1]}$, one constructs the Haar wavelet $\psi_h(t)$ (see (3.5.1)). Haar wavelet is a discontinuous function. From the scaling function $\varphi_s(t) = \frac{\sin(\pi t)}{\pi t}$, one constructs the Shannon wavelet

$$\psi(t) = \frac{\sin 2\pi(t - \frac{1}{2}) - \sin \pi(t - \frac{1}{2})}{\pi(t - \frac{1}{2})}.$$

Shannon wavelet is infinitely many times differentiable but it decays very slowly. A lot of good wavelets, such as Meyer wavelets, Battle–Lemarie wavelets, and Daubechies wavelets, have been constructed.

Meyer wavelets and Battle–Lemarie wavelets have good smoothness and decay fast. Fourier transform of Meyer wavelets $\psi(t)$ is

$$\hat{\psi}(\omega) = \begin{cases} e^{-i\frac{\omega}{2}} \sin\left(\frac{\pi}{2}\nu(\frac{3}{2\pi}|\omega| - 1)\right), & \frac{2}{3}\pi \le |\omega| \le \frac{4}{3}\pi, \\ e^{-i\frac{\omega}{2}} \cos\left(\frac{\pi}{2}\nu(\frac{3}{4\pi}|\omega| - 1)\right), & \frac{4}{3}\pi \le |\omega| \le \frac{8}{3}\pi, \end{cases}$$

where $\nu(x)$ is an n-degree differentiable real-valued function and

$$\nu(x) = \begin{cases} 0 & (x \le 0), \\ 1 & (x \ge 1) \end{cases} \qquad \nu(x) + \nu(1 - x) = 1 \quad (x \in \mathbb{R}).$$

Fourier transform of the Battle–Lemarie wavelet $\psi_k(t)$ of degree k is

$$\hat{\psi}_k(\omega) = \left(\frac{4}{i\omega}\right)^k e^{-i\frac{\omega}{2}} \sin^{2k}\frac{\omega}{4} \left(\frac{F_k(\frac{\omega}{2} + \pi)}{F_k(\frac{\omega}{2})F_k(\omega)}\right)^{\frac{1}{2}},$$

where $F_k(2\omega) = -\frac{\sin^{2k}\omega}{(2k-1)!}\frac{d^{2k-1}(\cot\omega)}{d\omega^{2k-1}}$. Especially, $F_1(\omega) = 1$ and $F_2(\omega) = \frac{1}{3}\sin^2\frac{\omega}{2} + \cos^2\frac{\omega}{2}$. Daubechies wavelet has good smoothness and is compactly supported. It is introduced in Section 6.6.

Let $\{\psi_{mn}\}_{m,n\in\mathbb{Z}}$ be a wavelet basis. Then $f \in L^2(\mathbb{R})$ can be expanded into a wavelet series

$$f(t) = \sum_{m,n\in\mathbb{Z}} d_{mn}\psi_{mn}(t)$$

in L^2-sense, where

$$d_{mn} = (f, \psi_{mn}) = \int_{\mathbb{R}} f(t)\bar{\psi}_{mn}(t)\,dt$$

are called *wavelet coefficients*. Another wavelet expansion formula is that for $M \in \mathbb{Z}$,

$$f(t) = \sum_{n\in\mathbb{Z}} c_{Mn}\varphi_{Mn}(t) + \sum_{m=M}^{\infty}\sum_{n\in\mathbb{Z}} d_{mn}\psi_{mn}(t),$$

where φ is the scaling function and

$$\varphi_{Mn}(t) = 2^{\frac{M}{2}}\varphi(2^M t - n),$$
$$c_{Mn} = (f, \varphi_{Mn}).$$

Moreover, $\{\varphi_{Mn}(t)\}_{n\in\mathbb{Z}}$ and $\{\psi_{mn}\}_{m=M,M+1,\dots;\,n\in\mathbb{Z}}$ are a normal orthogonal basis for $L^2(\mathbb{R})$.

If f has few nonnegligible wavelet coefficients d_{mn} and scale coefficients c_{Mn}, then f can be approximated by the sum of few terms in these expansions. This is very useful in data compression and noise removal.

We say ψ has p *vanishing moments* if $\int_{\mathbb{R}} t^k\psi(t)\,dt = 0$ $(k = 0, \dots, p)$. If f is smooth and the wavelet ψ has a high vanishing moment, the Taylor theorem shows that wavelet coefficients of the fine-scale are small.

3.5.3 High-dimensional wavelets

The notion of multiresolution analyses (MRA) is generalized easily to the high-dimensional space $L^2(\mathbb{R}^d)$.

An MRA consists of a sequence $\{V_m\}_{m\in\mathbb{Z}}$ of closed subspaces in $L^2(\mathbb{R}^d)$ satisfying

(a) $V_m \subset V_{m+1}$, $\overline{\bigcup_{m\in\mathbb{Z}} V_m} = L^2(\mathbb{R}^d)$, $\bigcap_{m\in\mathbb{Z}} V_m = \{0\}$;

(b) $f \in V_m$ if and only if $f(2\cdot) \in V_{m+1}$ $(m \in \mathbb{Z})$;

(c) there exists a $\varphi \in V_0$ such that $\{\varphi(\mathbf{t} - \mathbf{n})\}_{\mathbf{n}\in\mathbb{Z}^d}$ is a normal orthogonal basis of V_0, where φ is called a *scaling function*.

One may use a high-dimensional MRA to construct a high-dimensional wavelet basis as follows.

As in the one-dimensional case, the scaling function φ satisfies

$$\hat{\varphi}(2\omega) = H(\omega)\hat{\varphi}(\omega) \quad (\omega \in \mathbb{R}^d),$$

where H is a 2π-periodic function in $\mathbb{R}^d$ and $\int_{[-\pi,\pi]^d} |H(\omega)|^2 \, d\omega < \infty$. Let $H_{\mathbf{0}} = H$. Denote by $\{0, 1\}^d$ the set of vertices of $[0, 1]^d$. For example, $\{0, 1\}^2 = \{(0, 0), (0, 1), (1, 0), (1, 1)\}$. Take $2^d - 1$ functions $\{H_\mu\}_{\mu \in \{0,1\}^d \setminus \{\mathbf{0}\}}$ with period 2π such that the $2^d \times 2^d$ matrix $(H_\mu(\omega + \pi\nu))_{\mu,\nu \in \{0,1\}^d}$ is a unitary matrix. Let $\{\psi_\mu\}_{\mu \in (\{0,1\}^d \setminus \{\mathbf{0}\})}$ be such that

$$\hat{\psi}_\mu(2\omega) = H_\mu(\omega)\hat{\varphi}(\omega) \quad (\mu \in (\{0, 1\}^d \setminus \{\mathbf{0}\})),$$
$$\psi_{\mu m \mathbf{n}}(\mathbf{t}) = 2^{\frac{md}{2}} \psi_\mu(2^m \mathbf{t} - \mathbf{n}).$$

Then $\{\psi_{\mu m \mathbf{n}}\}_{\mu \in (\{0,1\}^d \setminus \{\mathbf{0}\}), m \in \mathbb{Z}, \mathbf{n} \in \mathbb{Z}^d}$ form an *normal orthogonal basis* for $L^2(\mathbb{R}^d)$.

The following is the simplest method for constructing a high-dimensional wavelet basis.

Suppose that φ is a univariate scaling function and ψ is the corresponding wavelet. Denote $\psi^0 = \varphi$ and $\psi^1 = \psi$. For each nonzero vertex $\mathbf{e} = (e_1, \ldots, e_d) \in \{0, 1\}^d \setminus \{\mathbf{0}\}$, define a d-variate function

$$\psi^{\mathbf{e}}(t_1, \ldots, t_d) = \psi^{e_1}(t_1) \cdots \psi^{e_d}(t_d).$$

The set $\Psi = \{\psi^{\mathbf{e}}, \mathbf{e} \in \{0, 1\}^d \setminus \{\mathbf{0}\}\}$ is a d-variate wavelet and

$$\psi^{\mathbf{e}}_{m\mathbf{n}}(\mathbf{t}) = 2^{\frac{md}{2}} \psi^{\mathbf{e}}(2^m \mathbf{t} - \mathbf{n}) \quad (m \in \mathbb{Z}, \mathbf{n} \in \mathbb{Z}^d)$$

is a d-dimensional wavelet basis, where $\mathbf{t} = (t_1, \ldots, t_d)$. Each $f \in L^2(\mathbb{R}^d)$ can be expanded into a wavelet series

$$f = \sum_{\mathbf{e} \in \{0,1\}^d \setminus \{\mathbf{0}\}} \sum_{m \in \mathbb{Z}} \sum_{\mathbf{n} \in \mathbb{Z}^d} c^{\mathbf{e}}_{m\mathbf{n}} \psi^{\mathbf{e}}_{m\mathbf{n}},$$

where $c^{\mathbf{e}}_{m\mathbf{n}} = (f, \psi^{\mathbf{e}}_{m\mathbf{n}}) = \int_{\mathbb{R}^d} f(\mathbf{t}) \overline{\psi^{\mathbf{e}}_{m,\mathbf{n}}(\mathbf{t})} \, d\mathbf{t}$.

Especially, the two-dimensional wavelets are

$$\psi^{(1)}(\mathbf{t}) = \varphi(t_1)\psi(t_2),$$
$$\psi^{(2)}(\mathbf{t}) = \psi(t_1)\varphi(t_2),$$
$$\psi^{(3)}(\mathbf{t}) = \psi(t_1)\psi(t_2),$$

where $\mathbf{t} = (t_1, t_2) \in \mathbb{R}^2$, and

$$\psi^{(k)}_{m\mathbf{n}}(\mathbf{t}) = 2^m \psi^{(k)}(2^m \mathbf{t} - \mathbf{n}), \quad m \in \mathbb{Z}, \ \mathbf{n} \in \mathbb{Z}^2 \quad (k = 1, 2, 3)$$

is a bivariate wavelet basis.

3.5.4 Wavelet packet

The emergence of the concept of the wavelet packet enables us to construct many normal orthogonal bases from an MRA. Wavelet packet is defined as follows.

Let $\{V_m\}$ be a d-dimensional MRA with scaling function φ and 2^d transfer functions $\{H_\mu\}_{\mu\in\{0,1\}^d}$. The corresponding wavelets $\{\psi_\mu\}$ satisfy

$$\hat{\psi}_\mu(\omega) = H_\mu\left(\frac{\omega}{2}\right)\hat{\varphi}\left(\frac{\omega}{2}\right), \quad \mu \in (\{0,1\}^d \setminus \{\mathbf{0}\}).$$

Define $\psi_{\nu_1,\dots,\nu_j}$ such that

$$\hat{\psi}_{\nu_1,\dots,\nu_j}(\omega) = \hat{\varphi}\left(\frac{\omega}{2^j}\right)\prod_1^j H_{\nu_i}\left(\frac{\omega}{2^i}\right),$$

where each $\nu_i \in \{0,1\}^d$ $(i = 1,\dots,j)$. Let ν_i be the $\tilde{\nu}_{l_i}$-th vertex in $\{0,1\}^d$. Denote $l = \tilde{\nu}_{l_1} + 2^d\tilde{\nu}_{l_2} + \cdots + 2^{(j-1)d}\tilde{\nu}_{l_j}$, where $\tilde{\nu}_1, \tilde{\nu}_2, \dots, \tilde{\nu}_{2^d}$ are an arbitrary serial number of $\{0,1\}^d$. Then l corresponds to the set $\nu_1,\dots,\nu_j$ one-to-one. Denote $\omega_l(\mathbf{t}) = \psi_{\nu_1,\dots,\nu_j}(\mathbf{t})$. The system

$$\{2^{\frac{md}{2}}\omega_l(2^m\mathbf{t}-\mathbf{k})\}_{l,m\in\mathbb{Z}_+^*,\mathbf{k}\in\mathbb{Z}^d}$$

is called a *wavelet packet*, where $\mathbb{Z}_+^* = \mathbb{Z}_+\bigcup\{0\}$.

Let S be the set of pairs (l,m) $(l,m\in\mathbb{Z}_+^*)$. From a wavelet packet the question is how to choose S such that $\{2^{\frac{md}{2}}\omega_l(2^m t-\mathbf{k})\}_{(l,m)\in S,\mathbf{k}\in\mathbb{Z}^d}$ form an normal orthogonal basis for $L^2(\mathbb{R}^d)$? The answer is as follows. Denote

$$I_{lm} = \{\tau\in\mathbb{Z}_+^*:\ 2^{md}l \le \tau < 2^{md}(l+1)\}, \quad \text{where } l\in\mathbb{Z}_+^*,\ m\in\mathbb{Z}_+^*.$$

Then

$$\{2^{\frac{md}{2}}\omega_l(2^m t-\mathbf{k})\}_{(l,m)\in S,\mathbf{k}\in\mathbb{Z}^d}$$

is an normal orthogonal basis for $L^2(\mathbb{R}^d)$ if and only if the set $S\in\mathbb{Z}_+^*\times\mathbb{Z}_+^*$ is such that $\{I_{lm}\}_{(l,m)\in S}$ forms a partition of $\mathbb{Z}_+^*$.

For example, let $\tau\in\mathbb{Z}_+^*$ and

$$S = \left\{(l,m):\ (l,m)\in\left([0,2^{\tau d})\times\{0\}\right)\bigcup\left([2^{\tau d},2^{(\tau+1)d})\times\mathbb{Z}_+^*\right)\right\}.$$

Then $\{I_{lm}\}_{(l,m)\in S}$ is a disjoint covering of $\mathbb{Z}_+^*$. In fact, since I_{lm} $((l,m)\in S)$ are disjoint, from

$$\bigcup_{2^{\tau d}\le l<2^{(\tau+1)d}} I_{lm} = [2^{(\tau+m)d}, 2^{(\tau+m+1)d})\bigcap\mathbb{Z}_+^*,$$

$$\bigcup_{m\in\mathbb{Z}_+^*}\left(\bigcup_{2^{\tau d}\le l<2^{(\tau+1)d}} I_{lm}\right) = [2^{\tau d},\infty)\bigcap\mathbb{Z}_+^*,$$

it follows that $\bigcup_{(l,m)\in S} I_{lm} = \mathbb{Z}_+^*$, and so

$$\{2^{\frac{md}{2}}\omega_l(2^m\mathbf{t}-\mathbf{k}),\quad \mathbf{k}\in\mathbb{Z}^d, (l,m)\in S\}$$

is a normal orthogonal basis for $L^2(\mathbb{R}^d)$. When $\tau = 0$, $S = \{(l, m) : (l, m) \in ([0, 1] \times \{0\}) \bigcup ([1, 2^d) \times \mathbb{Z}_+^*)\}$, and so

$$\{\varphi(\boldsymbol{t} - \boldsymbol{k}),\ 2^{\frac{md}{2}} \omega_l(2^m \boldsymbol{t} - \boldsymbol{k}),\ 1 \le l < 2^d,\ m = 0, 1, \dots, k \in \mathbb{Z}^d\}$$

is a normal orthogonal basis for $L^2(\mathbb{R}^d)$.

For example, let $S = \mathbb{Z}_+^* \times \{0\}$. Then $I_{l0} = \{l\}$ and $\bigcup_{l \in \mathbb{Z}_+^*} I_{l0} = \mathbb{Z}_+^*$. So $\{\omega_l(\boldsymbol{t} - \boldsymbol{k})\}_{l \in \mathbb{Z}_+^*, \boldsymbol{k} \in \mathbb{Z}^d}$ is a normal orthogonal basis for $L^2(\mathbb{R}^d)$.

3.5.5 Wavelet decomposition and pyramid algorithm

Let $\{V_m\}_{m \in \mathbb{Z}}$ be an MRA with scaling function φ and the corresponding wavelet ψ. Since $\overline{\bigcup_m V_m} = L^2(\mathbb{R})$, for $f \in L^2(\mathbb{R})$, one can choose a sufficiently large L such that

$$f \approx \operatorname{Proj}_{V_L} f =: f_L,$$

where $\operatorname{Proj}_{V_L} f$ is the projection of f on the space V_L. Since $V_m \subset V_{m+1}$ $(m \in \mathbb{Z})$,

$$V_{m+1} = V_m \bigoplus W_m \quad (m \in \mathbb{Z}), \tag{3.5.2}$$

where W_m is the orthogonal complement space of V_m on V_{m+1}. Moreover, $\{\varphi_{mn}\}_{n \in \mathbb{Z}}$ and $\{\psi_{mn}\}_{n \in \mathbb{Z}}$ are normal orthogonal bases of V_m and W_m, respectively. Let

$$\begin{aligned} f_m &= \operatorname{Proj}_{V_m} f, \\ g_m &= \operatorname{Proj}_{W_m} f. \end{aligned} \tag{3.5.3}$$

Then $f_{m+1} = f_m + g_m$ $(m \in \mathbb{Z})$ and

$$\begin{aligned} f_{m+1}(t) &= \sum_{n \in \mathbb{Z}} c_{m+1,n} \varphi_{m+1,n}(t), \\ f_m(t) &= \sum_{n \in \mathbb{Z}} c_{mn} \varphi_{mn}(t), \quad g_m(t) = \sum_{n \in \mathbb{Z}} d_{mn} \psi_{mn}(t). \end{aligned}$$

By (3.5.2) and (3.5.3), for $l < L$,

$$\begin{aligned} V_L &= V_l + \bigoplus_{m=l}^{L} W_m, \\ f_L &= f_l + \sum_{m=l}^{L} g_m. \end{aligned}$$

This gives a *wavelet decomposition formula*: for the sufficiently large L,

$$\begin{aligned} f(t) \approx f_L(t) &= \sum_{n \in \mathbb{Z}} c_{Ln} \varphi_{Ln}(t) \\ &= \sum_{n \in \mathbb{Z}} c_{ln} \varphi_{ln}(t) + \sum_{m=l}^{L-1} \sum_{n \in \mathbb{Z}} d_{mn} \psi_{mn}(t) \quad \text{for some } l < L, \end{aligned}$$

where $c_{mn} = (f, \varphi_{mn})$ and $d_{mn} = (f, \psi_{mn})$. In applications, a lot of coefficients are negligible. So the series $\sum_{n\in\mathbb{Z}} c_{Ln}\varphi_{Ln}(t)$ has only finitely many terms. From this decomposition formula, one uses the following pyramid algorithm to fast compute coefficients c_{ln}, d_{mn} $(m = l, \ldots, L-1)$ from finite many coefficients c_{Ln}, while there is no need to compute each coefficient by integrals.

Pyramid algorithm provides a fast algorithm that can compute coefficients $\{c_{mn}\}$ and $\{d_{mn}\}$. From the bi-scale equation and the wavelet formula, it follows that

$$\varphi(t) = \sum_{k\in\mathbb{Z}} p_k \varphi(2t-k),$$
$$\psi(t) = \sum_{k\in\mathbb{Z}} q_k \varphi(2t-k),$$

where $p_k = 2c_k$ and $q_k = 2(-1)^{k+1}\bar{c}_{1-k}$, and c_k are the bi-scale coefficients. The decomposition and synthesis formulas in the pyramid algorithm are, respectively,

$$c_{mn} = \sum_{l\in\mathbb{Z}} \bar{p}_{l-2n} c_{m+1,l},$$
$$d_{mn} = \sum_{l\in\mathbb{Z}} \bar{q}_{l-2n} c_{m+1,l}$$

and

$$c_{m+1,n} = \sum_{l\in\mathbb{Z}} p_{n-2l} c_{ml} + \sum_{l\in\mathbb{Z}} q_{n-2l} d_{ml}.$$

These formulas can fast compute $\{c_{mn}\}_{n\in\mathbb{Z}}$, $\{d_{mn}\}_{n\in\mathbb{Z}}$, and $\{c_{m+1,n}\}_{n\in\mathbb{Z}}$.

Consider the two-dimensional case.

Let φ and ψ be the univariate scaling function and the corresponding univariate wavelet. Then $\varphi(t_1)\varphi(t_2)$ is a bivariate scaling function and $\varphi(t_1)\psi(t_2)$, $\psi(t_1)\varphi(t_2)$, $\psi(t_1)\psi(t_2)$ are the corresponding bivariate wavelets. For $f \in L^2(\mathbb{R}^2)$ and sufficiently large L,

$$\begin{aligned}
f(t_1,t_2) \approx (\mathrm{Proj}_{V_L} f)(t_1,t_2) &= \sum_{n_1\in\mathbb{Z}}\sum_{n_2\in\mathbb{Z}} c^L_{n_1n_2}\varphi_{Ln_1}(t_1)\varphi_{Ln_2}(t) \\
&= \sum_{n_1\in\mathbb{Z}}\sum_{n_2\in\mathbb{Z}} c^{(l)}_{n_1n_2}\varphi_{ln_1}(t_1)\varphi_{ln_2}(t_2) \\
&\quad + \sum_{m=l}^{L-1}\sum_{n_1\in\mathbb{Z}}\sum_{n_2\in\mathbb{Z}} d^{(1,m)}_{n_1n_2}\varphi_{mn_1}(t_1)\psi_{mn_2}(t_2) \\
&\quad + \sum_{m=l}^{L-1}\sum_{n_1\in\mathbb{Z}}\sum_{n_2\in\mathbb{Z}} d^{(2,m)}_{n_1n_2}\psi_{mn_1}(t_1)\varphi_{mn_2}(t_2) \\
&\quad + \sum_{m=l}^{L-1}\sum_{n_1\in\mathbb{Z}}\sum_{n_2\in\mathbb{Z}} d^{(3,m)}_{n_1n_2}\psi_{mn_1}(t_1)\psi_{mn_2}(t_2),
\end{aligned}$$

where

$$h_{mn}(\mathbf{t}) = 2^{\frac{m}{2}} h(2^m\mathbf{t} - \mathbf{n}) \quad (\mathbf{n} = (n_1, n_2))$$

and the coefficients are, respectively,

$$c_{n_1n_2}^{(m)} = \int_{\mathbb{R}^2} f(t_1,t_2)\bar{\varphi}_{Mn_1}(t_1)\bar{\varphi}_{Mn_2}(t_2)\,\mathrm{d}t_1\,\mathrm{d}t_2,$$

$$d_{n_1n_2}^{(1,m)} = \int_{\mathbb{R}^2} f(t_1,t_2)\bar{\varphi}_{mn_1}(t_1)\bar{\psi}_{mn_2}(t_2)\,\mathrm{d}t_1\,\mathrm{d}t_2,$$

$$d_{n_1n_2}^{(2,m)} = \int_{\mathbb{R}^2} f(t_1,t_2)\bar{\psi}_{mn_1}(t_1)\bar{\varphi}_{mn_2}(t_2)\,\mathrm{d}t_1\,\mathrm{d}t_2,$$

$$d_{n_1n_2}^{(3,m)} = \int_{\mathbb{R}^2} f(t_1,t_2)\bar{\psi}_{mn_1}(t_1)\bar{\psi}_{mn_2}(t_2)\,\mathrm{d}t_1\,\mathrm{d}t_2.$$

The decomposition formulas are

$$c_{n_1n_2}^{(m)} = \sum_{l_1,l_2\in\mathbb{Z}} \bar{p}_{l_1-2n_1}\bar{p}_{l_2-2n_2}c_{l_1l_2}^{(m+1)},$$

$$d_{n_1n_2}^{(1,m)} = \sum_{l_1,l_2\in\mathbb{Z}} \bar{p}_{l_1-2n_1}\bar{q}_{l_2-2n_2}c_{l_1l_2}^{(m+1)},$$

$$d_{n_1n_2}^{(2,m)} = \sum_{l_1,l_2\in\mathbb{Z}} \bar{q}_{l_1-2n_1}\bar{p}_{l_2-2n_2}c_{l_1l_2}^{(m+1)},$$

$$d_{n_1n_2}^{(3,m)} = \sum_{l_1,l_2\in\mathbb{Z}} \bar{q}_{l_1-2n_1}\bar{q}_{l_2-2n_2}c_{l_1l_2}^{(m+1)}$$

and the reconstruction formula is

$$4c_{n_1n_2}^{(m+1)} = \sum_{l_1,l_2\in\mathbb{Z}} \bar{p}_{l_1-2n_1}\bar{p}_{l_2-2n_2}c_{l_1l_2}^{(m)} + \sum_{l_1,l_2\in\mathbb{Z}} \bar{p}_{l_1-2n_1}\bar{q}_{l_2-2n_2}d_{l_1l_2}^{(1,m)}$$
$$+ \sum_{l_1,l_2\in\mathbb{Z}} \bar{q}_{l_1-2n_1}\bar{p}_{l_2-2n_2}d_{l_1l_2}^{(2,m)} + \sum_{l_1,l_2\in\mathbb{Z}} \bar{q}_{l_1-2n_1}\bar{q}_{l_2-2n_2}d_{l_1l_2}^{(3,m)}.$$

3.5.6 Multiresolution approximation

Let $\{V_m\}_{m\in\mathbb{Z}}$ be an MRA of $L^2(\mathbb{R}^d)$ with scaling function φ. Define the *best approximation* of $f \in L^2(\mathbb{R}^d)$ in the space V_m as

$$E_m(f)_2 = \inf_{s\in V_m} \|f - s\|_{L^2(\mathbb{R}^d)}.$$

Since $\overline{\bigcup_m V_m} = L^2(\mathbb{R}^d)$ and $V_m \subset V_{m+1}$ ($m \in \mathbb{Z}$), then $E_m(f)_2 \to 0$ as $m \to \infty$.

Suppose that $\varphi \in W^r(L^2(\mathbb{R}^d))$ satisfies the Strang–Fix condition of order $r \in \mathbb{Z}_+$

$$\hat{\varphi}(\mathbf{0}) = 0,$$
$$D^{\mathbf{j}}\hat{\varphi}(2\mathbf{k}\pi) = 0 \quad (\mathbf{k} \in \mathbb{Z}^d \setminus \{\mathbf{0}\},\ |\mathbf{j}| < r),$$

where

$$D^{\mathbf{j}} = \frac{\partial^{|\mathbf{j}|}}{\partial t_1^{j_1}\cdots\partial t_d^{j_d}}, \quad \mathbf{t} = (t_1,\ldots,t_d),\ \mathbf{j} = (j_1,\ldots,j_d),\ |\mathbf{j}| = j_1 + \cdots + j_d.$$

Then

$$E_m(f)_2 \le C2^{-mr} \sum_{|j|\le r} \left(\int_{\mathbb{R}^d} |D^j f(t)|^p \,\mathrm{d}t \right)^{\frac{1}{p}},$$

where C is a constant independent of m, f, and s. Since the best approximation of f in V_m is $\mathrm{Proj}_{V_m} f$,

$$\|f - \mathrm{Proj}_{V_m} f\|_{L^2(\mathbb{R}^d)} = O\left(\frac{1}{2^{mr}}\right),$$

where $\|g\|_{L^2(\mathbb{R}^d)} = \left(\int_{\mathbb{R}^d} |g(\boldsymbol{t})|^2 \,\mathrm{d}\boldsymbol{t}\right)^{\frac{1}{2}}$. Since $\varphi_{m\boldsymbol{n}}(\boldsymbol{t}) = 2^{\frac{md}{2}} \varphi(2^m \boldsymbol{t} - \boldsymbol{n})$ $(\boldsymbol{n} \in \mathbb{Z}^d)$ is an orthonormal basis of V_m, we expand $\mathrm{Proj}_{V_m} f$ into Fourier series

$$\mathrm{Proj}_{V_m} f = \sum_{\boldsymbol{n}\in\mathbb{Z}^d} c_{m\boldsymbol{n}} \varphi_{m\boldsymbol{n}}(\boldsymbol{t}),$$

where the coefficients

$$c_{m\boldsymbol{n}} = (\mathrm{Proj}_{V_m} f, \varphi_{m\boldsymbol{n}}) = (f, \varphi_{m\boldsymbol{n}}) = \int_{\mathbb{R}^d} f(t) \varphi_{m\boldsymbol{n}}(\boldsymbol{t}) \,\mathrm{d}\boldsymbol{t}.$$

If the scaling function $\varphi(\boldsymbol{t})$ decays fast as $t \to \infty$, then a lot of coefficients $c_{m\boldsymbol{n}} \approx 0$. Therefore, few coefficients $c_{m\boldsymbol{n}}$ are used for a good approximation.

The above approximation takes place in $L^2(\mathbb{R}^d)$. There is a similar result for approximation in $L^p(\mathbb{R}^d)$.

3.5.7 The *N*-term wavelet approximation and thresholding value methods

Wavelets on intervals have been constructed and extended to those on parallelepipeds and polyhedral domains in $\mathbb{R}^d$. However, constructing wavelets on general domains is very difficult. For wavelet approximation in general domains, after the target function is extended smoothly to the whole space, we consider the N-term wavelet approximation and thresholding value methods.

(a) The *N*-term approximation

In numerical analysis the target functions f are often defined on a finite domain Ω with boundary Lipschitz condition. If the wavelet $\psi^{\boldsymbol{e}}$ $(\boldsymbol{e} \in \{0, 1\}^d/(0, 0))$ is compactly supported, then there are a lot of $(m, \boldsymbol{n})$ such that $\psi^{\boldsymbol{e}}_{m\boldsymbol{n}}(\boldsymbol{t}) = 0$ $(\boldsymbol{t} \in \Omega)$. Thus, a lot of terms in the wavelet expansion of f can be deleted. The best N-term approximation for f can be obtained by choosing the N-terms in the wavelet expansion

$$\sum_{l=1}^{N} c^{e_l}_{m_l,\boldsymbol{n}_l} \psi^{e_l}_{m_l,\boldsymbol{n}_l},$$

where $|c^{e_1}_{m_1,\boldsymbol{n}_1}| \ge |c^{e_2}_{m_2,\boldsymbol{n}_2}| \ge \cdots \ge |c^{e_N}_{m_N,\boldsymbol{n}_N}|$ are largest in absolute value of all coefficients $\{c^{\boldsymbol{e}}_{m\boldsymbol{n}}\}$.

(b) Thresholding value methods

In numerical implementation one likes to use thresholding value methods. For convenience, we consider the one-dimensional case. Expand the target function f into a wavelet series

$$f = \sum_{mn} c_{mn}(f)\psi_{mn}.$$

Let $\Lambda_\eta(f)$ be the set of all pairs (m, n) satisfying $|c_{mn}(f)| > \eta$. Denote by $\#(\Lambda_\eta(f))$ the cardinality of the set $\Lambda_\eta(f)$. If f satisfies a condition that there exists a $\tau > 0$ such that for all $\eta > 0$,

$$\#(\Lambda_\eta(f)) \le M_f^\tau \eta^{-\tau},$$

where M_f is a constant depending only on f, then for a given $\varepsilon > 0$, the *thresholding operator*

$$T_\varepsilon(f) = \sum_{(mn)\in\Lambda_\varepsilon(f)} c_{mn}(f)\psi_{mn} = \sum_{|c_{mn}|>\varepsilon} c_{mn}(f)\psi_{mn}$$

satisfies

$$\|f - T_\varepsilon(f)\|_{L^2(R)} \le CM_f^{\frac{\tau}{2}}\varepsilon^{1-\frac{\tau}{2}},$$

where C is an absolute constant. When $\varepsilon = M_f N^{-\frac{1}{\tau}}$, $\#\Lambda_\varepsilon(f) \le N$, and so the N-term approximation is estimated by

$$\|f - T_\varepsilon(f)\|_{L^2(R)} \le CM_f N^{\frac{1}{2}-\frac{1}{\tau}}.$$

This result is easily extended to L^p-approximation.

The thresholding operator has an instability. To improve it, the *soft thresholding operator* $s_\varepsilon(t)$ is introduced by

$$s_\varepsilon(t) := \begin{cases} 0 & (|t| < \varepsilon), \\ 2(|t| - \varepsilon)\operatorname{sgn} t & (\varepsilon \le |t| \le 2\varepsilon), \\ t & (|t| > 2\varepsilon). \end{cases}$$

The soft thresholding operator $s_\varepsilon(t)$ has the same approximation properties as the thresholding operator T_ε.

(c) Data compression

In the lossy compression replacing the original data by an approximation, we choose a multivariate scaling function φ and then expand the original data I into a series

$$I(\boldsymbol{t}) \sim \sum_{\boldsymbol{n}\in\mathbb{Z}^d} c_{m\boldsymbol{n}}\varphi(2^m\boldsymbol{t} - \boldsymbol{n}).$$

Using the decomposition formula of the pyramid algorithm, we obtain a wavelet expansion

$$I(\boldsymbol{t}) \sim \sum_{\boldsymbol{n}\in\mathbb{Z}^d} c_{0\boldsymbol{n}}\varphi(\boldsymbol{t} - \boldsymbol{n}) + \sum_{k=0}^{m-1} \sum_{\boldsymbol{e}\in\{0,1\}^d\setminus\{\boldsymbol{0}\}} \sum_{\boldsymbol{n}\in\mathbb{Z}^d} d_{k\boldsymbol{n}}^{\boldsymbol{e}}\psi_{k\boldsymbol{n}}^{\boldsymbol{e}}(\boldsymbol{t})$$

and then we use thresholding to give a compressed file $\{\tilde{d}^{e}_{kn}\}$ of wavelet coefficients. The compressed coefficient file is further compressed using a lossless encoder. From the encoded compressed file of wavelet coefficients, we use a decoder and then use the synthesis formula of the pyramid algorithm to give the reconstructed data.

3.6 Greedy algorithms

In this section we regard Hilbert space $\mathcal{H}$ as $L^2(\mathbb{R}^d)$. An arbitrary subset of the Hilbert space is called a *dictionary*. In applications, one considers the N-term approximation from the dictionary D, i.e., the approximation to $f \in \mathcal{H}$ by a linear combination of N-terms in the dictionary D.

In time frequency analysis, the translations of Gabor functions $g_{\alpha,\beta}(t) := e^{i\alpha t}e^{-\beta t^2}$ generate a dictionary D in $L^2(\mathbb{R})$, i.e.,

$$D := \{g_{\alpha,\beta}(t-\gamma) : \alpha, \beta, \gamma \in \mathbb{R}\}.$$

In an MRA of $L^2(\mathbb{R}^d)$, the wavelet packet $\{\omega_l\}_{l=0,1,\ldots}$ has been constructed in Section 3.5. The integral translations and dyadic dilations of the wavelet packet generate a dictionary D in $L^2(\mathbb{R}^d)$, i.e.,

$$D := \{\omega_{l,m,\boldsymbol{n}}(\boldsymbol{t}) = 2^{\frac{m}{2}}\omega_l(2^m\boldsymbol{t}-\boldsymbol{n}),\ l = 0,1,\ldots;\ m, \boldsymbol{n} \in \mathbb{Z}\}.$$

For a general dictionary D and any $\tau > 0$, $M > 0$, define $K_\tau^0(D, M)$ as the set of functions f satisfying

$$f = \sum_{g\in\Lambda} c_g\, g \quad (\Lambda \subset D,\ \#\Lambda < \infty),$$

$$\sum_{g\in\Lambda} |c_g|^\tau \le M^\tau,$$

where $\#\Lambda$ represents the cardinality of Λ. Let $K_\tau(D, M)$ be the closure of $K_\tau^0(D, M)$ in $\mathcal{H}$, and let

$$K_\tau(D) = \bigcup_{M>0} K_\tau(D, M).$$

Define

$$|f|_{K_\tau(D)} = \inf\{M : f \in K_\tau(D, M)\}. \tag{3.6.1}$$

Greedy algorithms are often used in numerical analysis. Most often greedy algorithms are as follows.

3.6.1 Pure greedy algorithm

The advantage of this algorithm is its simplicity. Let $f \in \mathcal{H}$. The algorithm is as follow.

Let $h = h(f) \in D$ be such that

$$(f, h(f)) = \sup_{h\in D}(f, h)$$

where $(f,g) = \int_{\mathbb{R}^d} f(\mathbf{t})\bar{g}(\mathbf{t})\,\mathrm{d}\mathbf{t}$. Define

$$U_1(f) = (f, h(f))h(f), \qquad R_1(f) = f - U_1(f),$$
$$U_2(f) = U_1(f) + U_1(R_1(f)), \quad R_2(f) = f - U_2(f).$$

For each $m \geq 2$, inductively define

$$U_m(f) = U_{m-1}(f) + U_1(R_{m-1}(f)), \quad R_m(f) = f - U_m(f).$$

Because at each iteration of the algorithm it approximates to the residual $R_m(f)$ as best possible by a single function from the dictionary D, this algorithm is called the *pure greedy algorithm.*

If the dictionary D is generated by an orthonormal basis, then $U_m(f)$ is a best m-term approximation of f from D, i.e.,

$$\sigma_m(f; D) := \inf_{s \in \Sigma_m} \|f - s\| = \|f - U_m\|,$$

where Σ_m is the set of linear combinations $s = \sum_{h \in \Lambda} c_h\, h$ ($\Lambda \subset D$, $\#\Lambda \leq m$). However, for a general dictionary, the pure greedy algorithm only gives an estimate as follows. For $f \in K_1(D)$,

$$\|f - U_m(f)\| \leq |f|_{K_1(D)} m^{-\frac{1}{6}} \quad (m \in \mathbb{Z}_+),$$

where $|f|_{K_1(D)}$ is stated in (3.6.1).

3.6.2 Relaxed greedy algorithm

The relaxed greedy algorithm is an improvement of the pure greedy algorithm. Let $f \in \mathcal{H}$ and D be any dictionary of $\mathcal{H}$. The relaxed greedy algorithm is follows.

Define

$$U_0^{(r)}(f) = 0, \qquad R_0^{(r)}(f) = f,$$
$$U_1^{(r)}(f) = U_1(f), \quad R_1^{(r)}(f) = R_1(f),$$

where $U_1(f)$ and $R_1(f)$ are stated in Section 3.6.1.

For a function $g \in \mathcal{H}$, let $h = h(g) \in D$ be such that $(g,h) = \sup_{\tau \in D}(g, \tau)$. Inductively define

$$U_m^{(r)}(f) = \left(1 - \tfrac{1}{m}\right) U_{m-1}^{(r)}(f) + \tfrac{1}{m} h(R_{m-1}^{(r)}(f)),$$
$$R_m^{(r)}(f) = f - U_m^{(r)}(f).$$

The relaxed greedy algorithm give an estimate as follows. For $f \in K_1(D)$,

$$\|f - U_m^{(r)}(f)\| \leq C m^{-\frac{1}{2}} \quad (m \in \mathbb{Z}_+),$$

where C is a constant. In this estimate, the approximation order is $m^{-\frac{1}{2}}$. From this, it is seen that it has better approximation properties than the pure greedy algorithm.

3.6.3 Orthogonal greedy algorithm

The orthogonal greedy algorithm is also an improvement of the pure greedy algorithm.

Let $\mathcal{H}_0$ be a finite-dimensional subspace of the Hilbert space $\mathcal{H}$. The orthogonal greedy algorithm is as follows. Define

$$U_0^{(0)}(f) = 0, \quad R_0^{(0)}(f) = f.$$

For $m \geq 1$, inductively define

$$\begin{aligned}\mathcal{H}_m &= \operatorname{span}\{\, h(R_0^{(0)})(f), \dots, h(R_{m-1}^{(0)})(f)\,\},\\ U_m^{(0)}(f) &= P_{\mathcal{H}_m}(f),\\ R_m^{(0)}(f) &= f - U_m^{(0)}(f),\end{aligned}$$

where $h(g)$ is stated in Section 3.6.2 and $P_{\mathcal{H}_m}$ is the best approximation to f in $\mathcal{H}_m$. The orthogonal greedy algorithm has the following estimate:

$$\|f - U_m^{(0)}(f)\| \leq |f|_{K_1(D)} m^{-\frac{1}{2}} \quad (m \in \mathbb{Z}_+). \tag{3.6.2}$$

In this estimate, the approximation order is $m^{-\frac{1}{2}}$. From this, it is seen that it has better approximation properties than the pure greedy algorithm. But in this algorithm, the computation of the projection $P_{\mathcal{H}_m}$ is more expensive.

Based on the estimate (3.6.2), the m-term approximation from a dictionary has a good result as follows.

If $f \in K_\tau(D)$ and $\frac{1}{\tau} = \alpha + \frac{1}{2}$, and $\alpha \geq \frac{1}{2}$, then

$$\sigma_m(f; D) := \inf_{s\in\Sigma_m} \|f - s\| \leq C\,|f|_{K_\tau(D)} m^{-\alpha} \quad (m \in \mathbb{Z}_+),$$

where C depends on τ if τ is small and each $s \in \Sigma_m$ can be written in the form $s = \sum_{h\in\Lambda} c_h\, h$ $(\Lambda \subset D, \#\Lambda \leq m)$.

Further reading

[1] Alhasan A, White DJ, De Brabanterb K. Continuous wavelet analysis of pavement profiles. *Automation in Construction*. 2016(63):134–143.

[2] Bolten M, Huckle TK, Kravvaritis CD. Sparse matrix approximations for multigrid methods. *Linear Algebra and its Applications*. 2016(502):58–76.

[3] Fang J, Lin S, Xu Z. Learning and approximation capabilities of orthogonal super greedy algorithm. *Knowledge-Based Systems*. 2016(95):86–98.

[4] Liu D, Yan P, Wei Q. Data-based analysis of discrete-time linear systems in noisy environment: Controllability and observability. *Information Sciences*. 2014(288):314–329.

[5] Morkisz PM, Plaskota L. Approximation of piecewise Holder functions from inexact information. *Journal of Complexity*. 2016(32):122–136.

[6] Moteki N. Discrete dipole approximation for black carbon-containing aerosols in arbitrary mixing state: A hybrid discretization scheme. *Journal of Quantitative Spectroscopy and Radiative Transfer*. 2016(178):306–314.

[7] Safarinejadian B, Estahbanati ME. A novel distributed variational approximation method for density estimation in sensor networks. *Measurement*. 2016(89):78–86.
[8] Sharma V, Yang D, Walsh W, Reindl T. Short term solar irradiance forecasting using a mixed wavelet neural network. *Renewable Energy*. 2016(90):481–492.
[9] Yi H, Shu H. The improvement of the Morlet wavelet for multi-period analysis of climate data. *Comptes Rendus Geoscience*. 2012(344):483–497.
[10] Zhang Z. Fourier expansions with polynomial terms for random processes. *Journal of Function Spaces*. 2015:1–13. doi:10.1155/2015/763075.
[11] Zhang Z. Hyperbolic cross truncations for stochastic Fourier cosine series. *The Scientific World Journal*. 2014:1–13. doi:10.1155/2014/265031.
[12] Zhang Z, Saito N. PHLST with adaptive tiling and its application to Antarctic remote sensing image approximation. *Inverse Problems and Imaging*. 2014(8):321–337.
[13] Zhang Z, Jorgensen P. Modulated Haar wavelet analysis of climatic background noise. *Acta Appl Math*. 2015(140):71–93.
[14] Zhu X, Gisbrecht A, Schleif FM, Hammer B. Approximation techniques for clustering dissimilarity data. *Neurocomputing*. 2012(90):72–84.
[15] Zou CX, Shen XD, Li HY, Li XZ, Li ZJ. Wavelet analysis of spring climate characteristics in arid aeolian area of agro-pastoral ecotone in China. *Water Science and Engineering*. 2012(5):269–277.

4 Interpolation

Data records with equidistant time intervals are fundamental prerequisites for the development of environmental modeling, simulation and impact assessment. Usually long-term environmental time series contain missing data or data with different sampling intervals. Interpolation can be used to handle missing environmental data or fill the intervals between two grid points so that series of measurements with small intervals are kept. In this chapter we will discuss curve fitting, Lagrange and Hermite interpolations, spline interpolation, trigonometric interpolation, and bivariate interpolation.

4.1 Curve fitting

Given observation data (x_k, y_k) $(k = 1, \dots, M)$, we will find a polynomial $P(x)$ of degree N $(N < M)$ such that the sum $\sum_{k=1}^{M}(P(x_k) - y_k)^2$ attains the minimal value. This is the so-called *curve fitting problem.*

4.1.1 Polynomial fitting

Let (x_k, y_k) $(k = 1, \dots, M)$ be the observation data. Take a polynomial with unknown $a_0, a_1, \dots, a_N$

$$f(x) = a_0 + a_1 x + \cdots + a_N x^N \quad (N < M)$$

to fit these data. Denote

$$F(a_0, a_1, \dots, a_N) = \sum_{1}^{M} (f(x_k) - y_k)^2 = \sum_{1}^{M} (a_0 + a_1 x_k + \cdots + a_N x_k^N - y_k)^2 .$$

For $\nu = 0, \dots, N$, let $\frac{\partial F}{\partial a_\nu} = 0$. Then

$$\sum_{1}^{M} (a_0 + a_1 x_k + a_2 x_k^2 + \cdots + a_N x_k^N - y_k) x_k^\nu = 0 .$$

Denote $R_l = \sum_{k=1}^{M} x_k^l$ and $S_\nu = \sum_{k=1}^{M} y_k x_k^\nu$. Then

$$R_\nu a_0 + R_{\nu+1} a_1 + R_{\nu+2} a_2 + \cdots + R_{\nu+N} a_N = S_\nu \quad (\nu = 0, 1, \dots, N)$$

or

$$\sum_{i=0}^{N} R_{\nu+i} a_i = S_\nu \quad (\nu = 0, 1, \dots, N).$$

This is a system of $N + 1$ linear equations with $N + 1$ unknown, and so it has a unique solution $a_0^*, a_1^*, \dots, a_N^*$.

DOI 10.1515/9783110424904-005

Proposition 4.1.1. *If $a_0^*, a_1^*, \dots, a_N^*$ is the solution of the system of linear equations*

$$\sum_{i=0}^{N} R_{\nu+i} a_i = S_\nu \quad (\nu = 0, 1, \dots, N),$$

then the polynomial fitting of data $\{x_k, y_k\}_{k=1,\dots,M}$ is $f(x) = \sum_{\nu=0}^{N} a_\nu^* x^\nu$.

It is more convenient to choose a linear combination of orthogonal polynomials to fit data. For the given observation data (x_k, y_k) $(k = 1, \dots, M)$, assume that $\{x_k\}$ satisfy $-1 = x_1 < x_2 < \cdots < x_M = 1$ and are equally spaced. Choose the following linear combination of normal Legendre polynomials $\hat{P}_n(x)$ (see (3.3.1)) to fit the data

$$f(x) = a_0 \hat{P}_0(x) + a_1 \hat{P}_1(x) + \cdots + a_N \hat{P}_N(x).$$

Let

$$F(a_0, a_1, \dots, a_N) = \sum_{k=1}^{M} (f(x_k) - y_k)^2 = \sum_{k=1}^{M} \left(\sum_{\mu=0}^{N} a_\mu \hat{P}_\mu(x_k) - y_k \right)^2.$$

Then $\frac{\partial F}{\partial a_\nu} = 0$ $(\nu = 0, 1, \dots, N)$ is equivalent to

$$\sum_{k=1}^{M} \left(\sum_{\mu=0}^{N} a_\mu \hat{P}_\mu(x_k) - y_k \right) \hat{P}_\nu(x_k) = 0 \quad (\nu = 0, 1, \dots, N).$$

The left-hand side is equal to $\sum_{\mu=0}^{N} \left(\sum_{k=1}^{M} \hat{P}_\mu(x_k) \hat{P}_\nu(x_k) \right) a_\mu - \sum_{k=1}^{M} y_k \hat{P}_\nu(x_k)$. So

$$\sum_{\mu=0}^{N} \alpha_{\mu,\nu} a_\mu - \beta_\nu = 0 \quad (\nu = 0, 1, \dots, N), \tag{4.1.1}$$

where

$$\alpha_{\mu,\nu} = \sum_{k=1}^{M} \hat{P}_\mu(x_k) \hat{P}_\nu(x_k), \quad \beta_\nu = \sum_{k=1}^{M} y_k \hat{P}_\nu(x_k). \tag{4.1.2}$$

Since $x_1, \dots, x_M$ are equally spaced nodes on $[-1, 1]$ and normal Legendre polynomials satisfy

$$\int_{-1}^{1} \hat{P}_\nu(x) \hat{P}_\mu(x) \, dx = \delta_{\nu\mu},$$

where $\delta_{\nu\mu}$ is the Kronecker delta, we get

$$\alpha_{\mu,\nu} = \sum_{k=1}^{M} \hat{P}_\mu(x_k) \hat{P}_\nu(x_k) \approx 0 \quad (\mu \neq \nu),$$

$$\alpha_{\nu,\nu} = \sum_{k=1}^{M} \hat{P}_\nu^2(x_k) \approx \frac{M-1}{2}.$$

From this and (4.1.1), $a_\nu \approx \frac{2\beta_\nu}{M-1}$ $(\nu = 0, 1, \dots, N)$, where β_ν are stated in (4.1.2).

Proposition 4.1.2. *Let $-1 = x_0 < x_1 < \cdots < x_M = 1$ and let them be equally spaced, and data (x_k, y_k) $(k = 0, 1, \dots, M)$ be given. Then the polynomial fitting data is $f(x) = \sum_{\nu=0}^{N} \beta_\nu \hat{P}_\nu(x)$, where $\beta_\nu \approx \frac{2}{M-1} \sum_{k=1}^{M} y_k \hat{P}_\nu(x_k)$ and $\hat{P}_\nu(x)$ is the ν-th normal Legendre polynomial.*

Now we consider the orthogonal polynomial with weight function to fit the given data.

Given data (x_k, y_k) $(k = 1, \dots, M)$ satisfying $a = x_1 < x_2 < \cdots < x_M = b$, the normal orthogonal polynomials $P_\nu^{(\rho)}(x)$ $(\nu = 0, 1, \dots)$ on $[a, b]$ with weight function $\rho(x)$ satisfy

$$\int_a^b P_\mu^{(\rho)}(x) P_\nu^{(\rho)}(x) \rho(x)\, dx = \delta_{\mu,\nu},$$

where $\delta_{\mu,\nu}$ is the Kronecker delta. Then the polynomial fitting data is

$$f(x) = \sum_{\mu=0}^{N} \beta_\mu^{(\rho)} P_\mu^{(\rho)}(x),$$

where

$$\beta_\mu^{(\rho)} \approx \frac{2}{M-1} \sum_{k=1}^{M} y_k P_\mu^{(\rho)}(x_k) \rho(x_k).$$

4.1.2 Orthogonality method

In fitting data using the orthogonality method, the BC-decomposition of matrices is crucial.

BC-decomposition

Let A be an $M \times N$ matrix with rank r, where $M \geq N$. Then the matrix A can be decomposed into $A = BC$, where B is an $M \times r$ matrix, C is an $r \times N$ matrix, and ranks of B and C are both r.

In fact, let $A = (\alpha_{ij})_{M \times N}$ and $\boldsymbol{a}_j = (\alpha_{1j}, \dots, \alpha_{Mj})^{\mathrm{T}}$ $(j = 1, \dots, N)$ be its j-th column vector. Since $\operatorname{rank}(A) = r$, there are r linearly independent column vectors. Say, $\boldsymbol{a}_1, \boldsymbol{a}_2, \dots, \boldsymbol{a}_r$ are the r linearly independent column vectors. We construct an orthonormal basis $\boldsymbol{e}_1, \boldsymbol{e}_2, \dots, \boldsymbol{e}_M$ on $\mathbb{R}^M$ such that for any $s = 2, \dots, M$,

$$\boldsymbol{e}_s \perp \boldsymbol{a}_j \quad (j = 1, \dots, s-1).$$

Let $P = (\boldsymbol{e}_1 | \boldsymbol{e}_2 | \cdots | \boldsymbol{e}_M)$. Then P is an orthogonal matrix of order M. Define $U = P^{\mathrm{T}} A$. Since $P^{\mathrm{T}} = P^{-1}$,

$$A = PU.$$

Let $U = (u_{kl})_{M \times N}$, where $u_{kl} = (\boldsymbol{e}_k, \boldsymbol{a}_l)$. Since $\boldsymbol{a}_1, \boldsymbol{a}_2, \dots, \boldsymbol{a}_r$ are linearly independent, $\boldsymbol{a}_l = \sum_{j=1}^{r} c_{jl} \boldsymbol{a}_j$ $(l > r)$. From this and $\boldsymbol{e}_s \perp \boldsymbol{a}_j$, it follows that $u_{kl} = (\boldsymbol{e}_k, \boldsymbol{a}_l) =$

$\sum_{j=1}^{r} c_{jl}(\boldsymbol{e}_k, \boldsymbol{a}_j) = 0$ $(k > r)$. Denote $\tilde{B} = (u_{kl})_{k,l=1,\dots,r}$ and $\tilde{C} = (u_{kl})_{k=1,\dots,r;\, l=r+1,\dots,N}$. So

$$U = \begin{pmatrix} \tilde{B} & \tilde{C} \\ 0 & 0 \end{pmatrix}$$

and the product PU only depends on the first r columns of P and the first r rows of U, and $A = PU = BC$, where $B = (\boldsymbol{e}_1|\boldsymbol{e}_2|\cdots|\boldsymbol{e}_r)$ and $C = (\tilde{B}|\tilde{C})$.

Consider a general system of linear independent functions $\varphi_1(x), \varphi_2(x), \dots, \varphi_N(x)$. We use their linear combination $F(x)$ to fit observation data (x_i, y_i) $(i = 1, \dots, M)$, where $M \gg N$, such that

$$y^2 := \sum_{i=1}^{M} \left(\sum_{j=1}^{N} \beta_j \varphi_j(x_i) - y_i \right)^2$$

attains the minimal value. Some often used function systems are the power function system $\{x^i\}$, the trigonometric function system $\sin(ix)$, and the exponential function system $\{ e^{\lambda_i x} \}$.

Let $\frac{\partial F}{\partial \beta_j} = 0$. Then

$$\sum_{i=1}^{M} \varphi_j(x_i) \left(\sum_{j=1}^{N} \beta_j \varphi_j(x_i) - y_i \right) = 0 \quad (i = 1, \dots, M). \tag{4.1.3}$$

This is a system of linear equations. The matrix form is $A^{\mathrm{T}}(A\beta - \boldsymbol{y}) = 0$, where

$$\begin{aligned} A &= (\alpha_{ij})_{M\times N}, \\ \beta &= (\beta_1, \dots, \beta_N)^{\mathrm{T}}, \\ \boldsymbol{y} &= (y_1, \dots, y_M)^{\mathrm{T}}, \end{aligned}$$

and $\alpha_{ij} = \varphi_j(x_i)$ $(i = 1, \dots, M; j = 1, \dots, N)$. Denote by $\beta^* = (\beta_1^*, \dots, \beta_N^*)$ the solution of (4.1.3). Then the combination fitting data is $\sum_{j=1}^{N} \beta_j^* \varphi_j(x)$. We solve out $\beta = \beta^*$ below.

Replacing A by its BC-decomposition in the matrix form of (4.1.3),

$$C^{\mathrm{T}} B^{\mathrm{T}} B C \beta = C^{\mathrm{T}} B^{\mathrm{T}} \boldsymbol{y}.$$

Multiplying both sides by C,

$$(CC^{\mathrm{T}})(B^{\mathrm{T}} B) C\beta = (CC^{\mathrm{T}}) B^{\mathrm{T}} \boldsymbol{y}.$$

Both CC^{T} and $B^{\mathrm{T}}B$ are $r \times r$ nonsingular matrices and $\operatorname{rank}(B) = \operatorname{rank}(C) = r$, so

$$C\beta = W, \quad \text{where } W = (B^{\mathrm{T}}B)^{-1} B^{\mathrm{T}} \boldsymbol{y}.$$

This implies that $C^{\mathrm{T}}(CC^{\mathrm{T}})^{-1} C\beta = C^{\mathrm{T}}(CC^{\mathrm{T}})^{-1} W$. Note that $C^{\mathrm{T}}(CC^{\mathrm{T}})^{-1} C = I$. The desired solution is

$$\beta^* = C^{\mathrm{T}}(CC^{\mathrm{T}})^{-1} W = C^{\mathrm{T}}(CC^{\mathrm{T}})^{-1} (B^{\mathrm{T}}B)^{-1} B^{\mathrm{T}} \boldsymbol{y}.$$

Write $\beta^* = (\beta_1^*, \dots, \beta_N^*)$. So the combination $F(x)$ fitting data is $\sum_{j=1}^{N} \beta_j^* \varphi_j(x)$.

4.2 Lagrange interpolation

Given a real sequence y_k $(k = 1, \dots, n)$ and nodes x_k $(k = 1, \dots, n)$, where $x_1 < x_2 < \cdots < x_n$, we construct a Lagrange interpolation polynomial $L_n(x)$ of degree $n-1$ such that $L_n(x_k) = y_k$ $(k = 1, \dots, n)$. Moreover, we introduce the uniform convergence and mean convergence of the Lagrange interpolation polynomial sequences.

4.2.1 Fundamental polynomials

Let $\omega_n(x)$ be the product of n factors $(x - x_k)$ $(k = 1, \dots, n)$, i.e.,

$$\omega_n(x) = (x - x_1)(x - x_2)\cdots(x - x_n).$$

Then $\omega_n(x)$ is a polynomial of degree n and $\omega_n(x_k) = 0$ $(k = 1, \dots, n)$, and

$$\omega_n'(x_k) = (x_k - x_1)\cdots(x_k - x_{k-1})(x_k - x_{k+1})\cdots(x_k - x_n) \quad (k = 1, \dots, n).$$

Define fundamental polynomials as

$$l_k(x) = \frac{\omega_n(x)}{\omega_n'(x_k)(x - x_k)} \quad (k = 1, \dots, n). \tag{4.2.1}$$

Then $l_k(x)$ is a polynomial of degree $n-1$ and $l_k(x_j) = \delta_{jk}$ for $j, k = 1, \dots, n$, where δ_{jk} is the Kronecker delta.

Let $P(x)$ be any polynomial of degree $n-1$. Then $P(x) = \sum_{k=1}^n P(x_k) l_k(x)$. In fact, let

$$Q(x) = \sum_1^n P(x_k) l_k(x).$$

Then $Q(x)$ is a polynomial of degree $n-1$ and $Q(x_k) = P(x_k)$ $(k = 1, \dots, n)$. These n pairs of values determine that $P(x) = Q(x)$, i.e., $P(x) = \sum_{k=1}^n P(x_k) l_k(x)$.

4.2.2 Lagrange interpolation polynomials

Lagrange interpolation polynomial of degree $n-1$ is defined as

$$L_n(x) = \sum_1^n y_k l_k(x) = \sum_1^n y_k \frac{\omega_n(x)}{\omega_n'(x_k)(x - x_k)}. \tag{4.2.2}$$

Clearly, $L_n(x_j) = y_j$ $(j = 1, \dots, n)$. Formula (4.2.2) is called the *Lagrange interpolation formula*. For convenience of computation, it is rewritten in the form

$$\begin{aligned} L_n(x) = {} & c_0 + c_1(x - x_1) + c_2(x - x_1)(x - x_2) + \cdots \\ & + c_{n-1}(x - x_1)(x - x_2)\cdots(x - x_{n-1}). \end{aligned} \tag{4.2.3}$$

This form is called the *Newton interpolation formula*. The coefficients $\{c_k\}$ are computed as follows.

$$\begin{aligned}
c_0 &= y_1, \\
c_1 &= \frac{y_2 - c_0}{x_2 - x_1}, \\
&\vdots \\
c_{k-1} &= \frac{y_k - c_0 - \sum_{l=1}^{k-2} c_l (x_k - x_1)\cdots(x_k - x_l)}{(x_k - x_1)\cdots(x_k - x_{k-1})} \quad (k = 3, \ldots, n).
\end{aligned}$$

The combination of Lagrange and Newton interpolation formulas gives

$$c_{k-1} = \sum_{\nu=1}^{k} \frac{y_\nu}{\omega_k'(x_\nu)} \quad (k = 1, \ldots, n). \tag{4.2.4}$$

When we add a node, if we use the Lagrange interpolation formula, this again necessitates computing each fundamental polynomial $l_i(x)$; if we use the Newton interpolation formula, the coefficients already computed do not have to be changed. Therefore, in numerical computations, it is best to use the Newton interpolation formula.

Let $f \in C^n([a, b])$ and $a \le x_1 < \cdots < x_n \le b$, and $L_n(x)$ be the Lagrange interpolation polynomial of degree $n - 1$. Then the error between the interpolation polynomial and the original function is

$$f(x) - L_n(x) = \frac{1}{n!} f^{(n)}(\xi_x)\omega_n(x) \quad (a < \xi_x < b), \tag{4.2.5}$$

and so

$$\max_{a\le x\le b} |f(x) - L_n(x)| \le \frac{1}{n!} \max_{a\le x\le b} |f^{(n)}(x)|(b - a)^n.$$

4.2.3 Equally spaced nodes and Chebyshev nodes

Consider equally spaced nodes

$$\begin{aligned}
x_1 &= a, \\
x_2 &= a + h, \\
&\vdots \\
x_n &= a + (n - 1)h.
\end{aligned}$$

Then

$$\begin{aligned}
\omega_k'(x_\nu) &= (x_\nu - x_1)\cdots(x_\nu - x_{\nu-1})(x_\nu - x_{\nu+1})\cdots(x_\nu - x_k) \\
&= (-1)^{k-\nu} h^{k-1} (\nu - 1)!(k - \nu)!\,.
\end{aligned}$$

From this and (4.2.4),

$$c_{k-1} = \frac{1}{h^{k-1}} \sum_{\nu=1}^{k} \frac{(-1)^{k-\nu}}{(\nu-1)!(k-\nu)!} y_\nu = \frac{\Delta^{k-1} y_1}{h^{k-1}(k-1)!},$$

where $\Delta^{k-1} y_1$ is the $(k-1)$-th difference. From this and (4.2.3), it follows that

$$L_n(x) = y_1 + \sum_{\nu=1}^{n-1} \frac{(x-a)(x-a-h)\cdots(x-a-(\nu-1)h)}{\nu!\, h^\nu} \Delta^\nu y_1.$$

This formula is called the *Newton interpolation formula with equally spaced nodes.*

Consider the equally spaced nodes $x_k^{(n)} = -1 + \frac{2k}{n}$ $(k = 0, 1, \ldots, n)$. The Lagrange interpolation polynomials of $f(x) = |x|$ $(-1 \le x \le 1)$ do not converge to $f(x)$ as $n \to \infty$ except $x = -1, 0, 1$. Therefore, we need to look for other nodes.

Consider Chebyshev nodes $x_k = \cos \frac{(2k-1)\pi}{2n}$ $(k = 1, \ldots, n)$, i.e., the zeros of Chebyshev polynomial $T_n(x) = \cos(n \arccos x)$. The term of highest degree of $T_n(x)$ is $2^{n-1}x^n$. So

$$\omega_n(x) = \frac{1}{2^{n-1}} T_n(x) = \frac{1}{2^{n-1}} \cos(n \arccos x), \tag{4.2.6}$$

$$\omega_n'(x) = \frac{n}{2^{n-1}} \frac{\sqrt{1 - T_n^2(x)}}{\sqrt{1 - x^2}},$$

$$\omega_n'(x_k) = \frac{n}{2^{n-1}\sqrt{1 - x_k^2}}.$$

By (4.2.1),

$$l_k(x) = \frac{T_n(x)}{n(x - x_k)} \sqrt{1 - x_k^2}$$

and the Lagrange interpolation formula with Chebyshev nodes is

$$L_n(x) = \frac{T_n(x)}{n} \sum_{1}^{n} \frac{y_k}{x - x_k} \sqrt{1 - x_k^2}.$$

By (4.2.6), $|\omega_n(x)| \le \frac{1}{2^{n-1}}$ $(|x| \le 1)$. From this and the error formula (4.2.5), it follows that for the Chebyshev nodes $x_1, \ldots, x_n$, the interpolation error is

$$|f(x) - L_n(x)| = \left| \frac{1}{n!} f^{(n)}(\xi_x) \omega_n(x) \right| \le \frac{1}{2^{n-1} n!} \max_{|t| \le 1} |f^{(n)}(t)|.$$

From this, it is seen that Chebyshev nodes are optimal nodes since $|\omega_n(x)| \le \frac{1}{2^{n-1}}$ $(|x| \le 1)$ holds if and only if the nodes $x_1, \ldots, x_n$ are Chebyshev nodes.

4.2.4 Convergence of interpolation polynomials

Given a continuous function f, if P_n is an interpolation polynomial of f with n nodes, we expect that $\{P_n\}$ converges to f as $n \to \infty$. Consider the triangular matrix of nodes on $[a, b]$

$$
\begin{array}{llll}
x_1^{(1)} & & & \\
x_1^{(2)}, & x_2^{(2)} & & \\
\vdots & \vdots & & \\
x_1^{(n)}, & x_2^{(n)}, & \dots, & x_n^{(n)}.
\end{array}
$$

The sequence of Lagrange interpolation polynomials of $f \in C([a, b])$ is

$$L_n(x) = \sum_{k=1}^{n} f(x_k^{(n)}) l_k^{(n)}(x) \quad (n \in \mathbb{Z}_+),$$

where $l_k^{(n)}(x)$ $(k = 1, \dots, n)$ are fundamental polynomials based on nodes $x_k^{(n)}$ $(k = 1, \dots, n)$

$$l_k^{(n)}(x) = \frac{\omega_n(x)}{\omega_n'(x_k^{(n)})(x - x_k^{(n)})},$$

where $\omega_n(x) = \prod_{k=1}^{n} (x - x_k^{(n)})$.

Proposition 4.2.1. *Let $f \in C([a, b])$ and its best approximation by polynomials of degree n be E_n, and let*

$$\lambda_n(x) = \sum_{k=1}^{n} |l_k^{(n)}(x)|,$$

$$\lambda_n = \max_{a \le x \le b} |\lambda_n(x)|.$$

If $\lim_{n\to\infty} \lambda_n E_{n-1} = 0$, then the Lagrange interpolation polynomial $L_n(x)$ converges to $f(x)$ on $[a, b]$ uniformly.

For Chebyshev nodes $x_k^{(n)} = \cos \frac{(2k-1)\pi}{2n}$ $(k = 1, \dots, n)$,

$$\lambda_n = \max_{0 \le x \le 1} \lambda_n(x) \le 8 + \frac{4}{\pi} \log n.$$

If $f \in C([-1, 1])$ and f is a piecewise differentiable function on $[-1, 1]$, then $f \in \operatorname{lip} 1$. From this and the Jackson inequality in Section 3.1, $E_n = O(\frac{1}{n})$, and so

$$\lim_{n\to\infty} \lambda_n E_{n-1} = 0.$$

By Proposition 4.2.1, the interpolation polynomial $L_n(x)$ with Chebyshev nodes converges uniformly to $f(x)$ on $[-1, 1]$.

There exists a continuous function $f(x)$ on $[-1, 1]$ such that its Lagrange interpolation polynomial with Chebyshev nodes diverges everywhere. But, in fact, for any triangular matrix $\{x_k^{(n)}\}_{k=1,\dots,n}$, there exists a continuous function such that its Lagrange interpolation polynomials do not converge to itself uniformly.

4.2.5 Mean convergence

From Section 3.3, we see that Legendre polynomials are orthogonal polynomials with weight 1. They have n different zeros on $[-1,1]$. It is convenient to choose the zeros of Legendre polynomials as nodes for mean convergence, and the following proposition holds.

Let nodes $\{x_k^{(n)}\}_{k=1,\dots,n}$ be zeros of the n-th Legendre polynomial. If $f \in C([-1,1])$, then its Lagrange interpolation polynomial $L_n(x)$ converges to $f(x)$ in the mean square sense, i.e.,

$$\lim_{n\to\infty}\int_{-1}^{1}(L_n(x)-f(x))^2\,\mathrm{d}x = 0.$$

For orthogonal polynomials on $[a,b]$ with weight $\rho(x)$, their zeros $\{x_k^{(n)}\}_{k=1,\dots,n}$ are all simple zeros and $a < x_1^{(n)} < x_2^{(n)} < \cdots < x_n^{(n)} = b$, and can be estimated precisely. The product

$$\omega_n(x) = (x-x_1^{(n)})(x-x_2^{(n)})\cdots(x-x_n^{(n)})$$

is an orthogonal polynomial with the first term coefficient 1 and has a simple and clear representation, and its derivative $\omega_n'(x_k^{(n)})$ is estimated easily. Take these zeros $x_k^{(n)}$ $(k=1,\dots,n)$ as nodes. For $f \in C([a,b])$, the Lagrange interpolation polynomials $L_n(x)$ with nodes $x_k^{(n)}$ $(k=1,\dots,n)$ converges to $f(x)$ in the mean square sense

$$\lim_{n\to\infty}\int_{a}^{b}(L_n(x)-f(x))^2\rho(x)\,\mathrm{d}x = 0.$$

Especially, for $f \in C([-1,1])$, the Lagrange interpolation polynomial with Chebyshev nodes $x_k = \cos\frac{(2k-1)\pi}{2n}$ $(k=1,\dots,n)$ satisfies

$$\lim_{n\to\infty}\int_{-1}^{1}(L_n(x)-f(x))^2\frac{1}{\sqrt{1-x^2}}\,\mathrm{d}x = 0.$$

4.3 Hermite interpolation

For nodes x_k $(k=1,\dots,n)$ satisfying $x_1 < x_2 < \cdots < x_n$, we will find the lowest polynomial $H(x)$ such that $H^{(l)}(x_k) = y_k^{(l)}$ $(k=1,\dots,n;\ l=0,1,\dots,\alpha_k-1)$, i.e.,

$$\begin{cases} H(x_k) = y_k, \\ H'(x_k) = y_k', \\ \vdots \\ H^{(\alpha_k-1)}(x_k) = y_k^{(\alpha_k-1)} \quad (k=1,\dots,n). \end{cases}$$

Such a polynomial exists and is unique. The polynomial $H(x)$ is called the *Hermite interpolation polynomial.*

If $\alpha_1 = \alpha_2 = \cdots = \alpha_n = 1$, the Hermite interpolation polynomial is reduced to the Lagrange interpolation polynomial. If $n = 1$, the Hermite interpolation polynomial is reduced to the Taylor polynomial

$$H(x) = y_1 + \frac{y_1'}{1!}(x - x_1) + \cdots + \frac{y_1^{\alpha_1 - 1}}{(\alpha_1 - 1)!}(x - x_1)^{\alpha_1 - 1}.$$

4.3.1 Hermite interpolation formula with remainder term

Assume that $f \in C^m([a, b])$, where $m = \alpha_1 + \cdots + \alpha_n$, and nodes $x_k \subset [a, b]$ $(k = 1, \ldots, n)$, and its Hermite interpolation polynomial $H(x)$ satisfying $H^{(l)}(x_k) = f^{(l)}(x_k)$ $(k = 1, \ldots, n;\ l = 0, 1, \ldots, \alpha_k - 1)$. Then the error formula is as follows:

$$f(x) = H(x) + \frac{f^{(m)}(\xi)}{m!}\Omega(x) \quad (a < \xi < b),$$

where $\Omega(x) = (x - x_1)^{\alpha_1}(x - x_2)^{\alpha_2} \cdots (x - x_n)^{\alpha_n}$ and $m = \alpha_1 + \alpha_2 + \cdots + \alpha_n$.

4.3.2 Interpolation polynomial with double points

When $\alpha_1 = \cdots = \alpha_n = 2$, the problem is reduced to finding a polynomial $H(x)$ of degree $2n - 1$ satisfying

$$\begin{aligned} H(x_k) &= y_k, \\ H'(x_k) &= y_k' \quad (k = 1, \ldots, n). \end{aligned}$$

The polynomial $H(x)$ is called the *interpolation polynomial with double points.*

Let

$$\begin{aligned} P_{2n-1}(x) &= \sum_1^n y_k \left(1 - \frac{\omega_n''(x_k)}{\omega_n'(x_k)}(x - x_k)\right) l_k^2(x), \\ Q_{2n-1}(x) &= \sum_1^n y_k'(x - x_k) l_k^2(x). \end{aligned}$$

where $\omega_n(x) = (x - x_1) \cdots (x - x_n)$ and $l_k(x)$ are fundamental polynomials stated in Section 4.2. Then

$$\begin{aligned} P_{2n-1}(x_k) &= y_k, \quad P_{2n-1}'(x_k) = 0, \\ Q_{2n-1}(x_k) &= 0, \quad Q_{2n-1}'(x_k) = y_k' \quad (k = 1, \ldots, n). \end{aligned}$$

Both $P_{2n-1}(x)$ and $Q_{2n-1}(x)$ are polynomials of degree $2n - 1$. The following proposition holds.

Given two real number sequences y_k, y_k' $(k = 1, \ldots, n)$ and nodes x_k $(k = 1, \ldots, n)$, where $x_1 < \cdots < x_n$, the Hermite interpolation polynomial $H_{2n-1}(x)$ satisfying

$$\begin{aligned} H_{2n-1}(x_k) &= y_k, \\ H_{2n-1}'(x_k) &= y_k' \quad (k = 1, \ldots, n) \end{aligned}$$

can be decomposed in the form

$$H_{2n-1}(x) = P_{2n-1}(x) + Q_{2n-1}(x).$$

Especially, for the case $n = 2$, given real numbers y_1, y_2 and y_1', y_2', and nodes x_1, x_2, the Hermite interpolation polynomial $H_3(x)$ satisfying

$$\begin{aligned} H_3(x_1) &= y_1, \quad H_3(x_2) = y_2, \\ H_3'(x_1) &= y_1', \quad H_3'(x_2) = y_2' \end{aligned}$$

can be decomposed into

$$H_3(x) = P_3(x) + Q_3(x),$$

where

$$\begin{aligned} P_3(x) &= y_1 \left(1 - \frac{2}{(x_1 - x_2)}(x - x_1)\right)\left(\frac{x - x_2}{x_1 - x_2}\right)^2 \\ &\quad + y_2 \left(1 - \frac{2}{(x_2 - x_1)}(x - x_2)\right)\left(\frac{x - x_1}{x_2 - x_1}\right)^2, \\ Q_3(x) &= y_1'(x - x_1)\left(\frac{x - x_2}{x_1 - x_2}\right)^2 + y_2'(x - x_2)\left(\frac{x - x_1}{x_2 - x_1}\right)^2. \end{aligned} \tag{4.3.1}$$

Let $f \in C([-1, 1])$ and $x_k^{(n)} = \cos \frac{(2k-1)\pi}{2n}$ $(k = 1, \ldots, n)$ be Chebyshev nodes. If $H_{2n-1}(x)$ is a polynomial of degree $2n - 1$ satisfying $H_{2n-1}(x_k^{(n)}) = f(x_k^{(n)})$ and $H_{2n-1}'(x_k^{(n)}) = 0$, then the polynomial $H_{2n-1}(x)$ converges to $f(x)$ on $[-1, 1]$ uniformly. Such an interpolation method is called the *Féjer interpolation method.*

4.4 Spline interpolation

Spline interpolation is to replace polynomials by piecewise polynomials as interpolation functions and requires that piecewise polynomials are smooth at each node.

4.4.1 Spline functions

Given nodes $a = x_0 < x_1 < \cdots < x_n = b$, if $s(x)$ is a constant on each interval $[x_k, x_{k+1}]$ $(k = 0, \ldots, n - 1)$, then $s(x)$ is called a spline function of degree 0 with nodes x_k $(k = 0, \ldots, n)$. If $s(x)$ is a liner function on each interval $[x_k, x_{k+1}]$ and $s \in C([a, b])$, then $s(x)$ is called a spline function of degree 1. In general, if $s(x)$ is a polynomial of degree m on each interval $[x_k, x_{k+1}]$ and $s \in C^{m-1}([a,b])$, then $s(x)$ is called a *spline function of degree* m.

For a spline function $s(x)$ of degree m, let $d_k = s^{(m)}(x_k+0) - s^{(m)}(x_k-0)$, and let s_{mk} be the restriction of $s(x)$ on $[x_k, x_{k+1}]$, i.e., $s_{mk} = s|_{[x_k,x_{k+1})}$. Then

$$s_{mk}(x) = s_{m,k-1}(x) + R(x),$$

where $R(x)$ is a polynomial of degree m. From $s(x) \in C^{m-1}([a,b])$, it follows that

$$R^{(l)}(x_k) = 0 \quad (l = 0, 1, \dots, m-1),$$
$$R^{(m)}(x_k) = d_k.$$

So

$$s_{mk}(x) = s_{m,k-1}(x) + \frac{d_k}{(m-1)!}(x - x_k)^{m-1}.$$

Introduce two notations $x_+ = \max(0, x)$ and $x_+^{m-1} = (x_+)^{m-1}$ $(m \geq 2)$. Then

$$s(x) = s|_{[x_0,x_1)}(x) + \sum_{k=1}^{n-1} \frac{d_k}{(m-1)!}(x - x_k)_+^{m-1}.$$

The general structure of spline functions is as follows.

Proposition 4.4.1. *Let $s(x)$ be a spline function of degree m with nodes $x_0 < x_1 < \dots < x_n$. Then there exist $c_0, \dots, c_m$ and $d_1, \dots, d_{n-1}$ such that*

$$s(x) = c_0 + c_1 x + \dots + c_m x^m + \sum_{k=1}^{n-1} \frac{d_k}{(m-1)!}(x - x_k)_+^{m-1}.$$

4.4.2 Spline interpolation

Given nodes $a = x_0 < x_1 < \dots < x_n = b$ and numerical values $y_0, y_1, \dots, y_n$, if a function $s(x)$ satisfies

(a) $s(x)$ is a polynomial of degree ≤ 3 on each subinterval $[x_{k-1}, x_k)$ $(k = 1, \dots, n)$,

(b) $s(x_k) = y_k$ $(k = 0, 1, \dots, n)$ and $s(x) \in C^2([a,b])$,

then $s(x)$ is called a *cubic spline function* on $[a, b]$.

Let $s(x)$ be a cubic spline function on $[a, b]$. Then $s(x)$ satisfies on each interval $[x_k, x_{k+1})$

$$s(x_k) = y_k, \quad s(x_{k+1}) = y_{k+1}.$$

Denote

$$s'(x_k) = \mu_k, \quad s'(x_{k+1}) = \mu_{k+1},$$

where $\{\mu_k\}_{k=1,\ldots,n}$ are unknown. By (4.3.1),

$$s(x) = \left(1 + 2\frac{x - x_k}{\delta_k}\right)\left(\frac{x - x_{k+1}}{\delta_k}\right)^2 y_k + \left(1 - 2\frac{x - x_{k+1}}{\delta_k}\right)\left(\frac{x - x_k}{\delta_k}\right)^2 y_{k+1}$$
$$+ (x - x_k)\left(\frac{x - x_{k+1}}{\delta_k}\right)^2 \mu_k + (x - x_{k+1})\left(\frac{x - x_k}{\delta_k}\right)^2 \mu_{k+1}, \qquad (4.4.1)$$

where $\delta_k = x_{k+1} - x_k$. This implies that

$$s''(x) = \left(\frac{6}{\delta_k^2} - \frac{12}{\delta_k^3}(x_{k+1} - x)\right) y_k + \left(\frac{6}{\delta_k^2} - \frac{12}{\delta_k^3}(x - x_k)\right) y_{k+1}$$
$$+ \left(\frac{6}{\delta_k} - \frac{6}{\delta_k}(x_{k+1} - x)\right)\mu_k - \left(\frac{2}{\delta_k} - \frac{6}{\delta_k^2}(x - x_k)\right)\mu_{k+1}. \qquad (4.4.2)$$

Thus, the right-derivative and the left-derivative are, respectively,

$$s''(x_k^+) = -\frac{6}{\delta_k^2} y_k + \frac{6}{\delta_k^2} y_{k+1} - \frac{4}{\delta_k}\mu_k - \frac{2}{\delta_k}\mu_{k+1} \qquad (k = 0, 1, \ldots, n-1),$$
$$s''(x_k^-) = \frac{6}{\delta_{k-1}^2} y_{k-1} - \frac{6}{\delta_{k-1}^2} y_k + \frac{2}{\delta_{k-1}}\mu_{k-1} + \frac{4}{\delta_{k-1}}\mu_k \quad (k = 0, 1, \ldots, n-1).$$

Since $s''(x_k^+) = s''(x_k^-)$ $(k = 1, \ldots, n-1)$,

$$\frac{2}{\delta_{k-1}}\mu_{k-1} + 4\left(\frac{1}{\delta_{k-1}} + \frac{1}{\delta_k}\right)\mu_k + \frac{2}{\delta_k}\mu_{k+1}$$
$$= -\frac{6}{\delta_{k-1}^2} y_{k-1} + \left(\frac{6}{\delta_{k-1}} - \frac{6}{\delta_k^2}\right) y_k + \frac{6}{\delta_k^2} y_{k+1} \quad (k = 1, \ldots, n-1).$$

Finding unknown $\mu_0, \mu_1, \ldots, \mu_n$ meets three kinds of boundary conditions. The first kind is

$$s'(x_0) = \mu_0,$$
$$s'(x_n) = \mu_n.$$

The second kind is

$$s''(x_0) = 0,$$
$$s''(x_n) = 0.$$

From this and (4.4.2), it follows that

$$2\mu_0 + \mu_1 = \frac{3}{\delta_0}(y_1 - y_0),$$
$$\mu_{n-1} + 2\mu_n = \frac{3}{\delta_{n-1}}(y_n - y_{n-1}).$$

The third kind is that $s(x)$ is a periodic function with period $x_n - x_0$, in this case, $y_0 = y_n$, and

$$s'(x_0) = s'(x_n),$$
$$s''(x_0) = s''(x_0).$$

From this and (4.4.2), it follows that $\mu_0 = \mu_n$ and

$$\frac{3}{\delta_0^2}(y_1 - y_0) - \frac{1}{\delta_0}(2\mu_0 + \mu_1) = \frac{3}{\delta_{n-1}^2}(y_{n-1} - y_n) + \frac{1}{\delta_{n-1}}(\mu_{n-1} - 2\mu_n).$$

Each kind of boundary condition gives $n+1$ linear equations with $n+1$ unknown numbers. Solve these equations to find $\mu_0, \mu_1, \dots, \mu_n$.

Proposition 4.4.2. *Let $f \in C^4([a,b])$, and let $s(x)$ be the cubic spline function on $[a,b]$ corresponding to nodes $a = x_0 < x_1 < \cdots < x_n = b$. Denote $\lambda = \max_{0 \le k \le n+1} |x_{k+1} - x_k|$. Then*

$$|f^{(i)}(x) - s^{(i)}(x)| \le C\lambda^{4-i} \quad (i = 1, 2, 3),$$

where C is a constant.

4.4.3 B-splines

Let $N_1(x) = \chi_{[0,1]}(x)$, where $\chi_{[0,1]}(x)$ is the characteristic function of $[0,1]$ and

$$N_2(x) = \int_0^1 N_1(x-t)\,\mathrm{d}t,$$

$$\vdots$$

$$N_m(x) = \int_0^1 N_{m-1}(x-t)\,\mathrm{d}t.$$

The $N_m(x)$ is called the *B-spline of degree* $m-1$. B-spline N_m has the following properties:

$$N_m(x) > 0 \quad (0 < x < m),$$
$$\operatorname{supp} N_m(x) = [0, m],$$
$$\sum_{l \in \mathbb{Z}} N_m(x - l) = 1 \quad (x \in \mathbb{R}).$$

Let $s_m \in L^2(\mathbb{R})$ be a spline function of degree m with nodes $\mathbb{Z}$. Then

$$s_m(x) = \sum_{k \in \mathbb{Z}} c_k N_m(x-k)$$

in the $L^2(\mathbb{R})$ sense. Especially, if $\operatorname{supp} s_m(x) = [N_1, N_2]$ $(N_1, N_2 \in \mathbb{Z})$, then $s_m(x)$ is a sum of finitely many terms.

4.5 Trigonometric interpolation and fast Fourier transform

Given N points $x_0, x_1, \dots, x_{N-1}$, define the *discrete Fourier transform* as

$$X_k = \frac{1}{N}\sum_{n=0}^{N-1} x_n \,\mathrm{e}^{-\mathrm{i}n\frac{2\pi k}{N}} \quad (k = 0, \dots, N-1).$$

Using the formula

$$\sum_{j=0}^{N-1} \mathrm{e}^{-\mathrm{i}k\frac{2\pi j}{N}} \mathrm{e}^{\mathrm{i}m\frac{2\pi j}{N}} = \delta_{km} \quad (0 \le k, m \le N-1),$$

where δ_{km} is the Kronecker delta, the *inverse discrete Fourier transform* is

$$x_n = \sum_{k=0}^{n} X_k \,\mathrm{e}^{\mathrm{i}k\frac{2\pi n}{N}} \quad (n = 0, 1, \dots, N-1).$$

This implies the following formula.

Trigonometric interpolation formula

Let $f(x)$ be defined on $[0, 1]$ and $x_j = \frac{j}{N}$ $(j = 0, 1, \dots, N-1)$ be N nodes. Denote

$$c_k = \frac{1}{N}\sum_{j=0}^{N-1} f(x_j)\,\mathrm{e}^{-2\pi \mathrm{i}kx_j} \quad (k = 0, \dots, N-1).$$

Then the trigonometric polynomial $P(x) = \sum_{k=0}^{N-1} c_k \,\mathrm{e}^{2\pi \mathrm{i}kx}$ satisfies $P(x_k) = f(x_k)$.

Fast Fourier transform

Fast Fourier transform is a fast algorithm computing discrete Fourier transform by the halving trick.

Given a 2^N-point time series $\boldsymbol{x} = (x_0, x_1, \dots, x_{2^N-1})$, its discrete Fourier transform is

$$X_k = \frac{1}{2^N}\sum_{n=0}^{2^N-1} x_n \mathrm{e}^{-\mathrm{i}n\frac{2\pi k}{2^N}} \quad (k = 0, 1, \dots, 2^N-1). \tag{4.5.1}$$

Now we halve X_k by the halving trick.

First, we compute the first half $X_0, X_1, \dots, X_{2^{N-1}-1}$. Decompose the given 2^N-point time series $\boldsymbol{x}$ into two 2^{N-1}-point time series

$$\boldsymbol{u} = (x_0, x_2, \dots, x_{2^N-2}) =: (u_0, u_1, \dots, u_{2^{N-1}-1}),$$
$$\boldsymbol{v} = (x_1, x_3, \dots, x_{2^N-1}) =: (v_0, v_1, \dots, v_{2^{N-1}-1}),$$

i.e., $\boldsymbol{u}$ is an even sample and $\boldsymbol{v}$ is an odd sample of $\boldsymbol{x}$. From (4.5.1), it follows that

$$X_k = \frac{1}{2^N} \sum_{n=0}^{2^{N-1}-1} u_n e^{-2n\mathrm{i}\frac{2\pi k}{2^N}} + \frac{1}{2^N} \sum_{n=0}^{2^{N-1}-1} v_n e^{-(2n+1)\mathrm{i}\frac{2\pi k}{2^N}},$$

and so

$$X_k = \frac{1}{2}(U_k + e^{-\mathrm{i}\frac{2\pi k}{2^N}} V_k) \quad (k = 0, 1, \ldots, 2^{N-1} - 1),$$

where

$$U_k = \frac{1}{2^{N-1}} \sum_{n=0}^{2^{N-1}-1} u_n e^{-\mathrm{i}n\frac{2\pi k}{2^{N-1}}},$$

$$V_k = \frac{1}{2^{N-1}} \sum_{n=0}^{2^{N-1}-1} v_n e^{-\mathrm{i}n\frac{2\pi k}{2^{N-1}}} \quad (k = 0, 1, \ldots, 2^{N-1} - 1).$$

Similarly, the second half $X_{2^{N-1}}, X_{2^{N-1}+1}, \ldots, X_{2^N-1}$ is computed as follows:

$$X_{k+2^{N-1}} = \frac{1}{2}(U_k - e^{-\mathrm{i}\frac{2\pi k}{2^N}} V_k) \quad (k = 0, 1, \ldots, 2^{N-1} - 1).$$

where U_k and V_k are stated as above.

We continue to halve the obtained U_k and V_k by the halving trick.

Halving U_k $(k = 0, 1, \ldots, 2^{N-1} - 1)$, this gives that

$$U_k = \frac{1}{2}\left(U'_k + e^{-\mathrm{i}\frac{2\pi k}{2^{N-1}}} U''_k\right),$$
$$U_{k+2^{N-2}} = \frac{1}{2}\left(U'_k - e^{-\mathrm{i}\frac{2\pi k}{2^{N-1}}} U''_k\right) \quad (k = 0, 1, \ldots, 2^{N-2} - 1),$$

where U'_k and U''_k are the discrete Fourier transforms of two 2^{N-2}-point time series which consist of even samples and odd samples of $\boldsymbol{u}$, respectively.

Halving V_k $(k = 0, 1, \ldots, 2^{N-1} - 1)$, this gives that

$$V_k = \frac{1}{2}\left(V'_k + e^{-\mathrm{i}\frac{2\pi k}{2^{N-1}}} V''_k\right),$$
$$V_{k+2^{N-2}} = \frac{1}{2}\left(V'_k - e^{-\mathrm{i}\frac{2\pi k}{2^{N-1}}} V''_k\right) \quad (k = 0, 1, \ldots, 2^{N-2} - 1),$$

where V'_k and V''_k are the discrete Fourier transforms of two 2^{N-2}-point time series which consist of even samples and odd samples of $\boldsymbol{v}$, respectively.

Continue this procedure until a one-point time series.

Using the fast Fourier transform algorithm, the total number of multiplication operations is equal to $N2^{N-1}$. While using the original discrete Fourier transform algorithm, the total number of multiplication operations is equal to 2^{2N}. This means that the fast Fourier transform has better computationally efficiency.

4.6 Bivariate interpolation

Given a set of nodes in the xy-plane $(x_1, y_1), (x_2, y_2), \ldots, (x_n, y_n)$, where each node (x_i, y_i) is associated with a real number c_i, we find a smooth and easily computed function F such that $F(x_i, y_i) = c_i$ $(1 \le i \le n)$.

4.6.1 Cartesian product and grids

Assume that the set of nodes is a Cartesian product $N = \{x_1, x_2, \ldots, x_p\} \times \{y_1, y_2, \ldots, y_q\}$, i.e., grids

$$N = \{(x_i, y_j) : 1 \le i \le p, 1 \le j \le q\}.$$

Let $u_i(x)$ $(i = 1, \ldots, p)$ be univariate real functions such that $u_i(x_j) = \delta_{ij}$ $(1 \le i, j \le p)$, where δ_{ij} is the Kronecker delta. For example, $u_i(x)$ may be the fundamental polynomials in the Lagrange interpolation formula (see (4.2.1)). Let f be a bivariate function. Define an operator P as

$$(Pf)(x, y) = \sum_1^p f(x_i, y) u_i(x).$$

The operator Pf is a bivariate function that interpolates f on vertical lines L_i: $\{(x_i, y) : -\infty < y < \infty\}$ $(i = 1, \ldots, p)$. Define another operator Q as

$$(Qf)(x, y) = \sum_1^q f(x, y_j) v_j(y),$$

where $v_j(y_i) = \delta_{ij}$ $(1 \le i, j \le q)$. The operator Qf is a bivariate function that interpolates f on horizontal lines L^j: $\{(x, y_j) : -\infty < x < \infty\}$ $(j = 1, \ldots, q)$. Then $P(Qf)$ is a function that interpolates f at the nodes (x_i, y_j) $(i = 1, \ldots, p; j = 1, \ldots, q)$.

In fact, from

$$P(Qf)(x, y) = P\left(\sum_1^q f(x, y_j) v_j(y)\right) = \sum_{i=1}^p \sum_{j=1}^q f(x_i, y_j) u_i(x) v_j(y),$$

it follows that $P(Qf)(x_i, y_j) = f(x_i, y_j)$ $(i = 1, \ldots, p; j = 1, \ldots, q)$.

4.6.2 Tensor product

A function $\sum_{0 \le i+j \le k} c_{ij} x^i y^j$ is called a *bivariate polynomial* of degree $\le k$, where c_{ij} are constants. The space of all bivariate polynomials of degree at most k is denoted by $\prod_k(\mathbb{R}^2)$. The set $\{x^i y^j\}_{0 \le i+j \le k}$ is the *basis* of $\prod_k(\mathbb{R}^2)$. So the *dimension* of $\prod_k(\mathbb{R}^2)$ is $\frac{1}{2}(k+1)(k+2)$.

For a node set N, we ask whether the interpolation is possible on N. The following results are known:

(a) Assume that the set N consists of nodes $(x_1, y_1), (x_2, y_2), \dots, (x_{\tilde{k}}, y_{\tilde{k}})$, where $\tilde{k} = \frac{1}{2}(k+1)(k+2)$, and these nodes lie on lines $L_0, L_1, \dots, L_k$, and for each i, the line L_i contains just $i+1$ nodes. Then, for arbitrary associated data $c_1, c_2, \dots, c_{\tilde{k}}$, there is a polynomial $p \in \prod_k(\mathbb{R}^2)$ such that $p(x_l, y_l) = c_l$ $(l = 1, 2, \dots, \tilde{k})$.

(b) For any set of $k+1$ distinct nodes $(x_1, y_1), (x_2, y_2), \dots, (x_{k+1}, y_{k+1})$ and data $c_1, c_2, \dots, c_{k+1}$, there is a $p \in \prod_k(\mathbb{R}^2)$ such that $p(x_l, y_l) = c_l$ $(l = 1, \dots, k+1)$.

4.6.3 Shepard interpolation

Let n nodes $p_i = (x_i, y_i)$ $(i = 1, \dots, n)$. We select a real-valued function φ on $\mathbb{R}^2 \times \mathbb{R}^2$ such that $\varphi(p, q) = 0$ if and only if $p = q$. Define cardinality functions

$$u_i(p) = \prod_{\substack{j=1,\dots,n \\ j \neq i}} \frac{\varphi(p, p_j)}{\varphi(p_i, p_j)} \quad (1 \le i \le n).$$

This leads to an *interpolation formula* of a function f at the given nodes p_i as follows:

$$F = \sum_1^n f(p_i) u_i.$$

Especially, if

$$\varphi(p, p_j) = \|p - p_j\|^2 = (x - x_j)^2 + (y - y_j)^2,$$

where $p = (x, y)$ and $p_j = (x_j, y_j)$, then the interpolation formula is

$$F(x, y) = \sum_{i=1}^n f(x_i, y_i) \prod_{\substack{j=1,\dots,n \\ j \neq i}} \frac{(x - x_j)^2 + (y - y_j)^2}{(x_i - x_j)^2 + (y_i - y_j)^2}.$$

4.6.4 Triangulation

Triangulation is another method for interpolation problems.

If a set of triangles $\Delta_1, \Delta_2, \dots, \Delta_m$ satisfies the following three conditions:

- each node is the vertex of some triangle Δ_s;
- each vertex of a triangle in the set is a node;
- if a node belongs to a triangle, then the node must be a vertex of that triangle,

then this set is called a *triangulation*.

In a triangle Δ_s, define a linear function as

$$l_s(x, y) = \alpha_s x + \beta_s y + \gamma_s, \quad (x, y) \in \Delta_s.$$

Choose α_s, β_s, and γ_s such that each linear function $l_s(x, y)$ $(s = 1, 2, \ldots, m)$ takes the prescribed values c_i, c_j, and c_k at vertices (x_i, y_i), (x_j, y_j), and (x_k, y_k), i.e.,

$$\begin{aligned} l_s(x_i, y_i) &= c_i, \\ l_s(x_j, y_j) &= c_j, \\ l_s(x_k, y_k) &= c_k \quad (s = 1, 2, \ldots, m). \end{aligned}$$

Then the obtained pieces linear function is continuous on the union $\bigcup_{s=1}^{m} \Delta_s$.

Further reading

[1] Agudelo OM, Viaene P, De Moor B. Improving the PM10 estimates of the air quality model AURORA by using Optimal Interpolation. *IFAC-PapersOnLine*. 2015(48):1154–1159.

[2] Arun PV. A comparative analysis of different DEM interpolation methods. *The Egyptian Journal of Remote Sensing and Space Science*. 2013(16):133–139.

[3] Chaplot V, Darboux F, Bourennane H, Legedois S, Silvera N, Phachomphon K. Accuracy of interpolation techniques for the derivation of digital elevation models in relation to landform types and data density. *Geomorphology*. 2006(77):126–141.

[4] Chudinov AV, Gao W, Huang Z, Cai W, Zhou Z, Raznikov VV, Kozlovski VI, Sulimenkov IV. Interpolational and smoothing cubic spline for mass spectrometry data analysis. *International Journal of Mass Spectrometry*. 2016(396):42–47.

[5] Duran-Rosal AM, Heras-Martinez C, Tallon-Ballesteros AJ, Martinez-Estudillo AC, Salcedo-Sanz S. Massive missing data reconstruction in ocean buoys with evolutionary product unit neural networks. *Ocean Engineering*. 2016(117):292–301.

[6] Janssen S, Dumont G, Fierens F, Mensink C. Spatial interpolation of air pollution measurements using CORINE land cover data. *Atmospheric Environment* 2008(42):4884–4903.

[7] Jeong SYY, Choi YJ, Park P. Parametric interpolation using sampled data. *Computer-Aided Design*. 2006(38):39–47.

[8] Kilibarda M, Tadi MP, Hengl T, Lukovic J, Bajat B. Global geographic and feature space coverage of temperature data in the context of spatio-temporal interpolation. *Spatial Statistics*. 2015(14):22–38.

[9] Krivoruchko K, Gribov A, Krause E. Multivariate areal interpolation for continuous and count data. *Procedia Environmental Sciences*. 2011(3):14–19.

[10] Li Q, Dehler SA. Inverse spatial principal component analysis for geophysical survey data interpolation. *Journal of Applied Geophysics*. 2015(115):79–91.

[11] Lin J, Cromley PG. Evaluating geo-located Twitter data as a control layer for areal interpolation of population. *Applied Geography*. 2015(58):41–47.

[12] Liu R, Chen Y, Sun C, Zhang P, Wang J, Yu W, Shen Z. Uncertainty analysis of total phosphorus spatial-temporal variations in the Yangtze River Estuary using different interpolation methods. *Marine Pollution Bulletin*. 2014(86):68–75.

[13] Liu S, Wang CCL. Quasi-interpolation for surface reconstruction from scattered data with radial basis function. *Computer Aided Geometric Design*. 2015(29):435–447.

[14] Mendez D, Labrador M, Ramachandran K. Data interpolation for participatory sensing systems. *Pervasive and Mobile Computing*. 2013(9):132–148.

[15] Nardelli BB, Droghei R, Santoleri R. Multi-dimensional interpolation of SMOS sea surface salinity with surface temperature and in situ salinity data. *Remote Sensing of Environment*. 2016: in press.

[16] Nardelli BB, Pisano A, Tronconi C, Santoleri R. Evaluation of different covariance models for the operational interpolation of high resolution satellite Sea Surface Temperature data over the Mediterranean Sea. *Remote Sensing of Environment*. 2015(164):334–343.

[17] Plouffe CCF, Robertson C, Chandrapala L. Comparing interpolation techniques for monthly rainfall mapping using multiple evaluation criteria and auxiliary data sources: A case study of Sri Lanka. *Environmental Modelling & Software*. 2015(67):57–71.

[18] Scudiero E, Corwin DL, Morari F, Anderson RG, Skaggs TH. Spatial interpolation quality assessment for soil sensor transect datasets. *Computers and Electronics in Agriculture*. 2016(123): 74–79.

[19] Singh SK, McMillan H, Bardossy A. Use of the data depth function to differentiate between case of interpolation and extrapolation in hydrological model prediction. *Journal of Hydrology*. 2013(477):213–228.

[20] Slattery SR. Mesh-free data transfer algorithms for partitioned multiphysics problems: Conservation, accuracy, and parallelism. *Journal of Computational Physics*. 2016(307):164–188.

[21] Steinbuch L, Brus DJ, van Bussel LGJ, Heuvelink GBM. Geostatistical interpolation and aggregation of crop growth model outputs. *European Journal of Agronomy*. 2016(77):111–121.

[22] Su T, Cao Z, Lv Z, Liu C, Li X. Multi-dimensional visualization of large-scale marine hydrological environmental data. *Advances in Engineering Software*. 2016(95):7–15.

[23] Tokumitsu M, Hasegawa K, Ishida Y. Toward resilient sensor networks with spatiotemporal interpolation of missing data: An example of space weather forecasting. *Procedia Computer Science*. 2015(60):1585–1594.

[24] Urquhart EA, Hoffman MJ, Murphy RR, Zaitchik BF. Geospatial interpolation of MODIS-derived salinity and temperature in the Chesapeake Bay. *Remote Sensing of Environment*. 2013(135): 167–177.

[25] Wagner PD, Fiener P, Wilken F, Kumar S, Schneider K. Comparison and evaluation of spatial interpolation schemes for daily rainfall in data scarce regions. *Journal of Hydrology*. 2012(464–465):388–400.

[26] Wang Q, Shi W, Atkinson PM. Sub-pixel mapping of remote sensing images based on radial basis function interpolation. *ISPRS Journal of Photogrammetry and Remote Sensing*. 2014(92): 1–15.

[27] Yang Z, Liu Y, Li C. Interpolation of missing wind data based on ANFIS. *Renewable Energy*. 2011(36):993–998.

5 Statistical methods

In order to handle various environmental issues well, it is crucial to apply statistical methodology to ensure well-conducted data collection, analyze and interpret environmental data, and describe environmental changes with sound and validated models. In this chapter, we will provides comprehensive coverage of the methodology used in the statistical investigation of environmental issues, including regression analysis, principal component analysis, discriminant analysis, cluster analysis, factor analysis, and canonical correlation analysis. All these statistical methods can be easily implemented by popular software packages such as SPSS, SAS, and R.

5.1 Linear regression

Linear regression analysis is the most widely used among all statistical techniques. It is used to model the relationship between two variables by fitting a linear equation to observed data. One variable is considered to be an explanatory variable, and the other is considered to be a dependent variable. The simplest linear regression model is

$$Y = \beta_0 + \beta_1 x + \varepsilon \quad (\varepsilon \sim N(0, \sigma^2)), \tag{5.1.1}$$

where Y is a dependent variable, x is an explanatory variable, and ε is the random error. If the data is a sample $\{ Y_i, x_i \}_{i=1,\dots,n}$, by (5.1.1),

$$Y_i = \beta_0 + \beta_1 x_i + \varepsilon_i \quad (i = 1, \dots, n). \tag{5.1.2}$$

Common assumptions are that $\varepsilon_1, \dots, \varepsilon_n$ are independent, $E[\varepsilon_i] = 0$ $(i = 1, \dots, n)$ and $\mathrm{Var}(\varepsilon_i) = \sigma^2$ $(i = 1, \dots, n)$, and $\varepsilon_1, \dots, \varepsilon_n$ are normal distributed.

5.1.1 Estimate of regression coefficients β_0 and β_1

Let $\hat{Y}_i$ be the estimate of Y_i, i.e.,

$$\hat{Y}_i = \beta_0 + \beta_1 x_i \quad (i = 1, \dots, n). \tag{5.1.3}$$

Use the method of least squares to estimate β_0 and β_1.

Let

$$Q(\beta_0, \beta_1) := \sum_1^n (Y_i - \hat{Y}_i)^2 = \sum_1^n (Y_i - \beta_0 - \beta_1 x_i)^2. \tag{5.1.4}$$

We need to choose $\hat{\beta}_0, \hat{\beta}_1$ such that $Q(\hat{\beta}_0, \hat{\beta}_1)$ attains the minimal value. This means solving two equations

$$\frac{\partial Q}{\partial \beta_0} = 0, \quad \frac{\partial Q}{\partial \beta_1} = 0.$$

DOI 10.1515/9783110424904-006

Let $\bar{x} = \frac{1}{n}\sum_{i=1}^n x_i$ and $\bar{Y} = \frac{1}{n}\sum_{i=1}^n Y_i$. By (5.1.4),

$$\frac{\partial Q}{\partial \beta_0} = -2\sum_1^n (Y_i - \beta_0 - \beta_1 x_i) = -2(n\bar{Y} - n\beta_0 - n\bar{x}\beta_1).$$

Since $\frac{\partial Q}{\partial \beta_0} = 0$,

$$\beta_0 = \bar{Y} - \beta_1 \bar{x}. \tag{5.1.5}$$

Denote $\tilde{x}_i = x_i - \bar{x}$ and $\tilde{Y}_i = Y_i - \bar{Y}$. By (5.1.4) and (5.1.5), we get

$$\begin{aligned}\frac{\partial Q}{\partial \beta_1} &= -2\sum_1^n (Y_i - (\beta_0 + \beta_1 x_i))x_i = -2\sum_1^n (\tilde{Y}_i + \bar{Y} - (\beta_0 + \beta_1 x_i))x_i \\ &= -2\sum_1^n (\tilde{Y}_i + (\beta_0 + \beta_1 \bar{x}) - (\beta_0 + \beta_1 x_i))x_i \\ &= -2\sum_1^n (\tilde{Y}_i - \beta_1 \tilde{x}_i)\tilde{x}_i = -2S_{xY} + 2\beta_1 S_{xx},\end{aligned}$$

where $S_{xY} = \sum_{i=1}^n \tilde{x}_i \tilde{Y}_i$ and $S_{xx} = \sum_{i=1}^n (\tilde{x}_i)^2$. Since $\frac{\partial Q}{\partial \beta_1} = 0$,

$$\beta_1 S_{xx} = S_{xY}.$$

From this and (5.1.5), the regression coefficients β_0, β_1 are estimated

$$\begin{aligned}\hat{\beta}_1 &= \frac{S_{xY}}{S_{xx}} = \frac{\sum_1^n \tilde{x}_i \tilde{Y}_i}{\sum_1^n (\tilde{x}_i)^2} = \frac{\sum_1^n (x_i - \bar{x})(Y_i - \bar{Y})}{\sum_1^n (x_i - \bar{x})^2}, \\ \hat{\beta}_0 &= \bar{Y} - \hat{\beta}_1 \bar{x}.\end{aligned}$$

The equation $\hat{Y} = \hat{\beta}_0 + \hat{\beta}_1 x$ is called the regression equation.

5.1.2 Estimate of σ^2

By (5.1.1), $E[(Y - (\beta_0 + \beta_1 x))^2] = E[\varepsilon^2] = \sigma^2$. Now we estimate σ^2 using the sample $(Y_i, x_i)_{i=1,\dots,n}$.

By $\hat{\beta}_1 = S_{xY}/S_{xx}$, the sum of squares

$$\begin{aligned}\sum_{i=1}^n (Y_i - \hat{\beta}_0 - \hat{\beta}_1 x_i)^2 &= \sum_1^n (Y_i - \bar{Y} - \hat{\beta}_1(x_i - \bar{x}))^2 \\ &= \sum_1^n (Y_i - \bar{Y})^2 - 2\hat{\beta}_1 \sum_1^n (Y_i - \bar{Y})(x_i - \bar{x}) + (\hat{\beta}_1)^2 \sum_1^n (x_i - \bar{x})^2 \\ &= S_{YY} - \hat{\beta}_1 S_{xY}.\end{aligned}$$

The corresponding statistic is $Q = S_{YY} - \hat{\beta}_1 S_{xY}$. It can be proved that $\frac{Q}{\sigma^2} \sim \chi^2(n-2)$. So $E[Q/(n-2)] = \sigma^2$. This implies that the unbiased estimator

$$\hat{\sigma}^2 = \frac{Q}{n-2} = \frac{S_{YY} - \hat{\beta}_1 S_{xY}}{n-2}.$$

5.1.3 Decomposition formula

Since

$$\sum_1^n (Y_i - \bar{Y})^2 = \sum_1^n (Y_i - \hat{Y}_i + \hat{Y}_i - \bar{Y})^2$$
$$= \sum_1^n (Y_i - \hat{Y})^2 + 2\sum_1^n (Y_i - \hat{Y}_i)(\hat{Y}_i - \bar{Y}) + \sum_1^n (\hat{Y}_i - \bar{Y})^2.$$

From $\tilde{Y}_i = Y_i - \bar{Y}$ and $\hat{Y}_i = \bar{Y} + \hat{\beta}_1 \tilde{x}_i$ and $\tilde{x}_i = x_i - \bar{x}$, and (5.1.5), the second term on the right-hand side

$$\sum_{i=1}^n (Y_i - \hat{Y}_i)(\hat{Y}_i - \bar{Y}) = \sum_{i=1}^n (\tilde{Y}_i - \hat{\beta}_1 \tilde{x}_i)\hat{\beta}_1 \tilde{x}_i = \hat{\beta}_0 \hat{\beta}_1 \sum_1^n \tilde{x}_i = 0.$$

Thus the sum of squares of deviation $Y_i - \bar{Y}$ is decomposed as follows:

$$\sum_1^n (Y_i - \bar{Y})^2 = \sum_1^n (Y_i - \hat{Y}_i)^2 + \sum_1^n (\hat{Y}_i - \bar{Y})^2, \tag{5.1.6}$$

where $\hat{Y}_i = \hat{\beta}_0 + \hat{\beta}_1 x_i$.

In (5.1.6), the sum $Q := \sum_{i=1}^n (Y_i - \hat{Y}_i)^2$ is called the *residual sum of squares*; the sum $U := \sum_{i=1}^n (\hat{Y}_i - \bar{Y})^2$ is called the *regression sum of squares*. So (5.1.6) can be rewritten as $S_{YY} = Q + U$. Since residuals $Y_i - \hat{Y}_i$ represent the differences between the fitted values and the observed values, they can be used to check if the assumed model fits the data well or not. Define

$$R^2 = \frac{U}{S_{YY}} = 1 - \frac{Q}{S_{YY}} = 1 - \frac{\sum_1^n (\hat{Y}_i - Y_i)^2}{\sum_1^n (Y_i - \bar{Y})^2}.$$

Here R^2 is called the *coefficient of determination* which indicates the usefulness of the regression model. If R approximates 1, the regression model is very useful.

5.1.4 Extensions

Many more complicated regressions may be reduced to linear regression problems. Here we give some examples which may be reduced to linear regression.

For the model $G(Y) = \beta_0 + \beta_1 g(x) + \varepsilon$ and $\varepsilon \sim N(0, \sigma^2)$, where β_0, β_1, and σ^2 are independent of x, its samples are $\{x_i, Y_i\}_{i=1,\ldots,n}$. Let $Y^* = G(Y)$ and $x^* = g(x)$. Then

$$Y^* = \beta_0 + \beta_1 x^* + \varepsilon,$$
$$\varepsilon \sim N(0, \sigma^2)$$

is a linear regression model and the samples are $\{g(x_i), G(Y_i)\}_{i=1,\ldots,n}$. By using the least square method, the regression coefficients $\hat{\beta}_0$ and $\hat{\beta}_1$ can be estimated and the regression model is $G(Y) = \hat{\beta}_0 + \hat{\beta}_1 g(x) + \varepsilon$.

For the model $Y = ae^{bx}\varepsilon$ and $\log\varepsilon \sim N(0, \sigma^2)$ and the model $Y = ax^b\varepsilon$ and $\log\varepsilon \sim N(0, \sigma^2)$, where a, b, and ε are independent of x, taking logarithms on both sides, these two models are transformed into two linear regression models, respectively.

5.1.5 Nonlinear regression models

In general, the univariate regression model is

$$\begin{gathered} Y = \mu(x;\beta_1,\dots,\beta_p) + \varepsilon, \\ \varepsilon \sim N(0,\sigma^2), \end{gathered}$$

where $\beta_1, \dots, \beta_p$, and σ^2 are independent of x. If μ is a linear function of $\beta_1, \dots, \beta_p$, then μ is called a *linear regression model*, otherwise, it is called a *nonlinear regression model.*

Let $\{x_i, Y_i\}_{i=1,\dots,n}$ $(n > p)$ be a known sample. A general nonlinear regression model can be written as

$$Y_i = \mu(x_i;\beta_1,\dots,\beta_p) + \varepsilon_i \quad (i = 1,\dots,n).$$

For a nonlinear regression model, the assumption of random errors ε_i's are the same as those for a linear model. Use least squares to estimate $\beta_1, \dots, \beta_p$. Let

$$Q(\beta_1,\dots,\beta_p) = \sum_1^n \left(Y_i - \mu(x_i;\beta_1,\dots,\beta_p)\right)^2 .$$

Differentiating both sides of the equation with respect to β_k gives

$$\frac{\partial Q}{\partial \beta_k} = 2\sum_1^n (Y_i - \mu(x_i;\beta_1,\dots,\beta_p))\frac{\partial \mu}{\partial \beta_k}(x_i;\beta_1,\dots,\beta_p) = 0 \quad (k = 1,\dots,p).$$

One often solves these equations using the Newton–Rapson iterative algorithm (see Chapter 7) and obtains the estimates $\hat{\beta}_k$'s of β_k's $(k = 1, \dots, p)$. Finally, the estimate of Y is $Y \approx \mu(x;\hat{\beta}_1,\dots,\hat{\beta}_p)$.

5.2 Multiple regression

In this section, we discuss multivariate linear regression models with one or multiple dependent variables. The multivariate regression model incorporates the correlation between the responses. So it may provide more efficient inference than separate univariate regression models but it is more complicated and is more different to handle.

5.2.1 A dependent variable case

In multivariate analysis, we choose Y as a response and variables $x_1, \ldots, x_p$ as parameters, and a sample $\{ Y_i; x_{i1}, \ldots, x_{ip} \}_{i=1,\ldots,n}$. Multivariate linear regression models attempt to find an approximate relationship

$$Y = \sum_{0}^{p} \beta_k x_k + \varepsilon \qquad (\varepsilon \sim N(\mathbf{0}, \sigma^2 I)), \quad x_0 = \mathbf{1}. \tag{5.2.1}$$

This implies that

$$Y_i = \sum_{k=0}^{p} \beta_k x_{ik} + \varepsilon_i \qquad (\varepsilon_i \sim N(0, \sigma^2)), \quad x_{i0} = 1 \quad (i = 1, \ldots, n)$$

and $\{\varepsilon_i\}$ are independent. The matrix form is $Y = X\beta + \varepsilon$ where $Y = (Y_1, \ldots, Y_n)^{\mathrm{T}}$, $\varepsilon = (\varepsilon_1, \ldots, \varepsilon_n)^{\mathrm{T}}$, $\beta = (\beta_0, \beta_1, \ldots, \beta_p)^{\mathrm{T}}$, and

$$X = \begin{pmatrix} 1 & x_{11} & \cdots & x_{1p} \\ \vdots & \vdots & \ddots & \vdots \\ 1 & x_{n1} & \cdots & x_{np} \end{pmatrix}.$$

(a) Estimate of regression coefficients $\boldsymbol{\beta} = (\boldsymbol{\beta}_0, \ldots, \boldsymbol{\beta}_p)$

Choose $\{\beta_k\}_{k=1,\ldots,p}$ such that the following sum of squares attains the minimal value

$$Q(\beta) = \sum_{i=1}^{n} (Y_i - \sum_{k=0}^{p} \beta_k x_{ik})^2 = (Y - X\beta)^{\mathrm{T}}(Y - X\beta). \tag{5.2.2}$$

Let $\frac{\partial Q}{\partial \beta_j} = 0$ $(j = 1, \ldots, n)$. Then $\sum_{i=1}^{n} (Y_i - \sum_{k=0}^{p} \beta_k x_{ik}) x_{ij} = 0$. This is equivalent to

$$\sum_{k=0}^{p} \left(\sum_{i=1}^{n} x_{ik} x_{ij} \right) \beta_k = \sum_{i=1}^{n} Y_i x_{ij} \quad (j = 1, \ldots, n).$$

Its matrix form is $X^{\mathrm{T}} X \beta = X^{\mathrm{T}} Y$. So $\hat{\beta} = (X^{\mathrm{T}} X)^{-1} X^{\mathrm{T}} Y$ is the least square estimate of β. Moreover, it is an unbiased estimate of β. It can be proved that $\hat{\beta} \sim N(\beta, \sigma^2 (X^{\mathrm{T}} X)^{-1})$. From this and (5.2.2), it follows that

$$Q(\hat{\beta}) = \sum_{i=1}^{n} \left(Y_i - \sum_{k=0}^{p} \hat{\beta}_k x_{ik} \right)^2 = Y^{\mathrm{T}} Y - Y^{\mathrm{T}} X \hat{\beta},$$

$$\hat{\sigma}^2 = \frac{Q(\hat{\beta})}{n - p - 1} = \frac{1}{n - p - 1} \sum_{1}^{n} (Y_i - \hat{Y}_i)^2,$$

where $\hat{Y}_i = \sum_{k=0}^{p} \hat{\beta}_k x_{ik}$ $(i = 1, \ldots, n)$ and $\hat{\sigma}^2$ is a unbiased estimate of σ^2. The following proposition holds.

For the model (5.2.1),

$$\hat{\beta} \sim N(\beta, \sigma^2(X^{\mathrm{T}}X)^{-1}),$$

$$\frac{Q(\hat{\beta})}{\sigma^2} \sim \chi^2_{n-p-1}.$$

If $\beta_1 = \cdots = \beta_p = 0$, then $V/\sigma^2 \sim \chi^2_p$, where $V = \sum_{i=1}^n (\hat{Y}_i - \bar{Y})^2$.

Define a test statistic $F = \frac{V/p}{Q/(n-p-1)}$. If $\beta_1 = \cdots = \beta_p = 0$, then $F \sim F(p, n-p-1)$. By the known sample, we compute the value of F and significant probability $\tilde{p}$. If $\tilde{p}$ is less than significance level α, then we reject hypothesis H_0. Otherwise, the hypothesis H_0 holds.

(b) Model selection

If in a regression model, some variable does not have a significant effect on the response, then we may remove it from this model. In general, a simple model is better than a complex model.

In a linear regression model, if the first model chosen by us is

$$Y = \sum_0^p \beta_k x_k + \varepsilon_1 \quad (\beta_0 = 1),$$

$$\hat{Y}_i^{(1)} = \sum_0^p \hat{\beta}_k x_k \qquad (\hat{\beta}_0 = 1)$$

and the second model chosen by us is

$$Y = \sum_0^q b_k x_k + \varepsilon_2 \quad (b_0 = 1),$$

$$\hat{Y}_i^{(2)} = \sum_0^q \hat{b}_k x_k \qquad (\hat{b}_0 = 1),$$

where $p > q$. Denote

$$R_1 = \sum_1^n (Y_i^{(1)} - \hat{Y}_i^{(1)})^2,$$

$$R_2 = \sum_1^n (Y_i^{(2)} - \hat{Y}_i^{(2)})^2.$$

If

$$F = \frac{(R_2 - R_1)/(p-q)}{R_1/(n-p)} > F_\alpha(p-q, n-p),$$

then we choose the first model. Otherwise, we choose the second model.

5.2.2 Multiple dependent variable case

Suppose that there are p dependent variables $Y_1, \dots, Y_p$ and m independent variables $x_1, \dots, x_m$, and data matrixes are $X = (x_{ij})_{n\times m}$ and $Y = (Y_{ij})_{n\times p}$. If

$$Y_{ij} = \beta_{0j} + \sum_{k=1}^{m} \beta_{kj} x_{ik} + \varepsilon_{ij} \quad (i = 1, \dots, n;\ j = 1, \dots, p)$$

then $Y = (I_n|X)\beta + E = C\beta + E$, where

$$I_n = (1, 1, \dots, 1)^{\mathrm{T}}, \quad C = (I_n|X),$$
$$\beta = (\beta_{ij})_{m\times p}, \qquad E = (\varepsilon_{ij})_{n\times p}.$$

Assume that $\varepsilon_i = (\varepsilon_{i1}, \dots, \varepsilon_{ip})^{\mathrm{T}}$ $(i = 1, \dots, n)$ are independent, $E[\varepsilon_i] = 0$, ε_i's have the same covariance matrix Σ, and $\varepsilon_i \sim N_p(0, \Sigma)$ $(i = 1, \dots, n)$. The model $Y = (I_n|X)\beta + E = C\beta + E$ is called a *multivariate linear regression model* of multiple dependent variables, where Y, E are random matrices and $C = (I_n|X)$.

Similar to the dependent variable case, the parameter matrix β and covariance matrix Σ are estimated using the least square method as follows:

$$\hat{\beta} = (C^{\mathrm{T}}C)^{-1}C^{\mathrm{T}}Y =: (b_{ij})_{(m+1)\times p}.$$

So the regression equation with p dependent variables is $\hat{Y}_j = b_{0j} + b_{1j}x_1 + \cdots + b_{mj}x_m$ $(j = 1, \dots, p)$.

5.3 Case study: Tree-ring-based climate reconstructions

Due to the lack of reliable long-term meteorological records, it is hard to understand past climate change. Reconstructions of past climate conditions may be derived from various paleoclimatology proxies. Compared with other proxies, tree-ring-based reconstructions have many advantages, including wide spatial distribution, high climate sensitivity, high annual resolution and calendar-exact dating. Currently, the world's longest tree-ring chronology extends over more than 7000 years.

5.3.1 Collection of tree-ring data

Since tree radial growth is always subject to climatic influences, one can use various parameters from sample tree-rings to reconstruct historical temperature and precipitation. In order to reconstruct past temperature, the sample sites of tree-ring data should be in upper-elevation tree-line locations and cold mountain valley environments. In order to reconstruct past precipitation, sample sites should be in a steep rocky, south facing slope. In each sample site, after removing a cylinder of wood 5 mm in diameter along the radius of a tree, core samples are collected at breast height from trees.

Core samples are air dried and polished, then tree-ring widths are measured with a precision of 0.01 mm by using the LINTAB system or a similar system. After eliminating age-related growth trends, tree-ring width data need be standardized using the program ARSTAN in order to get tree-ring width chronologies. Except for tree-ring widths, tree-ring isotopic data can also be used to reconstruct past climate. Each ring of tree core samples is cut by using a scalpel blade and cellulose is extracted, $\delta^{18}O$ and $\delta^{13}C$ values of cellulose are measured by using a stable isotope radio mass spectrometer.

5.3.2 Tree-ring-based climate reconstruction

Main tree-ring related parameters include
a_1 tree-ring width;
a_2 ($\delta^{18}O$);
a_3 ($\delta^{13}C$),
a_4 mean latewood density.

Main reconstructed climate parameters include
b_1 temperature,
b_2 precipitation,
b_3 runoff,
b_4 drought,
b_5 CO_2 concentration C_a.

Correlation analysis is used to examine the relationship between tree-ring related parameters a_i and climate parameters b_j. Suppose that $X_i^{(1)}, \dots, X_i^{(n)}$ is the tree-ring chronology for parameter a_i and $Y_j^{(1)}, \dots, Y_j^{(n)}$ is the observation of climate parameter b_j. Then the estimate of the correlation coefficient is

$$r_{ij} = \frac{\sum_{l=1}^n (X_i^{(l)} - \bar{X}_i)(Y_j^{(l)} - \bar{Y}_j)}{\sqrt{\sum_{l=1}^n (X_i^{(l)} - \bar{X}_i)^2} \sqrt{\sum_{l=1}^n (Y_j^{(l)} - \bar{Y}_j)^2}},$$

where $\bar{X}_i = \frac{1}{n}\sum_{l=1}^n X_i^{(l)}$ and $\bar{Y}_j = \frac{1}{n}\sum_{l=1}^n Y_j^{(l)}$. If $\{r_{i,j}\}$ is large, we can use the tree-ring parameter a_i to reconstruct the climate parameter b_j. Finally, with the help of linear regression formula, we can reconstruct the climate factor b_j from climate parameter a_i as follows:

$$b_j = \beta + \beta_1 a_i + \varepsilon_{i,j} \quad (\varepsilon_{i,j} \sim N(0, \sigma^2)),$$

where β, β_1, and σ^2 are estimated well by samples of a_i and b_j (see Section 5.1).

Below we give some results on tree-ring-based reconstruction of climate.

The Hulunbuir region between 47–53° N and 115–126° E is extremely sensitive to climate changes. Therefore, it is an ideal region to carry out tree-ring research. Y. Liu et al. (2009) collected tree core samples and found that tree-ring width (TRW) is highly correlated with precipitation P_{76} for previous July to current June with the correlation coefficient $r = 0.711$. They give the following linear regression formula:

$$P_{76} = 222.408\,\mathrm{TRW} - 133.115$$

to reconstruct P_{76} since 1865. During the calibration period 1952–2003, total precipitation reconstructed by Liu et al. tracked the observation very well. In 2012, G. Bao et al. further reconstructed April–September mean maximum temperature (MMT_{49}) from 1868 to 2008 by tree-ring width chronologies by using the following linear regression formula:

$$\mathrm{MMT}_{49} = -2.807\,\mathrm{TRW} + 21.288.$$

Moreover, Bao et al. (2012) further found significant correlations between the reconstructed MMT and Pacific Decadal Oscillation/Nino. This explains the influences of large-scale atmospheric-oceanic variability on regional temperature and droughts in the Hulubuir grassland.

The Luoshan Mountains are in the south part of the Tengger Desert of China and are surrounded by land subject to desertification. Palmer Drought Severity Index (PDSI) is a standardized measure of surface moisture conditions. In 2013, Y. Wang et al. used tree-ring width to reconstruct annual PDSI in Tengger Desert for the period 1897–2007 as follows:

$$\mathrm{PDSI} = 4.90 + 4.15\,\mathrm{TRW}.$$

In the West Tianmu Mountains of China, X. Zhao et al. (2006) collected tree-ring samples and reconstructed atmospheric CO_2 concentration C_a by using tree-ring $\delta^{13}C$ values

$$C_a = 8598 + 810.922\,\delta^{13}\mathrm{C} + 19.748\,(\delta^{13}\mathrm{C})^2.$$

This is a curvilinear regression formula. Their results show that in 1685–1840, the evaluated atmospheric CO_2 concentration was stable, but after 1840 it exhibited a rapid increase.

In the Wuyi Mountains of China, F. Chen et al. (2012) used tree-ring width of current and pervious years to reconstruct July–October minimum temperature MT_{710} in 1803–2008 as follows:

$$\mathrm{MT}_{710}(t) = 22.662 - 150\,\mathrm{TRW}(t) - 1.567\,\mathrm{TRW}(t-1).$$

This is a nonlinear regression formula. From this, they showed that there is a strong relationship between the reconstruction and Summer Asian-Pacific Oscillation which suggest linkages of regional temperature variability with the Asian-Pacific climate system.

5.4 Covariance analysis

The analysis of covariance is used to adjust or control for differences between groups. Consider several independent random variables that have normal distribution with unknown means and unknown but common variance. A test of the equality of several means is called *analysis of variance*.

Suppose that $X_1, \dots, X_m$ are independent and each $X_k \sim N(\mu_k, \sigma^2)$ $(k = 1, \dots, m)$. Let $(X_{1k}, \dots, X_{nk})$ be a sample of each X_k. We test the hypothesis $H_0: \mu_1 = \mu_2 = \cdots = \mu_m = \mu$. Denote

$$\overline{X_{\cdot k}} = \frac{1}{n} \sum_{j=1}^{n} X_{jk},$$

$$\overline{X_{\cdot\cdot}} = \frac{1}{mn} \sum_{j=1}^{n} \sum_{k=1}^{m} X_{jk}.$$

It is easily deduced that the sum of squares

$$mnS^2 = \sum_{j=1}^{n} \sum_{k=1}^{m} (X_{jk} - \bar{X}_{\cdot\cdot})^2$$

has the decomposition formula

$$mnS^2 = \sum_{j=1}^{n} \sum_{k=1}^{m} (X_{jk} - \bar{X}_{\cdot k})^2 + n \sum_{k=1}^{m} (\bar{X}_{\cdot k} - \bar{X}_{\cdot\cdot})^2 =: Q_1 + Q_2.$$

Note that the sample variance of X_k: $S_k^2 = \frac{1}{n-1} \sum_{j=1}^{n} (X_{jk} - \bar{X}_{\cdot k})^2$ and $E[S_k^2] = \mathrm{Var}(X_k) = \sigma^2$. Then

$$E[Q_1] = \sum_{k=1}^{m} E[(n-1)S_k^2] = (n-1)m\sigma^2. \tag{5.4.1}$$

Denote $\mu = \frac{1}{m} \sum_{k=1}^{m} \mu_k$ and $\delta_k = \mu_k - \mu$ $(k = 1, \dots, m)$. By the independence of X_{jk} $(j = 1, \dots, m;\ k = 1, \dots, n)$,

$$\bar{X}_{\cdot k} \sim N\left(\mu_k, \frac{\sigma^2}{n}\right), \quad \bar{X}_{\cdot\cdot} \sim N(\mu, \frac{\sigma^2}{nm}),$$

and so the expectations of their squares are, respectively,

$$E[\bar{X}_{\cdot k}^2] = \mathrm{Var}(\bar{X}_{\cdot k}) + \mu_k^2 = \frac{\sigma^2}{n} + (\mu + \delta_k)^2,$$

$$E[\bar{X}_{\cdot\cdot}^2] = \mathrm{Var}(\bar{X}_{\cdot\cdot}) + \mu^2 = \frac{\sigma^2}{nm} + \mu^2.$$

Since $\bar{X}_{\cdot k} = \frac{1}{n} \sum_{j=1}^{n} X_{jk}$,

$$Q_2 = n \sum_{k=1}^{m} (\bar{X}_{\cdot k} - \bar{X}_{\cdot\cdot})^2 = n \sum_{k=1}^{m} (\bar{X}_{\cdot k}^2 + \bar{X}_{\cdot\cdot}^2 - 2\bar{X}_{\cdot k}\bar{X}_{\cdot\cdot}) = n \sum_{k=1}^{m} \bar{X}_{\cdot k}^2 - nm\bar{X}_{\cdot\cdot}^2.$$

Note that $\sum_{k=1}^m \delta_k = \sum_{k=1}^m \mu_k - m\mu = 0$. Then,

$$E[Q_2] = n\sum_1^m E\left[\bar{X}_{\cdot k}^2\right] - nmE\left[\bar{X}_{\cdot\cdot}^2\right] = n\sum_1^m \left(\frac{\sigma^2}{n} + (\mu+\delta_k)^2\right) - nm\left(\frac{\sigma^2}{nm} + \mu^2\right)$$
$$= (m-1)\sigma^2 + 2n\mu\sum_1^m \delta_k + n\sum_1^m \delta_k^2 = (m-1)\sigma^2 + n\sum_1^m \delta_k^2.$$

Comparing this with (5.4.1),

$$\frac{E[Q_2/(m-1)]}{E[Q_1/(m(n-1))]} \geq 1, \tag{5.4.2}$$

and the equality of (5.4.2) holds if and only if H_0 is true since $\delta_k = 0$ $(k = 1, \ldots, m)$.

The other test method is as follows. It is easy to prove that $\frac{Q_1}{\sigma^2} \sim \chi^2(m(n-1))$ and $\frac{Q_2}{\sigma^2} \sim \chi^2(m-1)$. Since Q_1 and Q_2 are independent,

$$\frac{Q_2/(m-1)}{Q_1/(m(n-1))} \sim F(m-1, m(n-1)).$$

This implies that H_0 is not true if and only if $\frac{Q_2/(m-1)}{Q_1/(m(n-1))} \geq F_\alpha(m-1, m(n-1))$, where α is a given significance level.

5.5 Discriminant analysis

Discriminant analysis is used to separate individuals into different populations based on given multivariate data. Suppose that there are k m-dimensional populations $G_1, \ldots, G_k$. For a given sample $X = (x_1, \ldots, x_m)^{\mathrm{T}}$, how to decide $X \in G_l$?

5.5.1 Mahalanobis distance method

Let G be an m-variate population with mean value vector $\mu = (\mu_1, \ldots, \mu_m)^{\mathrm{T}}$ and covariance matrix Σ. The *Mahalanobis distance* between an individual $X = (x_1, \ldots, x_m)^{\mathrm{T}}$ and the population G is defined as

$$d^2(X, G) = (X-\mu)^{\mathrm{T}}\Sigma^{-1}(X-\mu).$$

For example, let G be a univariate population with mean μ and variance σ^2, and let X be a sample. Then the Mahalanobis distance is

$$d^2(X, G) = (X-\mu)^2/\sigma^2.$$

Let G be a bivariate population with mean value vector μ and covariance matrix Σ

$$\mu = (\mu_1, \mu_2)^{\mathrm{T}},$$
$$\Sigma = \begin{pmatrix} \tau_{11} & \tau_{12} \\ \tau_{21} & \tau_{22} \end{pmatrix},$$

and let $X = (x_1, x_2)^{\mathrm{T}}$ be a sample. Then the Mahalanobis distance is

$$d^2(X, G) = (x_1 - \mu_1, x_2 - \mu_2) \begin{pmatrix} \tau_{11} & \tau_{12} \\ \tau_{21} & \tau_{22} \end{pmatrix}^{-1} \begin{pmatrix} x_1 - \mu_1 \\ x_2 - \mu_2 \end{pmatrix}.$$

Suppose that there are two populations G_i $(i = 1, 2)$ which have samples $X_1^{(i)}, \dots, X_{n_i}^{(i)}$ $(i = 1, 2)$. Then the estimates of the mean μ_i and covariance matrix Σ_i for G_i are, respectively,

$$\begin{aligned} \bar{X}_i &= \frac{1}{n_i} \sum_{k=1}^{n_i} X_k^{(i)}, & (i = 1, 2). \\ S_i &= \frac{1}{n_i - 1} \sum_{k=1}^{n_i} (X_k^{(i)} - \bar{X}_i)(X_k^{(i)} - \bar{X}_i)^{\mathrm{T}}, & (i = 1, 2). \end{aligned} \tag{5.5.1}$$

The Mahalanobis distance is estimated as

$$d^2(X, G_i) = (X - \bar{X}_i)^{\mathrm{T}} S_i^{-1} (X - \bar{X}_i) \quad (i = 1, 2).$$

We say $X \in G_1$ if $d^2(X, G_1) < d^2(X, G_2)$, and we say $X \in G_2$ if $d^2(X, G_1) \geq d^2(X, G_2)$.

For example, for two one-dimensional populations G_i with mean μ_i and variance $\sigma_i^2 (i = 1, 2)$. We say $x_0 \in G_1$ if $d^2(x, G_1) < d^2(x, G_2)$, i.e.,

$$\frac{(x_0 - \mu^{(1)})^2}{\sigma_1^2} < \frac{(x_0 - \mu^{(2)})^2}{\sigma_2^2},$$

which is equivalent to $\mu_* < x_0 < \mu^*$, and we say $x_0 \in G_2$ if $d^2(x, G_1) \geq d^2(x, G_2)$, i.e.,

$$\frac{(x_0 - \mu^{(1)})^2}{\sigma_1^2} \geq \frac{(x_0 - \mu^{(2)})^2}{\sigma_2^2},$$

which is equivalent to $x_0 \leq \mu_*$ or $x_0 \geq \mu^*$, where

$$\mu^* = \frac{\mu^{(1)}\sigma_2 - \mu^{(2)}\sigma_1}{\sigma_2 - \sigma_1}, \quad \mu_* = \frac{\mu^{(1)}\sigma_2 + \mu^{(2)}\sigma_1}{\sigma_2 + \sigma_1}.$$

When $\Sigma_1 = \Sigma_2 =: \Sigma$, the estimate of the covariance matrix Σ is

$$S = \frac{1}{n_1 + n_2 - 2} \sum_{i=1}^{2} \sum_{k=1}^{n_i} (X_k^{(i)} - \bar{X}_i)(X_k^{(i)} - \bar{X}_i)^{\mathrm{T}}. \tag{5.5.2}$$

The Mahalanobis distance is estimated as follows:

$$\begin{aligned} d^2(X, G_i) &= (X - \bar{X}_i)^{\mathrm{T}} S^{-1} (X - \bar{X}_i) \\ &= X^{\mathrm{T}} S^{-1} X - (\bar{X}_i)^{\mathrm{T}} S^{-1} X - X^{\mathrm{T}} S^{-1} \bar{X}_i + (\bar{X}_i)^{\mathrm{T}} S^{-1} \bar{X}_i \quad (i = 1, 2), \end{aligned}$$

where S is stated in (5.5.2) which is the estimate of the covariance matrix Σ. Since S is a symmetric matrix and $X^{\mathrm{T}} S^{-1} \bar{X}_i$ is a 1×1 matrix,

$$\begin{aligned} (X^{\mathrm{T}} S^{-1} \bar{X}_i)^{\mathrm{T}} &= (\bar{X}_i)^{\mathrm{T}} S^{-1} X, \\ d^2(X, G_i) &= X^{\mathrm{T}} S^{-1} X - 2T_i(X), \end{aligned}$$

where

$$T_i(X) = (\bar{X}_i)^T S^{-1} X - \tfrac{1}{2}(\bar{X}_i)^T S^{-1} \bar{X}_i \quad (i = 1, 2) \tag{5.5.3}$$

is a linear function. In fact, denote

$$(\bar{X}_i)^T S^{-1} = (\alpha_1^{(i)}, \ldots, \alpha_n^{(i)}),$$
$$(\bar{X}_i)^T S^{-1} \bar{X}_i = b_i.$$

Then, for $X = (x_1, \ldots, x_m)^T$,

$$(\bar{X}_i)^T S^{-1} X = \sum_{k=1}^{m} \alpha_k^{(i)} x_k,$$

and so

$$T_i(X) = \sum_{k=1}^{m} \alpha_k^{(i)} x_k - \frac{b_i}{2},$$

where $\alpha_k^{(i)}$ and b_i are constants, i.e., $T_i(x)$ is a linear function. The $T_i(X)$ is called a *linear discriminant function*. The difference of two Mahalanobis distances is

$$d^2(X, G_1) - d^2(X, G_2) = 2(T_2(X) - T_1(X)).$$

If $T_1(X) \ge T_2(X)$, then $x \in G_1$, otherwise $x \in G_2$.

More generally, suppose that there are k m-variate populations $\{G_i\}_{i=1,\ldots,k}$. For a given sample $X = (x_1, \ldots, x_m)$, if $d_l^2(X) = \min_{i=1,\ldots,k} \{ d_i^2(X) \}$, then $X \in G_l$.

5.5.2 Fisher method

Suppose that there are k m-dimensional populations $\{G_i\}_{i=1,\ldots,k}$ with means μ_i and the same covariance matrix Σ, and G_i has n_i samples $X_{i1}, \ldots, X_{in_i}$ $(i = 1, \ldots, k)$. Let

$$\hat{\mu}_i = \bar{X}_i = \frac{1}{n_i} \sum_{j=1}^{n_i} X_{ij},$$
$$\bar{X} = \frac{1}{n} \sum_{i=1}^{k} \sum_{j=1}^{n_i} X_{ij} = \frac{1}{n} \sum_{i=1}^{k} n_i \hat{\mu}_i,$$

where $n = n_1 + \cdots + n_k$.

Denote $C = (A + B)^{-1} A$, where

$$A = \sum_{i=1}^{k} n_i (\bar{X}_i - \bar{X})(\bar{X}_i - \bar{X})^T,$$
$$B = \sum_{i=1}^{k} \sum_{j=1}^{n_i} (X_{ij} - \bar{X}_i)(X_{ij} - \bar{X}_i)^T.$$

Let λ_1 be the largest eigenvalue of the matrix C and v_1 be the corresponding eigenvector. Then, for a given observation $X = (x_1, \ldots, x_m)^{\mathrm{T}}$, a linear discriminant function is $z_1 = v_1^{\mathrm{T}} X$.

Suppose that $\hat{\mu}_1, \ldots, \hat{\mu}_k$ lie on a straight line. We compute k distances in the straight line

$$d_i = |v_1^{\mathrm{T}} X - v_1^{\mathrm{T}} \hat{\mu}_i| \quad (i = 1, \ldots, k).$$

Let $d_l = \min_{i=1,\ldots,k}\{d_i\}$. Then $X \in G_l$.

Suppose that $\hat{\mu}_1, \ldots, \hat{\mu}_k$ do not lie on a straight line but on a plane. Let λ_2 be the second largest eigenvalue of the matrix C and v_2 be the corresponding eigenvector. Then, for a given observation $X = (x_1, \ldots, x_m)^{\mathrm{T}}$, two discriminant functions are $y_1 = v_1^{\mathrm{T}} X$ and $y_2 = v_2^{\mathrm{T}} X$. We compute k distances in the plane

$$d_i^2 = (v_1^{\mathrm{T}} X - v_1^{\mathrm{T}} \hat{\mu}_i)^2 + (v_2^{\mathrm{T}} X - v_2^{\mathrm{T}} \hat{\mu}_i)^2 \quad (i = 1, \ldots, k).$$

Let $d_l^2 = \min_{i=1,\ldots,k}\{ d_i^2 \}$. Then $X \in G_l$. If necessary, three discriminant functions are computed but usually two discriminant functions are sufficient by experience.

5.5.3 Bayes method

In the Mahalanobis distance method and Fisher method, we do not consider the prior probability and the losses of mistaken decisions. Bayes method solves these two problems.

Prior probability

Suppose that there are k populations $G_1, \ldots, G_k$ with given probabilities $q_1, \ldots, q_k$, where each $q_i > 0$ and $\sum_{i=1}^k q_i = 1$. Let $X = (x_1, \ldots, x_m)$ be an individual. Define a *generalized square distance* from X to a population G_i as

$$D^2(X, G_i) = d_i^2(X) + h_1(i) + h_2(i) \quad (i = 1, \ldots, k),$$

where $d_i^2(X)$ is the Mahalanobis distance between X and G_i and

$$h_1(i) = \begin{cases} 0 & \text{if } \Sigma_1 = \cdots = \Sigma_k, \\ \log|S_i| & \text{otherwise,} \end{cases}$$

$$h_2(i) = \begin{cases} 0 & \text{if } q_1 = \cdots = q_k, \\ -2\log|q_i| & \text{otherwise.} \end{cases}$$

Here Σ_i is the covariance matrix of G_i and S_i is an estimate of Σ_i.

The decision method of generalized square distance is $X \in G_l$ if $D^2(X, G_l) = \min_{i=1,\ldots,k} D^2(X, G_i)$.

Posterior probability

When an individual X is known, we compute the probability $P(X \in G_i)$

$$P(X \in G_i) = \frac{q_i f_i(x)}{\sum_1^k q_j f_j(x)} \quad (i = 1, \dots, k),$$

where $f_j(x)$ is the probability density of G_j and q_j is the given probability associated with G_j.

If each G_i is a normal population, then its density function is

$$f_i(x) = (2\pi)^{-\frac{m}{2}} |\Sigma_1|^{-\frac{1}{2}} e^{-\frac{1}{2}d_i^2(x)}$$

and

$$P(X \in G_i) = \frac{e^{-\frac{1}{2}d_i^2(x)}}{\sum_1^k e^{-\frac{1}{2}d_j^2(x)}},$$

where $d_i^2(x) = d^2(X, G_i)$. The decision is $X \in G_i$ if $P(X \in G_i) = \max_{j=1,\dots,k} P(x \in G_j)$.

Suppose that there are k populations $G_1, \dots, G_k$ with prior probabilities $q_1, \dots, q_k$. A discriminant criterion D means that a partition of R^m

$$R^m = D_1 \bigcup D_2 \bigcup \cdots \bigcup D_k$$

is given, where $D_1, \dots, D_k$ are mutually disjoint. For an individual $X \in G_i$, we use the criterion D to decide $X \in G_j$ $(j \neq i)$ and we may get a mistaken decision. Denote the probability of making such a mistaken decision by $P(j|i, D)$. Let the probability density function of the proposition G_i be $f_i(x_1, \dots, x_m)$. Then

$$P(j|i, D) = \int_{D_j} \cdots \int f_i(x_1, \dots, x_m)\, dx_1 \cdots dx_m = \int_{D_j} f_i(X)\, dX \quad (j \neq i). \tag{5.5.4}$$

Let $L(j|i, D)$ be the loss of a mistaken decision and $L(j|i, D)$ be determined by experience. Define the *mean loss* of mistaken decision as

$$g(D) = \sum_{l=1}^{k} q_l \sum_{j=1}^{k} P(j|l, D) L(j|l, D). \tag{5.5.5}$$

If a discriminant method D^* is such that $g(D^*) = \min_{\text{all } D} g(D)$, we say D^* conforms to the *Bayes criterion*.

Bayes criterion

Suppose that there are k populations with joint density function $f_1(X), \dots, f_k(X)$, prior probabilities $q_1, \dots, q_k$, and the losses of mistaken decisions $L(j|i, D)$. Then $D^* = (D_1^*, \dots, D_k^*)$ conforms to the Bayes criterion, where

$$\begin{aligned} D_i^* &= \{X \mid h_i(X) < h_j(X) \quad (j \neq i, j = 1, \dots, k)\}, \\ h_j(X) &= \sum_{l=1}^{k} q_l L(j|l, D) f_l(X). \end{aligned} \tag{5.5.6}$$

In fact, by (5.5.4), (5.5.5), and (5.5.6),

$$g(D^*) = \sum_{l=1}^{k} q_l \sum_{j=1}^{k} \left(\int_{D_j^*} f_l(X)\,\mathrm{d}X \right) L(j|l, D^*)$$

$$= \sum_{j=1}^{k} \int_{D_j^*} \left(\sum_{l=1}^{k} q_l f_l(X) L(j|l, D^*) \right) \mathrm{d}X = \sum_{j=1}^{k} \int_{D_j^*} h_j(X)\,\mathrm{d}X.$$

If $D = (D_1, \ldots, D_k)$ is any partition on $\mathbb{R}^m$, then the mean loss caused by them is $g(D) = \sum_{l=1}^{k} \int_{D_l} h_l(X)\mathrm{d}X$. So

$$g(D^*) - g(D) = \sum_{j=1}^{k} \int_{D_j^*} h_j(X)\,\mathrm{d}X - \sum_{l=1}^{k} \int_{D_l} h_l(X)\,\mathrm{d}X$$

$$= \sum_{l=1}^{k} \sum_{j=1}^{k} \int_{D_j^* \cap D_l} (h_j(X) - h_l(X))\,\mathrm{d}X.$$

By (5.5.6), $g(D^*) \le g(D)$, i.e., D^* conforms the Bayes criterion.

5.6 Cluster analysis

Cluster analysis is to partition all individuals into subgroups such that individuals in the same subgroup have similar characteristics. This necessitates devising a rule to measure the similarity between two individuals.

Let $X_1, \ldots, X_n$ be n individuals of m-dimensional random vectors. Ordinarily, one measures the similarity by the distance between X_i and X_j. When the distance is small, we say X_i and X_j are *similar*.

Now we define the distances between two individuals and between two subgroups.

Let $X_i = (x_{i1}, \ldots, x_{im})$ $(i = 1, \ldots, n)$. To ensure that all individuals have a similar scale, one uses the standardized data of individuals. Denote the individual mean vector by $\bar{X} = (\bar{X}_1, \ldots, \bar{X}_n)^{\mathrm{T}}$ and the individual covariance matrix by $S = (S_{ij})_{m\times m}$. Then the *standardized transform* is defined as

$$x_{ij}^* = \frac{x_{ij} - \bar{X}_j}{S_j} \quad (i = 1, \ldots, n;\ j = 1, \ldots, m),$$

where $S_j^2 = \frac{1}{n-1} \sum_{i=1}^{n} (x_{ij} - \bar{X}_j)^2$ $(j = 1, \ldots, m)$.

5.6.1 Distance between individuals X_i and X_j

- Minkowski distance is defined as

$$d_{ij}(p) = \left(\sum_{k=1}^{m} |x_{ik} - x_{jk}|^p \right)^{\frac{1}{p}} \quad (i, j = 1, \ldots, n;\ 0 \le p \le \infty).$$

Specially, the following distance measures are called
absolute value distance:

$$d_{ij}(1) = \sum_{k=1}^{m} |x_{ik} - x_{jk}|;$$

Euclidean distance:

$$d_{ij}(2) = \left(\sum_{k=1}^{m} |x_{ik} - x_{jk}|^2 \right)^{\frac{1}{2}};$$

Euclidean distance with variance weight:

$$d_{ij}^*(2) = \left(\sum_{k=1}^{m} \frac{|x_{ik} - x_{jk}|^2}{S_k} \right)^{\frac{1}{2}};$$

Chebyshev distance:

$$d_{ij}(\infty) = \max_{k=1,\dots,m} |x_{ik} - x_{jk}|.$$

- Mahalanobis distance is defined as

$$d_{ij}(M) = (X_i - X_j)^{\mathrm{T}} S^{-1} (X_i - X_j).$$

- The distance is defined as $d_{ij}^2 = 1 - c_{ij}^2$, where c_{ij} is equal to cosine of angle α_{ij} between m-dimensional vectors $(x_{i1}, \dots, x_{im})$ and $(x_{j1}, \dots, x_{jm})$

$$c_{ij} = \cos \alpha_{ij} = \frac{\sum_{k=1}^{m} x_{ik} x_{jk}}{\left(\sum_{k=1}^{m} x_{ik}^2\right)^{\frac{1}{2}} \left(\sum_{k=1}^{m} x_{jk}^2\right)^{\frac{1}{2}}} \quad (i, j = 1, \dots, n).$$

The c_{ij} is called the *similarity coefficient* between X_i and X_j.

5.6.2 Distance between subgroups G_p and G_q

Denote by D_{pq} the distance between G_p and G_q.

- Single linkage: $D_{pq} = \min_{X_i \in G_p, X_j \in G_q} d_{ij}$.
- Complete method: $D_{pq} = \max_{X_i \in G_p, X_j \in G_q} d_{ij}$.
- Gentraid method: Let $G_r = G_p \bigcup G_q$, and n_p and n_r be cardinal numbers of G_p and G_q, respectively, and $\bar{X}_p$ and $\bar{X}_q$ be mean values of G_p and G_q, respectively. Then the mean value of G_r is

$$\bar{X}_r = \frac{1}{n_r}(n_p \bar{X}_p + n_q \bar{X}_q) \quad (n_r = n_p + n_q).$$

  For a subgroup G_k $(k \neq p, q)$ with the mean value $\bar{X}_k$, the distance D_{rk} between G_r and G_k is $D_{rk} = d(\bar{X}_r, \bar{X}_k)$, where d is the Euclidean distance.
- Average linkage: Let n_p, n_q be a cardinal number of G_p, G_q, respectively. Define the distance as

$$D_{pq} = \frac{1}{n_p n_q} \sum_{X_i \in G_p,\, X_j \in G_q} d_{ij}^2.$$

5.6.3 Hierarchical cluster method

Suppose that there are n individuals and each individual has m indices.

Step 1. Start from n clusters with each cluster containing only one individual. Compute the distance between any two individuals to obtain a matrix of distances.

Step 2. Combine two nearest pair of clusters.

Step 3. Compute the distances between the newly formed clusters to obtain a new matrix of distances.

Step 4. Repeat steps 2 and 3 until there is one cluster left.

Step 5. Determine the number of clusters and members of each cluster.

5.7 Principal component analysis

In multivariate statistical analysis, many variables may be highly correlated and the covariance matrix is of high dimension. For statistical inference, there may be too many parameters. Therefore, one hopes to reduce the number of variables without much loss of information.

5.7.1 Principle component decomposition

The main idea of principal component analysis (PCA) is to transform the set of variables $X_1, \ldots, X_m$ with mean 0 and covariance matrix Σ into a smaller set of uncorrelated new variables $Y_1, \ldots, Y_k$ $(k < m)$ without much loss of information and $Y_1, \ldots, Y_k$ are a linear combination of original variables $X_1, \ldots, X_m$.

Consider a linear combination

$$Y_i = \sum_{j=1}^{m} a_{ij} X_j \quad (i = 1, \ldots, m).$$

The matrix form is $\boldsymbol{Y} = A\boldsymbol{X}$, where

$$A = (a_{ij})_{i,j=1,\ldots,m},$$
$$\boldsymbol{X} = (X_1, \ldots, X_m)^{\mathrm{T}},$$
$$\boldsymbol{Y} = (Y_1, \ldots, Y_m)^{\mathrm{T}}.$$

Denote the i-th row vector of A by $\boldsymbol{A}_i$, i.e., $\boldsymbol{A}_i = (A_{i1}, \ldots, A_{im})$. The covariance matrix of Y is

$$E[\boldsymbol{Y}\boldsymbol{Y}^{\mathrm{T}}] = E[A\boldsymbol{X}\boldsymbol{X}^{\mathrm{T}}A^{\mathrm{T}}] = AE[\boldsymbol{X}\boldsymbol{X}^{\mathrm{T}}]A^{\mathrm{T}}.$$

Since $\Sigma = E[\boldsymbol{X}\boldsymbol{X}^{\mathrm{T}}]$, $E[\boldsymbol{Y}\boldsymbol{Y}^{\mathrm{T}}] = A\Sigma A^{\mathrm{T}}$, and so

$$\operatorname{Var}(Y_i) = \boldsymbol{A}_i \Sigma \boldsymbol{A}_i^{\mathrm{T}},$$
$$\operatorname{Cov}(Y_i, Y_j) = \boldsymbol{A}_i \Sigma \boldsymbol{A}_j^{\mathrm{T}} \quad (i, j = 1, \ldots, m).$$

Since the covariance matrix Σ is a real symmetric matrix and is nonnegative definite, all eigenvalues are nonnegative real numbers satisfying $\lambda_1 \geq \lambda_2 \geq \cdots \geq \lambda_m \geq 0$ and the corresponding eigenvectors $\boldsymbol{a}_1^*, \ldots, \boldsymbol{a}_m^*$ with unit length are linear independent. Denote $\boldsymbol{a}_i^* = (a_{1i}^*, \ldots, a_{mi}^*)^{\mathrm{T}}$ $(i = 1, \ldots, m)$. Define

$$Y_i^* = \sum_{j=1}^{m} a_{ij}^* X_j. \tag{5.7.1}$$

Then

$$\begin{aligned} \operatorname{Var}(Y_i^*) &= (\boldsymbol{a}_i^*)^{\mathrm{T}} \Sigma \boldsymbol{a}_i^* = \lambda_i \quad && (i, j = 1, \ldots, m), \\ \operatorname{Cov}(Y_i^*, Y_j^*) &= 0 && (i \neq j;\ i, j = 1, \ldots, m). \end{aligned} \tag{5.7.2}$$

These new variables $Y_1^*, \ldots, Y_m^*$ are obtained by linear combination of original variables $X_1, \ldots, X_m$. These new variables are called *principal components*. The goal of principal component analysis is to replace the original set of variables by the first few principal components if the first few principal components can explain most variability.

We rewrite (5.7.1) into the matrix form $\boldsymbol{Y}^* = A^* \boldsymbol{X}$, where

$$\begin{aligned} \boldsymbol{Y}^* &= (Y_1^*, \ldots, Y_m^*)^{\mathrm{T}}, \\ \boldsymbol{X} &= (X_1, \ldots, X_m)^{\mathrm{T}}, \\ A^* &= (a_{ij}^*)_{i,j=1,\ldots,m} \end{aligned}$$

and A^* is an orthogonal matrix whose column vectors are m eigenvectors $A^* = (\boldsymbol{a}_1^*, \ldots, \boldsymbol{a}_m^*)$. So

$$\boldsymbol{X} = (A^*)^{-1} \boldsymbol{Y}^* = (A^*)^{\mathrm{T}} \boldsymbol{Y}^*.$$

Since $A^* = (\boldsymbol{a}_1^*, \ldots, \boldsymbol{a}_m^*)$ and $\boldsymbol{Y}^* = (Y_1^*, \ldots, Y_m^*)^{\mathrm{T}}$,

$$\boldsymbol{X} = \sum_{1}^{m} Y_i^* \boldsymbol{a}_i^*. \tag{5.7.3}$$

Since the total variation of $\boldsymbol{X} = (X_1, \ldots, X_m)^{\mathrm{T}}$ is equal to $\operatorname{Tr}(\Sigma)$ (i.e., the trace of covariance matrix Σ). The trace $\operatorname{Tr}(\Sigma)$ is equal to the sum of eigenvalues $\operatorname{Tr}(\Sigma) = \lambda_1 + \cdots + \lambda_m$. By (5.7.2), the importance of the j-th principal component can be measured by the ratio $\lambda_j / \operatorname{Tr}(\Sigma)$ $(j = 1, \ldots, m)$. So the importance of the first k principal components can be measured by the ratio

$$\sum_{1}^{k} \lambda_j / \operatorname{Tr}(\Sigma).$$

If we use two principal components to replace the original m variables without much loss of information, we can obtain better parameter estimates and better use of graphical tools.

In practice, the covariance matrix Σ of X is unknown. We should use samples to estimate it. Suppose that there is a sample $\{\boldsymbol{x}_j\}_{j=1,\dots,n}$, where $\boldsymbol{x}_j = (x_{1j}, \dots, x_{mj})^{\mathrm{T}}$ $(j = 1, \dots, n)$ are the m-dimensional vectors. The covariance matrix Σ has an estimate $\hat{\Sigma} = (\alpha_{ij})_{i,j=1,\dots,m}$, where $\alpha_{ij} = \frac{1}{n}\sum_{k=1}^{n} x_{ki}x_{kj}$ $(i, j = 1, \dots, m)$. The accuracy of these estimates depends on the sample size n. Since results of principal component analysis depend on units of variables, it is desirable that the original data has a similar scale. Therefore, one often performs PCA on the standardized data $z_{ij} = x_{ij}/\hat{\sigma}_j$, where $\hat{\sigma}_j$ is the estimate of variance of $\boldsymbol{x}_j$.

Let $\hat{\lambda}_1 \geq \hat{\lambda}_2 \geq \cdots \geq \hat{\lambda}_m$ be the eigenvalues of $\hat{\Sigma}$ and the corresponding eigenvectors with unit length be $\hat{\boldsymbol{a}}_1^*, \dots, \hat{\boldsymbol{a}}_m^*$. Denote $\hat{\boldsymbol{a}}_i^* = (\hat{a}_{1i}^*, \dots, \hat{a}_{mi}^*)^{\mathrm{T}}$ $(i = 1, \dots, m)$. Define

$$\hat{\boldsymbol{y}}_i^* = \sum_{j=1}^{m} \hat{a}_{ij}^* \boldsymbol{x}_j \quad (i = 1, \dots, m).$$

Let $\hat{\boldsymbol{y}}_i^* = (\hat{y}_{i1}^*, \dots, \hat{y}_{in}^*)^{\mathrm{T}}$. Then

$$\hat{y}_{il}^* = \sum_{j=1}^{m} \hat{a}_{ij}^* x_{jl} \quad (i = 1, \dots, m;\ l = 1, \dots, n).$$

These new data $\{\hat{y}_{il}^*\}_{m\times n}$ are called *principal component scores* and are used for further analysis.

5.7.2 Rotation of principal components

Assume that for m random variables $X_1, \dots, X_m$ with mean 0, we find k principal components $Y_1, \dots, Y_k$ without much loss of information. Each variable X_i has practical meaning but each Y_i does not always have practical meaning. Take a $k \times k$ orthogonal matrix G. Define

$$\boldsymbol{Z} = G\boldsymbol{Y},$$

where $\boldsymbol{Z} = (z_1, \dots, z_k)^{\mathrm{T}}$ and $\boldsymbol{Y} = (Y_1, \dots, Y_k)^{\mathrm{T}}$. The $z_1, \dots, z_k$ are called the *rotated principal components*. They are some new linear combination of $X_1, \dots, X_m$ and possess the following two properties:

- the total variance is invariant, i.e.,

$$\sum_{1}^{k} \mathrm{Var}(z_i) = \sum_{1}^{k} \mathrm{Var}(Y_i) = \sum_{1}^{k} \lambda_i,$$

  where $\lambda_1, \dots, \lambda_k$ are the first k eigenvalues of covariance matrix of X;
- $z_1, \dots, z_k$ are correlated, i.e., $\mathrm{Cov}(z_i, z_j) \neq 0$ $(i \neq j)$.

5.8 Canonical correlation analysis

Canonical correlation analysis is a kind of statistical method to study the relationship between two random vectors $\boldsymbol{X}$ and $\boldsymbol{Y}$, where $\boldsymbol{X} = (X_1, \ldots, X_p)^{\mathrm{T}}$ and $\boldsymbol{Y} = (Y_1, \ldots, Y_q)^{\mathrm{T}}$. Assume that the means of $\boldsymbol{X}$ and $\boldsymbol{Y}$ are both $\mathbf{0}$. We hope to choose a p-dimensional vector $\boldsymbol{f}_1 = (f_{11}, \ldots, f_{1p})^{\mathrm{T}}$ and a q-dimensional vector $\boldsymbol{g}_1 = (g_{11}, \ldots, g_{1q})^{\mathrm{T}}$ such that $\alpha_1 = \sum_{k=1}^{p} f_{1k} X_k$ and $\beta_1 = \sum_{l=1}^{q} g_{1l} Y_l$ satisfy the following two conditions:

(a) the correlation coefficients of α_1 and β_1 attain the maximal value;

(b) $\mathrm{Var}(\alpha_1) = \mathrm{Var}(\beta_1) = 1$.

To find f_1 and g_1, we first state the concepts of the square roots and singular values of matrices.

Let A be a real symmetric nonnegative definite matrix. Then there is an orthogonal matrix Γ such that

$$A = \Gamma \operatorname{diag}(\lambda_1, \ldots, \lambda_m) \Gamma^{\mathrm{T}},$$

where $\lambda_1 \geq \lambda_2 \geq \cdots \geq \lambda_m > 0$. Define the square root of A as

$$A^{\frac{1}{2}} = \Gamma \operatorname{diag}(\sqrt{\lambda_1}, \ldots, \sqrt{\lambda_m}) \Gamma^{\mathrm{T}}.$$

It is easy to prove that $A^{\frac{1}{2}} A^{\frac{1}{2}} = A$.

Let B be an $m \times n$ matrix with rank r. Then there exists an $m \times m$ orthogonal matrix C and an $n \times n$ orthogonal matrix D such that

$$B = C \begin{pmatrix} G & 0 \\ 0 & 0 \end{pmatrix} D^{\mathrm{T}},$$

where $G = \operatorname{diag}(\mu_1, \ldots, \mu_r)$ ($r \leq \min\{m, n\}$) and $\mu_1 \geq \mu_2 \geq \cdots \geq \mu_r \geq 0$. Each μ_k is called a *singular value*. The k-th column vector $\boldsymbol{C}_k$ of the matrix C is called the *left singular vector* of B. The k-th column vector $\boldsymbol{D}_k$ of the matrix D is called the *right singular vector* of D.

Using the Lagrange multiplier method, the following $\boldsymbol{f}_1$ and $\boldsymbol{g}_1$

$$\boldsymbol{f}_1 = \Sigma_{\boldsymbol{XX}}^{-\frac{1}{2}} \boldsymbol{C}_1, \quad \boldsymbol{g}_1 = \Sigma_{\boldsymbol{YY}}^{-\frac{1}{2}} \boldsymbol{D}_1$$

satisfy the above conditions (a) and (b), where $\Sigma_{\boldsymbol{XX}}$ and $\Sigma_{\boldsymbol{YY}}$ are covariance matrices of $\boldsymbol{X}$ and $\boldsymbol{Y}$, respectively, and $\boldsymbol{C}_1$ and $\boldsymbol{D}_1$ are the left and right singular vectors of matrix $S = \Sigma_{\boldsymbol{XX}}^{-\frac{1}{2}} \Sigma_{\boldsymbol{XY}} \Sigma_{\boldsymbol{YY}}^{-\frac{1}{2}}$ corresponding to the maximal singular values μ_1 of S, and $\Sigma_{\boldsymbol{XY}}$ is the covariance matrix of $\boldsymbol{X}$ and $\boldsymbol{Y}$. Moreover,

$$\mathrm{Cov}(\alpha_1, \beta_1) = \sqrt{\mu_1}.$$

Assume that S has r nonzero singular values $\mu_1 \geq \mu_2 \geq \cdots \geq \mu_r > 0$. We further find $\boldsymbol{f}_k$ and $\boldsymbol{g}_k$ $(k = 1, \ldots, r)$ such that $\alpha_k = \boldsymbol{f}_k^{\mathrm{T}} \boldsymbol{X}$ and $\beta_k = \boldsymbol{g}_k^{\mathrm{T}} \boldsymbol{Y}$ $(k = 1, \ldots, r)$ satisfy

$$\begin{aligned} &\operatorname{Var}(\alpha_k) = \operatorname{Var}(\beta_k) = 1, && \operatorname{Cov}(\alpha_k, \beta_k) = \sqrt{\mu_k}, \\ &(\boldsymbol{f}_k, \boldsymbol{f}_l) = (\boldsymbol{g}_k, \boldsymbol{g}_l) = 0 \quad (k \neq l), && (\boldsymbol{f}_k, \boldsymbol{g}_l) = \delta_{kl} \quad (k, l = 1, \ldots, r), \end{aligned}$$

where δ_{ij} is the Kronecker delta.

5.9 Factor analysis

Factor analysis is to remove redundancy from a set of correlated statistical variables and represent these variables with a smaller set of new statistical variables.

5.9.1 The factor analysis model

Let $\boldsymbol{X} = (X_1, \ldots, X_p)^{\mathrm{T}}$ be a random vector with mean 0. The *factor analysis model* can be written as

$$X_k = \sum_{i=1}^{m} \lambda_{ki} f_i + \varepsilon_k \quad (k = 1, \ldots, p, \ m \leq p),$$

where f_i's are called *factors* and λ_{ki} is called a *loading* of the k-th variable on the i-th factor, which reflects the relative importance of the i-th factor for the k-th variable, and ε_k's are random errors. The matrix form of the factor analysis model is

$$\boldsymbol{X} = \Lambda \boldsymbol{F} + \boldsymbol{\epsilon}, \tag{5.9.1}$$

where $\Lambda = (\lambda_{kl})_{p \times m}$ is the *factor loading matrix*, $\boldsymbol{F} = (f_1, \ldots, f_m)^{\mathrm{T}}$ is the *common factor vector*, and $\boldsymbol{\epsilon} = (\varepsilon_1, \ldots, \varepsilon_m)^{\mathrm{T}}$.

The assumptions for factor analysis are

- factors f_k's are independently and identically distributed with mean 0 and variance 1;
- random errors ε_k's are independent with mean $\mathbf{0}$ and variance ψ_k;
- f_k and ε_j are independent for any k and j.

5.9.2 The factor analysis equation

Let $\boldsymbol{X}$ be a random vector with mean $\mathbf{0}$. Then the covariance matrix of X is

$$\Sigma = E[\boldsymbol{X}\boldsymbol{X}^{\mathrm{T}}] = E[(\Lambda \boldsymbol{F})(\Lambda \boldsymbol{F})^{\mathrm{T}}] + E[\boldsymbol{\epsilon}(\Lambda \boldsymbol{F})^{\mathrm{T}}] + E[(\Lambda \boldsymbol{F})\boldsymbol{\epsilon}^{\mathrm{T}}] + E[\boldsymbol{\epsilon}\boldsymbol{\epsilon}^{\mathrm{T}}] = A + B + C + D$$

and by the assumption,

$$\begin{aligned} A &= E[\Lambda \boldsymbol{F}\boldsymbol{F}^{\mathrm{T}}\Lambda^{\mathrm{T}}] = \Lambda E[\boldsymbol{F}\boldsymbol{F}^{\mathrm{T}}]\Lambda^{\mathrm{T}} = \Lambda\Lambda^{\mathrm{T}}, \\ B &= E[\boldsymbol{\epsilon}\boldsymbol{F}^{\mathrm{T}}]\Lambda^{\mathrm{T}} = 0, \\ C &= \Lambda E[\boldsymbol{F}\boldsymbol{\epsilon}^{\mathrm{T}}] = 0, \\ D &= E[\boldsymbol{\epsilon}\boldsymbol{\epsilon}^{\mathrm{T}}] = \mathrm{diag}(\psi_1, \dots, \psi_p) =: \Psi. \end{aligned}$$

So

$$\Sigma = \Lambda\Lambda^{\mathrm{T}} + \Psi, \tag{5.9.2}$$

where $\Psi = \mathrm{diag}(\psi_1, \dots, \psi_p)$. This equation (5.9.2) is called a *factor analysis equation*.

Let $\Sigma = (\sigma_{kl})_{p\times m}$. Then the variance of X_k is

$$\sigma_{kk} = \sum_{j=1}^{m} \lambda_{kj}^2 + \psi_k \quad (k = 1, \dots, p).$$

Therefore, the proportion of variance of X_k explained by factors $\boldsymbol{f}_1, \dots, \boldsymbol{f}_m$ is $\sum_{j=1}^{m} \lambda_{kj}^2 / \sigma_{kk}$. The the factor loading matrix Λ is not unique.

5.9.3 Parameter estimate method

The goal of factor analysis is to describe the structure of covariance of p variables by only a few factors. So this needs estimate factor loadings $\{\lambda_{ki}\}$ and variances ψ_k of errors.

Maximal likelihood method

If factors f_k's and errors ϵ_k's follow normal distributions, the maximal likelihood estimates of the loading matrix and random error can be obtained. By the factor analysis equation, the *sample likelihood function* is

$$L(\bar{\boldsymbol{X}}, \Sigma) = L(\bar{\boldsymbol{X}}, \Lambda\Lambda^{\mathrm{T}} + \Psi) =: \varphi(\Lambda, \Psi).$$

We choose Λ_0 and Ψ_0 such that $\varphi(\Lambda, \Psi)$ attains the maximal value. It can be proved that Λ_0 and Ψ_0 satisfy the system of equations

$$\begin{cases} S\Psi_0^{-1}\Lambda_0 = \Lambda_0(I + \Lambda^{\mathrm{T}}\Psi_0^{-1}\Lambda_0), \\ \Lambda_0 = \mathrm{diag}(S - \Lambda_0\Lambda_0^{\mathrm{T}}), \end{cases} \tag{5.9.3}$$

where $S = \frac{1}{n}\sum_{k=1}^{n}(X_k - \bar{\boldsymbol{X}})(X_k - \bar{\boldsymbol{X}})^{\mathrm{T}}$ is the sample covariance matrix.

To ensure that the system of equations (5.9.3) has a unique solution, Bayes suggested to add a condition that $\Lambda_0^{\mathrm{T}}\Psi_0^{-1}\Lambda_0$ is a diagonal matrix. The system of equations is solved using the iteration method.

Principal component method

Let the eigenvalues of the sample covariance matrix Σ be $\lambda_1 \geq \cdots \geq \lambda_p \geq 0$ and the corresponding unit orthogonal eigenvectors be $\boldsymbol{l}_1, \boldsymbol{l}_2, \ldots, \boldsymbol{l}_p$, i.e.,

$$
\begin{aligned}
lSl^{\mathrm{T}} &= \operatorname{diag}(\lambda_1, \ldots, \lambda_p),\\
S &= l^{\mathrm{T}} \operatorname{diag}(\lambda_1, \ldots, \lambda_p) l = \sum_1^p \lambda_i \boldsymbol{l}_i \boldsymbol{l}_i^{\mathrm{T}} = \sum_1^m \lambda_i \boldsymbol{l}_i \boldsymbol{l}_i^{\mathrm{T}} + \sum_{m+1}^p \lambda_i \boldsymbol{l}_i \boldsymbol{l}_i^{\mathrm{T}},
\end{aligned}
$$

where $l = (\boldsymbol{l}_1, \ldots, \boldsymbol{l}_p)$. This formula is called *spectral decomposition*. When the eigenvalues $\lambda_{m+1}, \ldots, \lambda_p$ are small, S may decompose approximately into

$$
S \approx (\sqrt{\lambda_1}\, \boldsymbol{l}_1, \ldots, \sqrt{\lambda_m}\, \boldsymbol{l}_m) \begin{pmatrix} \sqrt{\lambda_1}\, \boldsymbol{l}_1^{\mathrm{T}} \\ \vdots \\ \sqrt{\lambda_m}\, \boldsymbol{l}_m^{\mathrm{T}} \end{pmatrix} + \operatorname{diag}(\psi_1, \ldots, \psi_p) = \Lambda\Lambda^{\mathrm{T}} + \Psi,
$$

where

$$
\begin{aligned}
\Lambda &= (\sqrt{\lambda_1}\, \boldsymbol{l}_1, \ldots, \sqrt{\lambda_m}\, \boldsymbol{l}_m) =: (\lambda_{ki})_{p\times m},\\
\Psi &= \operatorname{diag}(\psi_1, \ldots, \psi_p), \quad \psi_k = S_{kk} - \sum_{i=1}^m \lambda_{ki}^2 \quad (k = 1, \ldots, p).
\end{aligned}
$$

This gives a solution of the factor analysis equation. The j-th column of the loading matrix is $\sqrt{\lambda_j}$ times the j-th principal component of X. Denote the error

$$
S - (\Lambda\Lambda^{\mathrm{T}} + \Psi) = (\eta_{ij})_{p\times p}.
$$

It can be proved that $\sum_{i,j=1}^p \eta_{ij}^2 \leq \sum_{m+1}^p \lambda_j^2$. Therefore, we may choose m such that the error is very small. One often chooses m such that

$$
\frac{\sum_1^m \lambda_j}{\sum_1^p \lambda_j} \geq 0.7 .
$$

5.9.4 Rotation of a loading matrix

The goal of factor analysis is not only to find common factors but also to know the practical meaning of each common factor. In Section 5.9.3 we found the initial common factor using the maximal likelihood method and principal component method. In order to ensure that each common factor has practical meaning, we should choose an orthogonal matrix Q such that $\Lambda^* := \Lambda Q$ is also a loading matrix, and the new common factors have practical meaning.

In fact, since Q is an orthogonal matrix, $QQ^{\mathrm{T}} = I$ and $\Lambda^*(\Lambda^*)^{\mathrm{T}} = \Lambda QQ^{\mathrm{T}}\Lambda^t = \Lambda\Lambda^{\mathrm{T}}$, and so

$$
\begin{aligned}
\Sigma &= \Lambda^*(\Lambda^*)^{\mathrm{T}} + \Psi,\\
\boldsymbol{X} &= \Lambda \boldsymbol{F} + \boldsymbol{\epsilon} = \Lambda Q(Q^{\mathrm{T}}\boldsymbol{F}) + \boldsymbol{\epsilon} = \Lambda^* \boldsymbol{Z} + \boldsymbol{\epsilon},
\end{aligned}
$$

where $\boldsymbol{Z} = Q^{\mathrm{T}}\boldsymbol{F}$ and the covariance matrix of $\boldsymbol{Z}$ is $\Sigma_{\boldsymbol{Z}} = E[\boldsymbol{Z}\boldsymbol{Z}^{\mathrm{T}}] = E[Q^{\mathrm{T}}\boldsymbol{F}\boldsymbol{F}^{\mathrm{T}}Q] = I$ and

$$\mathrm{Cov}(\boldsymbol{Z}, \boldsymbol{\epsilon}) = \mathrm{Cov}(Q^{\mathrm{T}}\boldsymbol{F}, \boldsymbol{\epsilon}) = Q^{\mathrm{T}}\,\mathrm{Cov}(\boldsymbol{F}, \boldsymbol{\epsilon}) = 0.$$

It is seen that if $\boldsymbol{F}$ is the common factor vector of the factor model, then, for any orthogonal matrix Q, $\boldsymbol{Z} = Q^{\mathrm{T}}\boldsymbol{F}$ is also the common factor vector and ΛQ is the loading matrix of the common factor Q. We should use rotation repeatedly such that many factor loadings tend to zero and maximize other factor loadings. So this lets us focus on those factors with large loadings. This method is called the *orthogonal rotation of factor axes*.

Suppose that the factor analysis model is $\boldsymbol{X} = \Lambda\boldsymbol{F} + \boldsymbol{\epsilon}$ and $\Lambda = (\lambda_{ij})_{p\times m}$. Let $h_i^2 = \sum_{j=1}^{m}\lambda_{ij}^2$ $(i = 1, \dots, p)$. Then h_i^2 is called *communality* of the variable X_i. Let

$$d_{ij}^2 = \frac{\lambda_{ij}^2}{h_i^2} \quad (i = 1, \dots, p;\ j = 1, \dots, m),$$

$$D = \begin{pmatrix} d_{11} & \cdots & d_{1j} & \cdots & d_{1m} \\ \vdots & \ddots & \vdots & \ddots & \vdots \\ d_{i1} & \cdots & d_{ij} & \cdots & d_{im} \\ \vdots & \ddots & \vdots & \ddots & \vdots \\ d_{p1} & \cdots & d_{pj} & \cdots & d_{pn} \end{pmatrix}.$$

The *variance of p data* in the j-th column of D is defined as

$$V_j = \frac{1}{p}\sum_{i=1}^{p}(d_{ij}^2 - \bar{d}_j^2),$$

where $\bar{d}_j = \frac{1}{p}\sum_{i=1}^{p} d_{ij}^2$ $(j = 1, \dots, m)$. This implies that

$$V_j = \frac{1}{p}\sum_{i=1}^{p}\left(\frac{\lambda_{ij}^4}{h_i^4} - \frac{1}{p^2}\left(\sum_{\mu=1}^{p}\frac{\lambda_{\mu j}^2}{h_\mu^2}\right)^2\right).$$

The variance of the factor loading matrix Λ is

$$V = \sum_{j=1}^{m} V_j = \frac{1}{p^2}\sum_{j=1}^{m}\left(p\sum_{i=1}^{p}\frac{\lambda_{ij}^4}{h_i^4} - \left(\sum_{\mu=1}^{p}\frac{\lambda_{\mu j}^2}{h_\mu^2}\right)^2\right).$$

Through a rotation of a loading matrix, we expect that the variance V is large such that the loading values tend to 1 or tend to 0. In this way, the corresponding common factors have a simplified structure.

Further reading

[1] Bao G, Liu Y, Linderholm HW. April-September mean maximum temperature inferred from Hailar pine tree rings in the Hulunbuir region, Inner Mongolia, back to 1868 AD. *Palaeogeogr Palaeoclimatol Palaeoecol*. 2012(313–314):162–172.
[2] Chen F, Yuan Y, Wei W, Yu S, Zhang T. Reconstructed temperature for Yongan, Fujian, Southeast China: Linkages to the Pacific Ocean climate variability. *Global Planet Change*. 2012(86–87): 11–19.
[3] Gao J, Shi Z, Xu L, Yang X, Jia Z, Lu S, Feng C, Shang J. Precipitation variability in Hulunbuir, northeastern China since 1829 AD reconstructed from tree-rings and its linkage with remote oceans. *J Arid Environ*. 2013(95):14–21.
[4] Gregg D, Rolfe J. The value of environment across efficiency quantiles: A conditional regression quantiles analysis of rangelands beef production in North Eastern Australia. *Ecological Economics*. 2016(128):44–54.
[5] Fernandez P, Mourato S, Moreira M, Pereira L. A new approach for computing a flood vulnerability index using cluster analysis. *Physics and Chemistry of the Earth, Parts A/B/C*. 2016: in press.
[6] Jones M, Randell D, Ewans K, Jonathan P. Statistics of extreme ocean environments: Non-stationary inference for directionality and other covariate effects. *Ocean Engineering*. 2016(119):30–46.
[7] Liu Y, Bao G, Song H, Cai Q, Sun J. Precipitation reconstruction from Hailar pine tree rings in the Hailar region, Inner Mongolia, China back to 1865 AD. *Palaeogeogr Palaeoclimatol Palaeoecol*. 2009(282):81–87.
[8] Parinet J, Julien M, Nun P, Robins RJ, Remaud G, Hohener P. Predicting equilibrium vapour pressure isotope effects by using artificial neural networks or multi-linear regression – A quantitative structure property relationship approach. *Chemosphere*. 2015(134):521–527.
[9] Parente J, Pereira MG, Tonini M. Space-time clustering analysis of wildfires: The influence of dataset characteristics, fire prevention policy decisions, weather and climate. *Science of the Total Environment*. 2016(559):151–165.
[10] Roberts-Jones J, Bovis K, Martin MJ, McLaren A. Estimating background error covariance parameters and assessing their impact in the OSTIA system. *Remote Sensing of Environment*. 2016(176):117–138.
[11] Singh S, Prakash A, Chakraborty NR, Wheeler C, Agarwal PK, Ghosh A. Trait selection by path and principal component analysis in Jatropha curcas for enhanced oil yield. *Industrial Crops and Products*. 2016(86):173–179.
[12] Tan KC, Lim HS, Jafri MZM. Prediction of column ozone concentrations using multiple regression analysis and principal component analysis techniques: A case study in peninsular Malaysia. *Atmospheric Pollution Research*. 2016: in press.
[13] Wang Y, Lu R, Ma Y, Sang Y, Meng H, Gao S. Annual variation in PDSI since 1897 AD in the Tengger Desert, Inner Mongolia, China, as recorded by tree-ring data. *J Arid Environ*. 2013(98): 20–26.
[14] Weichenthal S, Ryswyk KV, Goldstein A, Bagg S, Shekkarizfard M, Hatzopoulou M. A land use regression model for ambient ultrafine particles in Montreal, Canada: A comparison of linear regression and a machine learning approach. *Environmental Research*. 2016(146):65–72.
[15] Zhang Z. Tree-Rings, a key ecological indicator of environment and climate change. *Ecological Indicators*. 2015(51):107–116.
[16] Zhao X, Qian J, Wang J, He Q, Wang Z, Chen C. Using a tree ring $\delta^{13}C$ annual series to reconstruct atmospheric CO2concentration over the past 300 years. *Pedosphere*. 2006(16):371–79.

6 Numerical methods

Numerical methods are widely used in each branch of environmental science. While environmental simulation, prediction and impact assessment depend on mathematical models, numerical methods provide a practical approach in quickly computing the solution of these models. In this chapter, we will introduce numerical integration and differentiation, numerical linear algebra, and numerical solution of ordinary differential equations and partial differential equations by using iterative method, difference method, finite element method, and wavelet method.

6.1 Numerical integration

Let $x_k \in [a, b]$ $(k = 1, \dots, n)$. If a function f is a continuous function on $[a, b]$ and $f(x_k)$ $(k = 1, \dots, n)$ are known, the *quadrature formula* for the integral $\int_a^b f(x)\,\mathrm{d}x$ is

$$\int_a^b f(x)\,\mathrm{d}x \doteq \sum_1^n A_k f(x_k),$$

where x_k are called *nodes* and A_k are called *quadrature coefficients*.

Throughout this chapter, the notation $A \doteq B$ means that A is equal approximately to B.

One wants to choose suitable nodes x_k and coefficients A_k such that the error

$$e_n(f) = \int_a^b f(x)\,\mathrm{d}x - \sum_{k=1}^n A_k f(x_k)$$

is as small as possible.

6.1.1 Interpolation-type quadrature formulas

Assume that f is a continuous function on $[a, b]$ and $a \le x_1 \le x_2 \le \cdots \le x_n \le b$. Let

$$\omega_n(x) = (x - x_1)(x - x_2)\cdots(x - x_n),$$
$$l_k(x) = \frac{\omega_n(x)}{\omega_n'(x_k)(x - x_k)} \quad (k = 1, \dots, n).$$

The Lagrange interpolation polynomial for f is $L(x) = \sum_{k=1}^n f(x_k) l_k(x)$ (see Section 4.2). Since $f(x) \doteq L(x)$,

$$\int_a^b f(x)\,\mathrm{d}x \doteq \int_a^b \Big(\sum_1^n f(x_k) l_k(x)\Big)\,\mathrm{d}x = \sum_1^n A_k f(x_k), \tag{6.1.1}$$

DOI 10.1515/9783110424904-007

where $A_k = \int_a^b l_k(x)\,dx$ $(k = 1, \dots, n)$. Assume further that $f \in C^n([a, b])$. Then the error is estimated by

$$|e_n(f)| = \left| \frac{1}{n!} \int_a^b f^{(n)}(\xi_x)\omega_n(x)\,dx \right| \leq \frac{C}{n!} \max_{a \leq x \leq b} |f^{(n)}(x)|,$$

where $C = \int_a^b |\omega_n(x)|\,dx$. If f is a polynomial of degree $\leq n - 1$, then $f(x) \equiv L(x)$. So (6.1.1) is exact, i.e.,

$$\int_a^b f(x)\,dx = \sum_1^n A_k f(x_k).$$

Let $n = 2$, $x_1 = a$, $x_2 = b$ in (6.1.1). The *trapezoidal formula* is

$$\int_a^b f(x)\,dx \doteq \frac{b-a}{2}(f(a) + f(b)).$$

Let $n = 3$, $x_1 = a$, $x_2 = \frac{a+b}{2}$, $x_3 = b$ in (6.1.1), then the *Simpson formula* is

$$\int_a^b f(x)\,dx \doteq \frac{b-a}{6}\left(f(a) + 4f\left(\frac{a+b}{2}\right) + f(b) \right).$$

Let $h = \frac{b-a}{n}$ and $x_k = a + kh (k = 0, \dots, n)$. Applying the trapezoidal formula on each interval $[x_k, x_{k+1}]$ $(k = 0, \dots, n-1)$, the *complicated trapezoidal formula* is given by

$$\int_a^b f(x)\,dx \doteq \frac{h}{2}\left(f(a) + 2\sum_1^{n-1} f(a + kh) + f(b) \right).$$

If $f \in C^2([a, b])$, the error is

$$e_n(f) = -\frac{(b-a)^3}{12n^2} f''(\eta_1) \quad (a \leq \eta_1 \leq b).$$

Let $h = \frac{b-a}{2m}$ and $x_k = a + kh$ $(k - 0, \dots, 2m)$. Applying the Simpson formula on each interval $[x_{2k-2}, x_{2k}]$ $(k = 1, \dots, m)$, the *complicated Simpson formula* is given by

$$\int_a^b f(x)\,dx \doteq \frac{h}{3}\left(f(a) + 4\sum_1^m f(x_{2k-1}) + 2\sum_1^{m-1} f(x_{2k}) + f(b) \right).$$

If $f \in C^4([a, b])$, the error is

$$e_n(f) = -\frac{(b-a)^5}{2880m^4} f^{(4)}(\eta_2) \quad (0 \leq \eta_2 \leq b).$$

6.1.2 Gauss quadrature formula

One expects to choose nodes such that the quadrature formula becomes more precise. The zeros of Legendre polynomials play such a role. It is well known that Legendre polynomials

$$P_n(x) = \frac{1}{2^n n!}\frac{\mathrm{d}^n(x^2-1)^n}{\mathrm{d}x^n} \quad (n = 0, 1, \dots)$$

are orthogonal polynomials on $[-1, 1]$ with weight 1. Its n zeros $x_k^{(n)}$ satisfy $-1 < x_1^{(n)} < \cdots < x_n^{(n)} < 1$. If the interpolation nodes are zeros of the Legendre polynomial, then the *Gauss–Legendre quadrature formula* is

$$\int_{-1}^{1} f(x)\,\mathrm{d}x \doteq \sum_{1}^{n} A_k^{(n)} f(x_k^{(n)}),$$

where

$$A_k^{(n)} = \frac{2(1-(x_k^{(n)})^2)}{n^2 P_{n-1}^2(x_k^{(n)})}.$$

This formula is also the *classical Gauss quadrature formula*. For $f \in C^{2n}([-1, 1])$, the error is

$$e_n(f) = \frac{2^{2n+1}(n!)^4}{(2n+1)((2n)!)^3} f^{(2n)}(\xi) \quad (-1 < \xi < 1).$$

If f is a polynomial of degree $\leq 2n-1$, then the Gauss–Legendre quadrature formula is exact, i.e.,

$$\int_{-1}^{1} f(x)\,\mathrm{d}x = \sum_{1}^{n} A_k^{(n)} f(x_k^{(n)}).$$

More generally, assume that $\{\omega_n(x)\}$ is a monic orthogonal polynomial system on $[a, b]$ with weight $\rho(x)$. Take zeros $\{x_k^{(n)}\}_{k=1,\dots,n}$ of $\omega_n(x)$ as nodes. The general Gauss quadrature formula is

$$\int_{a}^{b} f(x)\rho(x)\,\mathrm{d}x \doteq \sum_{1}^{n} A_k^{(n)} f(x_k^{(n)}),$$

where

$$A_k^{(n)} = \int_{a}^{b} \frac{\rho(x)\omega_n(x)}{(x-x_k^{(n)})\omega_n'(x_k^{(n)})}\,\mathrm{d}x.$$

Assume further that $f \in C^{2n}([a, b])$. The error is

$$e_n(f) = \int_{a}^{b} f(x)\rho(x)\,\mathrm{d}x - \sum_{1}^{n} A_k^{(n)} f(x_k^{(n)}) = \frac{f^{(2n)}(\xi)}{(2n)!}\int_{a}^{b} \rho(x)\omega_n^2(x)\,\mathrm{d}x \quad (a < \xi < b).$$

If f is a polynomial of degree $\leq 2n-1$, the general Gauss quadrature formula is exact, i.e.,

$$\int_a^b f(x)\rho(x)\,\mathrm{d}x = \sum_1^n A_k^{(n)} f(x_k^{(n)}).$$

As a special case, the monic Chebyshev polynomials

$$\tilde{T}_n(x) = \frac{\cos(n \arccos x)}{2^{n-1}} \quad (n = 0, 1, \dots)$$

are an orthogonal polynomial on $[-1, 1]$ with weight $\rho(x) = \frac{1}{\sqrt{1-x^2}}$. Taking its zeros $x_k = \cos\frac{2k-1}{2n}\pi$ $(k = 1, \dots, n)$ as nodes, the general Gauss quadrature formula is

$$\int_{-1}^{1} \frac{f(x)}{\sqrt{1-x^2}}\,\mathrm{d}x = \frac{\pi}{n}\sum_1^n f\left(\cos\frac{2k-1}{2n}\pi\right) + \frac{\pi}{(2n)!2^{2n-1}} f^{(2n)}(\xi) \quad (-1 < \xi \leq 1).$$

This formula is also called the *Hermite formula*.

6.1.3 Gauss quadrature formula on infinite intervals

Laguerre polynomials

$$L_n(x) = \mathrm{e}^x \frac{\mathrm{d}^n}{\mathrm{d}x^n}(\mathrm{e}^{-x}x^n) \quad (n = 0, 1, \dots)$$

are orthogonal polynomials on $[0, \infty)$ with weight $\rho(x) = \mathrm{e}^{-x}$. Its n zeros $x_k^{(n)}$ satisfy $0 < x_1^{(n)} < \cdots < x_n^{(n)} < \infty$. Take these zeros as interpolation nodes. The *Gauss–Laguerre quadrature formula* is

$$\int_0^\infty \mathrm{e}^{-x} f(x)\,\mathrm{d}x \doteq \sum_1^n A_k^{(n)} f(x_k^{(n)}),$$

where

$$A_k^{(n)} = \frac{(n!)^2}{x_k^{(n)}(L_n'(x_k^{(n)}))^2}.$$

Assume that $f \in C^{2n}([0, \infty))$. Then the error is

$$e_n(f) = \frac{(n!)^2}{(2n)!} f^{(2n)}(\xi) \quad (0 < \xi < \infty).$$

Hermite polynomials

$$H_n(x) = (-1)^n \mathrm{e}^{x^2} \frac{\mathrm{d}^n}{\mathrm{d}x^n}(\mathrm{e}^{-x^2}) \quad (n = 0, 1, \dots)$$

are orthogonal polynomials on $(-\infty, \infty)$ with weight $\rho(x) = e^{-x^2}$. Take its zeros as interpolation nodes. The Gauss–Hermite quadrature formula is

$$\int_{\mathbb{R}} e^{-x^2} f(x)\,dx \doteq \sum_{1}^{n} A_k^{(n)} f(x_k),$$

where

$$A_k^{(n)} = \frac{2^{n+1} n! \sqrt{\pi}}{(H_n'(x_k^{(n)}))^2}.$$

Assume that $f \in C^{2n}(\mathbb{R})$. The error is

$$e_n(f) = \frac{n!\sqrt{\pi}}{2^n (2n)!} f^{(2n)}(\xi) \quad (\xi \in \mathbb{R}).$$

If f is a polynomial of degree $\leq 2n-1$, then the Gauss–Laguerre quadrature formula and the Gauss–Hermite quadrature formula are both exact.

6.2 Numerical differentiation

Assume that a function is smooth on a closed interval and its values at some points on the closed interval are known. The numerical differentiation algorithm is to estimate its derivative by its values at these points.

6.2.1 Differentiation via polynomial interpolation

Let $f(x)$ be a smooth function on $[a, b]$ and $a =< x_1 < \cdots < x_n = b$. We use the derivative of the Lagrange interpolation polynomial $L_n(x)$ to estimate the derivative of f as follows:

$$f'(x_k) \doteq L_n'(x_k) \quad (k = 1, \ldots, n).$$

By (4.2.5),

$$f(x) - L_n(x) = \frac{1}{n!} f^{(n)}(\xi_x)\,\omega_n(x) \quad (a < \xi_x < b),$$

where $\omega_n(x) = (x - x_1)\cdots(x - x_n)$. Differentiating both sides gives

$$f'(x) - L_n'(x) = \frac{f^{(n)}(\xi_x)}{n!}\frac{d\omega_n(x)}{dx} + \frac{\omega_n(x)}{n!}\frac{df^{(n)}(\xi_x)}{dx}.$$

Let $x = x_k$. Note that $\omega_n(x_k) = 0$. Then

$$f'(x_k) - L_n'(x_k) = \frac{f^{(n)}(\xi_x)}{n!}\omega_n'(x_k).$$

This derives the numerical differentiation formulas often used as follows.

Let $x_{k+1} - x_k = h$ $(k = 1, \ldots, n-1)$ and $y_k = f(x_k)$ $(k = 1, \ldots, n)$.

(a) Two-point form

$$L_1(x) = \frac{x - x_1}{h} y_2 - \frac{x - x_2}{h} y_1$$

is the interpolation function with nodes x_1 and x_2. So

$$f'(x_1) \doteq L_1'(x_1) = \frac{y_2 - y_1}{h},$$
$$f'(x_2) \doteq L_1'(x_2) = \frac{y_2 - y_1}{h}.$$

(b) Three-point form

$$L_2(x) = \frac{(x - x_2)(x - x_3)}{2h^2} y_1 - \frac{(x - x_1)(x - x_3)}{h^2} y_2 + \frac{(x - x_1)(x - x_2)}{2h^2} y_3$$

is the interpolation function with nodes x_1, x_2, x_3. So the three-point form is

$$f'(x_1) \doteq L_2'(x_1) = \frac{-3y_1 + 4y_2 - y_3}{2h},$$
$$f'(x_2) \doteq L_2'(x_2) = \frac{y_3 - y_1}{2h},$$
$$f'(x_3) \doteq L_2'(x_3) = \frac{y_1 - 4y_2 + 3y_3}{2h}.$$

From $L_2(x)$, the second-order numerical differentiation formula is

$$f''(x_1) = f''(x_2) = f''(x_3) = \frac{y_1 - 2y_2 + y_3}{h^2}.$$

In general, when $L_n(x)$ converges to $f(x)$, $L_n'(x)$ does not necessarily converge to $f'(x)$. To avoid this problem, we may use the spline interpolation function to find differentiation.

6.2.2 Differentiation via spline interpolation

Let $f \in C^4([a, b])$ and $s(x)$ be its cube spline interpolation function associated with partition $a = x_0 < x_1 < \cdots < x_n = b$. When $\lambda = \max_{0 \le k \le n-1} |x_{k+1} - x_k| \to 0$, by Proposition 4.4.2,

$$s(x) \to f(x),$$
$$s'(x) \to f'(x),$$
$$s''(x) \to f''(x)$$

uniformly and the errors are estimated as

$$|f^{(i)}(x) - s^{(i)}(x)| \le C\lambda^{4-i} \quad (i = 0, 1, 2),$$

where C is independent of x, and the partition and the representation of $s(x)$ is stated in Section 4.4.2.

Find the derivative of first order on $[x_k, x_{k+1}]$ $(k = 0, \dots, n-1)$, by (4.4.1), we get

$$\begin{aligned} s'(x) = &\frac{6}{\delta_k^2}\left(\frac{1}{\delta_k}(x - x_{k+1})^2 + (x - x_{k+1})\right) y_{k+1} \\ &+ \frac{6}{\delta_k^2}\left((x - x_k) + \frac{1}{\delta_k}(x - x_k)^2\right) y_k \\ &+ \frac{1}{\delta_k}\left(\frac{1}{\delta_k^3}(x - x_{k+1})^2 + 2(x - x_{k+1})\right) \mu_k \\ &- \frac{1}{\delta_k}\left(2(x - x_{k+1}) - \frac{3}{\delta_k}(x - x_k)^2\right) \mu_{k+1}. \end{aligned}$$

So $f'(x) \approx s'(x)$. Especially, $f'(x_k) = \mu_k$ $(k = 0, \dots, n)$.

The representation of $s''(x)$ is stated as in (4.4.2). So $f''(x) \doteq s''(x)$.

6.2.3 Richardson extrapolation

By the Taylor formula, it follows that for small h,

$$f(x + h) - f(x - h) = 2hf'(x) + \frac{2}{3!}h^3 f'''(x) + \frac{2}{5!}h^5 f^{(5)}(x) + \cdots .$$

or

$$\frac{f(x + h) - f(x - h)}{2h} = f'(x) + \frac{1}{3!}h^2 f'''(x) + \frac{1}{5!}h^4 f^{(5)}(x) + \cdots .$$

Denote $a_2 = -\frac{f'''(x)}{3!}$, $a_4 = -\frac{f^{(5)}(x)}{5!}$,.... This equality becomes that for any small h,

$$f'(x) = \varphi(h) + a_2 h^2 + a_4 h^4 + a_6 h^6 + \cdots =: L(h).$$

where

$$\varphi(h) = \frac{f(x + h) - f(x - h)}{2h}. \tag{6.2.1}$$

This leads to

$$L(h) = L(h/2) = L(h/4) = \cdots .$$

So

$$4L(h) = 4L\left(\frac{h}{2}\right) = 4\varphi\left(\frac{h}{2}\right) + a_2 h^2 + a_4 \frac{h^4}{4} + a_6 \frac{h^6}{16} + \cdots ,$$

$$f'(x) = L(h) = \frac{4}{3}\varphi\left(\frac{h}{2}\right) - \frac{1}{3}\varphi(h) - \frac{1}{4}a_4 h^4 - \frac{5}{16}a_6 h^6 + \cdots , \tag{6.2.2}$$

and so

$$f'(x) = \frac{4}{3}\varphi\left(\frac{h}{2}\right) - \frac{1}{3}\varphi(h) + O(h^4),$$

$$L(h) = g(h) + b_4 h^4 + b_6 h^6 + \cdots .$$

where $b_4 = -\frac{a_4}{4}$, $b_6 = -\frac{5a_6}{16}$, and

$$g(h) = \frac{4}{3}\varphi\left(\frac{h}{2}\right) - \frac{1}{3}\varphi(h). \tag{6.2.3}$$

Note that $L(h) = L(\frac{h}{2})$. Then

$$16L(h) = 16g\left(\frac{h}{2}\right) + b_4h^4 + b_6\frac{h^6}{4} + \cdots .$$

The combination of this with (6.2.2) gives

$$f'(x) = L(h) = \frac{16}{15}g\left(\frac{h}{2}\right) - \frac{1}{15}g(h) - \frac{b_6h^6}{20} + \cdots ,$$

where $b_6 = -\frac{5}{16}a_6 = \frac{5}{16}\frac{f^{(7)}(x)}{7!}$. From this and (6.2.1)–(6.2.3), it is seen that a simple combination of $\varphi(h)$, $\varphi(\frac{h}{2})$, and $\varphi(\frac{h}{4})$ furnishes an estimate of $f'(x)$ with accuracy $O(h^6)$.

Continue this procedure to give the more precise results.

6.3 Iterative methods

Iterative methods are used to solve nonlinear equations, systems of linear equations, ordinary differential equations, and the matrix eigenvalue problem.

6.3.1 Fixed point principle

Given an equation $f(x) = 0$, where f is continuous, its equivalent form is $x = \varphi(x)$, where $\varphi(x) = x + f(x)$. Starting with an initial value x_0, we get the iterative sequence $\{x_n\}$ as follows:

$$\begin{aligned} x_1 &= \varphi(x_0), \\ x_2 &= \varphi(x_1), \\ &\vdots \\ x_n &= \varphi(x_{n-1}), \\ &\vdots \end{aligned}$$

If $\{x_n\}$ converges to ξ, then $\xi = \varphi(\xi)$. So ξ is a solution of $f(x) = 0$.

Fixed point principle

Assume that a function $\varphi(x)$ satisfies

$$|\varphi(y_1) - \varphi(y_2)| \le M\,|y_1 - y_2| \quad (y_1, y_2 \in [a, b]),$$

where $0 < M < 1$ is a constant independent of y_1 and y_2. Take an initial value $x_0 \in (a, b)$. Then

- the iterative sequence $\{x_n\}$ satisfying $x_n = \varphi(x_{n-1})$ converges.
  Denote $\xi = \lim_{n\to\infty} x_n$.
- $\xi = \varphi(\xi)$, i.e., the ξ is the fixed point of $\varphi(x)$; and
- the rate of convergence is $|x_n - \xi| \le \frac{M^n}{1-M}|x_1 - x_0|$.

In fact, by the assumption, for any $k \in \mathbb{Z}_+$,

$$|x_{k+1} - x_k| = |\varphi(x_k) - \varphi(x_{k+1})| \le M\,|x_k - x_{k+1}| \le \cdots \le M^k|x_1 - x_0|.$$

This implies that

$$\begin{aligned}|x_{n+p} - x_n| &\le |x_{n+p} - x_{n+p-1}| + |x_{n+p-1} - x_{n+p-2}| + \cdots + |x_{n+1} - x_n|\\ &\le (M^{n+p-1} + M^{n+p-2} + \cdots + M^n)\,|x_1 - x_0| = \frac{M^n(1 - M^p)}{1 - M}\,|x_1 - x_0|. \qquad (6.3.1)\end{aligned}$$

Since $0 < M < 1$, $\lim_{n\to\infty}|x_{n+p} - x_n| = 0$. Cauchy criterion indicates the iterative sequence $\{x_n\}$ converges. Denote $\xi = \lim_{n\to\infty} x_n$. From this and $x_n = \varphi(x_{n-1})$, it follows that $\xi = \varphi(\xi)$, i.e., ξ is the fixed point of $\varphi(x)$. Letting $p \to \infty$ in (6.3.1), since $x_{n+p} \to \xi$ and $M^p \to 0$, we get the desired rate of convergence.

6.3.2 Iterative methods of univariate nonlinear equations

Let $f \in C^2([a,b])$, ξ be a zero of f, and x be an approximation to ξ. By Taylor's formula,

$$0 = f(\xi) = f(x + h) = f(x) + hf'(x) + O(h^2) \approx f(x) + hf'(x),$$

where $h = \xi - x$. This implies $h \doteq -f'(x)/f(x)$. Since x is an approximation to ξ, the better approximation to ξ should be $x - f(x)/f'(x)$.

Starting with an estimate x_0 of ξ, Newton's iterative method defines inductively

$$x_{n+1} = x_n - f(x_n)/f'(x_n) \quad (n = 0, 1, \ldots).$$

If x_0 is close to a zero of f, then the sequence $\{x_n\}$ converges to ξ. Especially, if $f \in C^2(\mathbb{R})$ is an increasing and convex function and has only a zero, then the Newton iteration sequence $\{x_n\}$ converges to the zero of f from any stating point.

Since

$$f'(x_n) \doteq \frac{f(x_n) - f(x_{n-1})}{x_n - x_{n-1}},$$

replacing $f'(x_n)$ by

$$\frac{f(x_n) - f(x_{n-1})}{x_n - x_{n-1}}$$

in the above Newton formula, the iteration formula becomes

$$x_{n+1} = x_n - f(x_n)\left[\frac{(x_n - x_{n-1})}{f(x_n) - f(x_{n-1})}\right] \quad (n = 0, 1, \dots).$$

This iterative method is simpler than the Newton iterative method but its convergence rate is slower.

A polynomial of degree n

$$p(x) = a_n x^n + a_{n-1}x^{n-1} + \cdots + a_0$$

has exactly n complex roots and its all roots lie in the disk with center 0 and radius $\rho = 1 + |a_n|^{-1} \max_{0\le k\le n}|a_k|$.

Horner's algorithm can give the values of a polynomial and its derivatives simultaneously. Define α_k and β_k for $k = 0, 1, \dots, n$ as follows:

$$\begin{aligned} \alpha_n &= a_n, & \beta_n &= 0,\\ \alpha_{n-1} &= a_{n-1} + x\alpha_n, & \beta_{n-1} &= \alpha_n + x\beta_n,\\ \alpha_{n-2} &= a_{n-2} + x\alpha_{n-1}, & \beta_{n-2} &= \alpha_{n-1} + x\beta_{n-1},\\ &\vdots & &\vdots\\ \alpha_0 &= a_0 + x\alpha_1, & \beta_0 &= \alpha_1 + x\beta_1. \end{aligned}$$

Then $\alpha_0 = p(x)$ and $\beta_0 = p'(x)$.

Using Horner's algorithm, one can rapidly compute $p(x_k)/p'(x_k)$ to obtain $x_{k+1} = x_k - p(x_k)/p'(x_k)$ from x_k in Newton's method. This iterative process yields a root ξ of $p(x)$.

Let $q(x) = p(x)/(x - \xi)$. Then $q(x)$ is a polynomial of degree $n - 1$. Denote

$$q(x) = b_{n-1}x^{n-1} + b_{n-2}x^{n-2} + \cdots + b_0.$$

We will find $b_{n-1}, \dots, b_0$. From $p(x) = (x - \xi)q(x)$, it follows that

$$a_n x^n + a_{n-1}x^{n-1} + \cdots + a_0 = (x - \xi)(b_{n-1}x^{n-1} + b_{n-2}x^{n-2} + \cdots + b_0).$$

Comparing coefficients on both sides, we get

$$\begin{aligned} b_{n-1} &= a_n,\\ b_{n-2} &= a_{n-1} + \xi b_{n-1},\\ &\vdots\\ b_0 &= a_1 + \xi b_1. \end{aligned}$$

i.e., we get $q(x)$. All roots of $q(x)$ are the remaining $(n - 1)$ roots of $p(x)$.

Repeating the above procedure in which $p(x)$ is replaced by $q(x)$, we get another root of $p(x)$. Continuing the procedure, finally, we get all roots of $p(x)$.

6.3.3 Iterative method of systems of bivariate nonlinear equations

Consider a system of nonlinear equations

$$\begin{cases} f_1(x,y)=0, \\ f_2(x,y)=0, \end{cases}$$

where f_1 and f_2 are both continuously differentiable functions. A pair of initial values (x_0, y_0) are given. Expanding f_1, f_2 in the neighborhood of (x_0, y_0) into Taylor series using the Taylor theorem and taking their linear principal parts, respectively, we get the following system of equations:

$$\begin{cases} f_1(x_0,y_0)+\frac{\partial f_1}{\partial x}(x_0,y_0)(x-x_0)+\frac{\partial f_1}{\partial y}(x_0,y_0)(y-y_0)=0, \\ f_2(x_0,y_0)+\frac{\partial f_2}{\partial x}(x_0,y_0)(x-x_0)+\frac{\partial f_2}{\partial y}(x_0,y_0)(y-y_0)=0. \end{cases}$$

If the Jacobian determinant

$$J_0=\begin{vmatrix} \frac{\partial f_1}{\partial x}(x_0,y_0) & \frac{\partial f_1}{\partial y}(x_0,y_0) \\ \frac{\partial f_2}{\partial x}(x_0,y_0) & \frac{\partial f_2}{\partial y}(x_0,y_0) \end{vmatrix}\neq 0,$$

then the solution is

$$\begin{cases} x_1=x_0-J_0^{-1}\begin{vmatrix} \frac{\partial f_1}{\partial y}(x_0,y_0) & f_1(x_0,y_0) \\ \frac{\partial f_2}{\partial y}(x_0,y_0) & f_2(x_0,y_0) \end{vmatrix}, \\ y_1=y_0-J_0^{-1}\begin{vmatrix} f_1(x_0,y_0) & \frac{\partial f_1}{\partial x}(x_0,y_0) \\ f_2(x_0,y_0) & \frac{\partial f_2}{\partial x}(x_0,y_0) \end{vmatrix}. \end{cases}$$

Continue this procedure, in general, starting from (x_k, y_k), if

$$J_k=\begin{vmatrix} \frac{\partial f_1}{\partial x}(x_k,y_k) & \frac{\partial f_1}{\partial y}(x_k,y_k) \\ \frac{\partial f_2}{\partial x}(x_k,y_k) & \frac{\partial f_2}{\partial y}(x_k,y_k) \end{vmatrix}\neq 0 \quad (k\in\mathbb{Z}_+),$$

the solution is

$$\begin{cases} x_{k+1}=x_k-J_k^{-1}\begin{vmatrix} \frac{\partial f_1}{\partial y}(x_k,y_k) & f_1(x_k,y_k) \\ \frac{\partial f_2}{\partial y}(x_k,y_k) & f_2(x_k,y_k) \end{vmatrix}, \\ y_{k+1}=y_k-J_k^{-1}\begin{vmatrix} f_1(x_k,y_k) & \frac{\partial f_1}{\partial x}(x_k,y_k) \\ f_2(x_k,y_k) & \frac{\partial f_2}{\partial x}(x_k,y_k) \end{vmatrix}. \end{cases}$$

This gives the iterative process.

If $\max\{|x_{N+1}-x_N|, |y_{N+1}-y_N|\}<\varepsilon$. Then (x_{N+1}, y_{N+1}) is a desired solution.

6.3.4 Iterative method to solve systems of linear equations

Consider a system of linear equations

$$\sum_{j=1}^{n} a_{ij}x_j = b_i \quad (i = 1, \dots, n). \tag{6.3.2}$$

The matrix form is $A\boldsymbol{x} = \boldsymbol{b}$, where $A = (a_{ij})_{n\times n}$, $\boldsymbol{x} = (x_1, \dots, x_n)^{\mathrm{T}}$, and $\boldsymbol{b} = (b_1, \dots, b_n)^{\mathrm{T}}$.

(a) Jacobian iterative method

The system of linear equations (6.3.2) can be rewritten in the form

$$x_i = \sum_{j=1}^{n} c_{ij}x_j + d_i \quad (i = 1, \dots, n).$$

Take $x_1^{(0)}, \dots, x_n^{(0)}$ as initial values. Applying the iterative formula, we get

$$x_i^{(k)} = \sum_{j=1}^{n} c_{ij}x_j^{(k-1)} + d_i \quad (i = 1, \dots, n),$$

which is called the *Jacobian iterative formula*. We find the k-th approximate solution such that

$$\max_{1\le i\le n} |x_i^{(k)} - x_i^{(k-1)}| < \varepsilon,$$

where ε is a predictive error. So $(x_1^{(k)}, \dots, x_n^{(k)})$ is an approximate solution of (6.3.2) with predictive error ε.

Let $(x_1^*, \dots, x_n^*)$ be the exact solution. If $\mu = \max_{1\le i\le n} \sum_{j=1}^{n} |c_{ij}| < 1$, then the error is estimated by

$$\max_{1\le i\le n} |x_i^{(k)} - x_i^*| \le \frac{\mu^k}{1-\mu} \max_{1\le i\le n} |x_i^{(1)} - x_i^*|.$$

If $\nu = \max_{1\le j\le n} \sum_{i=1}^{n} |c_{ij}| < 1$, then the error is estimated by

$$\sum_{i=1}^{n} |x_i^{(k)} - x_i^*| \le \frac{\nu^k}{1-\nu} \sum_{i=1}^{n} |x_i^{(1)} - x_i^*|.$$

If $p = \sum_{i,j=1}^{n} c_{ij}^2 < 1$, then the error is estimated by

$$\left(\sum_{i=1}^{n} (x_i^{(k)} - x_i^*)^2\right)^{\frac{1}{2}} \le \frac{p^{\frac{k}{2}}}{1-p^{\frac{1}{2}}} \left(\sum_{i=1}^{n} (x_i^{(1)} - x_i^*)^2\right)^{\frac{1}{2}}.$$

(b) Seldel iterative method

The Jacobian iterative formula can be rewritten in the form

$$x_i^{(k)} = \sum_{j=1}^{i-1} c_{ij} x_j^{(k)} + \sum_{j=i+1}^{n} c_{ij} x_j^{(k-1)} + d_i,$$

which is called the *Seldel iterative formula*, then the convergence rate of the iterative process is quicker.

If $\max_{1\le i\le n} \sum_{j=1}^{n} |c_{ij}| < 1$, then the error is estimated by

$$\max_{1\le i\le n} |x_i^{(k)} - x_i^{(0)}| \le \frac{\mu^k}{1-\mu} \max_{1\le i\le n} |x_i^{(1)} - x_i^{(0)}|,$$

where

$$\mu = \max_{1\le i\le n} \left\{ \frac{\sum_{j=1}^{n} |c_{ij}|}{1 - \sum_{j=1}^{i-1} |c_{ij}|} \right\}.$$

(c) Relaxation iterative method

The Seldel iterative formula is rewritten in the form

$$x_i^{(k)} = \omega \left(\sum_{j=1}^{i-1} c_{ij} x_j^{(k)} + \sum_{j=i+1}^{n} c_{ij} x_j^{(k-1)} + d_i \right) + (1-\omega) x_i^{(k-1)},$$

where ω is a constant which is called a *relaxation factor*. This formula is called the *relaxation iterative formula*. We may choose ω such that the convergence rate of the iterative process becomes quicker.

6.3.5 Maximum eigenvalues for matrixes

Let A be an $n \times n$ matrix. Suppose that the eigenvalues $\lambda_1, \dots, \lambda_n$ of A satisfy $|\lambda_1| > \cdots > |\lambda_n|$ and the corresponding eigenvectors $\boldsymbol{v}_1, \dots, \boldsymbol{v}_n$ satisfy $\|\boldsymbol{v}_1\| = \cdots = \|\boldsymbol{v}_n\| = 1$. If these eigenvectors constitute a basis for the n-dimensional space, then an initial vector $\boldsymbol{x}_0$ can be expressed into a linear combination of eigenvectors, i.e.,

$$\boldsymbol{x}_0 = a_1 \boldsymbol{v}_1 + a_2 \boldsymbol{v}_2 + \cdots + a_n \boldsymbol{v}_n = \sum_1^n a_i \boldsymbol{v}_i.$$

Define $\boldsymbol{x}_k = A\boldsymbol{x}_{k-1}$ $(k \in \mathbb{Z}_+)$. Since $A\boldsymbol{v}_i = \lambda_i \boldsymbol{v}_i$,

$$\boldsymbol{x}_1 = A\boldsymbol{x}_0 = \sum_1^n a_i (A\boldsymbol{v}_i) = \sum_1^n \lambda_i a_i \boldsymbol{v}_i;$$

$$\boldsymbol{x}_2 = A\boldsymbol{x}_1 = \sum_1^n \lambda_i a_i (A\boldsymbol{v}_i) = \sum_1^n \lambda_i^2 a_i \boldsymbol{v}_i.$$

In general,

$$\boldsymbol{x}_k = A\boldsymbol{x}_{k-1} = \sum_1^n \lambda_i^k a_i \boldsymbol{v}_i \quad (k \in \mathbb{Z}_+).$$

For $a_1 \neq 0$, since $|\lambda_i/\lambda_1| < 1$ $(i \geq 2)$, when k is sufficiently large,

$$\boldsymbol{x}_k = \lambda_1^k \left(a_1\boldsymbol{v}_1 + a_2 \left(\frac{\lambda_2}{\lambda_1}\right)^k \boldsymbol{v}_2 + \cdots + a_n \left(\frac{\lambda_n}{\lambda_1}\right)^k \boldsymbol{v}_n \right) = \lambda_1^k (a_1\boldsymbol{v}_1 + \varepsilon_k),$$

where $\varepsilon_k = o(1)$, i.e.,

$$\boldsymbol{x}_k \doteq \lambda_1^k a_1 \boldsymbol{v}_1,$$
$$\boldsymbol{x}_{k+1} \doteq \lambda_1^{k+1} a_1 \boldsymbol{v}_1$$

Let $\boldsymbol{x}_k = (x_{k1}, \ldots, x_{kn})$ $(k \in \mathbb{Z}_+)$ and $\boldsymbol{v}_1 = (v_{11}, \ldots, v_{1n})$. Then

$$x_{ki} \doteq \lambda_1^k a_1 v_{1i},$$
$$x_{k+1,i} \doteq \lambda_1^{k+1} a_1 v_{1i}.$$

This implies that

$$\frac{x_{k+1,i}}{x_{ki}} \doteq \frac{\lambda_1^{k+1} a_1 v_{1i}}{\lambda_1^k a_1 v_{1i}} = \lambda_1.$$

6.3.6 Iterative method of ordinary differential equations

Consider the initial value problem of the ordinary differential equation of order 1

$$\begin{cases} y' = f(x, y), \\ y(x_0) = y_0. \end{cases} \tag{6.3.3}$$

In the numerical solution of differential equations, we want to find the values of $y(x)$ on a sequence of points on the interval $[a, b]$

$$x_n = a + nh \quad (n = 0, 1, \ldots, N),$$

where $h = \frac{b-a}{N}$ is called the *step length*. Denote $y(x_n) \doteq y_n$ $(n = 1, \ldots, N)$.

(a) Euler method

The derivative may be represented by the difference quotient, i.e., $y'(x) \doteq \frac{1}{h}(y(x+h) - y(x))$. So

$$y'(x_n) \doteq \frac{y(x_n + h) - y(x_n)}{h} = \frac{y(x_{n+1}) - y(x_n)}{h} \doteq \frac{y_{n+1} - y_n}{h}.$$

By (6.3.3),

$$y'(x_n) = f(x_n, y(x_n)) \doteq f(x_n, y_n),$$
$$y(x_0) = y_0.$$

These equalities give an iterative formula as follows:

$$\begin{aligned} y_0 &= y(x_0), \\ y_{n+1} &\doteq y_n + hf(x_n, y_n) \quad (n = 0, 1, \dots, N-1). \end{aligned} \tag{6.3.4}$$

This iterative process gives the values of y_n $(n = 0, 1, \dots, N)$.

If $f(x, y)$ is smooth on $[a, b]$, then the solution of (6.3.3) is smooth. By the Taylor formula

$$y(x_{n+1}) = y(x_n + h) = y(x_n) + hf(x_n, y(x_n)) + \frac{1}{2}h^2 y''(\xi) \doteq y_{n+1} + \frac{1}{2}h^2 y''(\xi),$$

where $x_n < \xi < x_{n+1}$. Denote $M = \max_{a \le x \le b} |y''(x)|$. The local truncation error is

$$|y(x_{n+1}) - y_{n+1}| \le \frac{M}{2} h^2 \quad (n = 0, 1, \dots, N-1).$$

The accumulation of all the local truncation errors gives the global truncation error $O(h)$.

(b) Trapezoidal method

Integrating both sides of (6.3.3) from x_n to x_{n+1},

$$y(x_{n+1}) - y(x_n) = \int_{x_n}^{x_{n+1}} f(t, y(t))\, dt.$$

The integral on the right-hand side is computed by trapezoidal formula as follows:

$$\int_{x_n}^{x_{n+1}} f(t, y(t))\, dt = \frac{h}{2}(f(x_n, y_n) + f(x_{n+1}, y_{n+1})).$$

Thus,

$$y_{n+1} = y_n + \frac{h}{2}(f(x_n, y_n) + f(x_{n+1}, y_{n+1})). \tag{6.3.5}$$

This is an implied format. We use (6.3.4) in the Euler method to find initial values $y_{n+1}^{(0)}$

$$y_{n+1}^{(0)} \doteq y_n + hf(x_n, y_n),$$

and then use (6.3.5) to give an implied iterative formula

$$y_{n+1}^{(k+1)} = y_0 + \frac{h}{2}(f(x_n, y_n) + f(x_{n+1}, y_{n+1}^{(k)})) \quad (k = 0, 1, \dots).$$

This implies that

$$|y_{n+1}^{(k+1)} - y_{n+1}^{(k)}| \le \frac{M_1 h}{2} |y_{n+1}^{(k)} - y_{n+1}^{(k-1)}|.$$

A similar argument of the fixed point principle implies that if $0 < \frac{M_1 h}{2} < 1$, then $\{y_{n+1}^{(k)}\}$ converges as $k \to \infty$.

For the trapezoidal method, the global truncation error is $O(h^2)$.

(c) System of ordinary differential equations

$$\begin{cases} y'(x) = f(x, y, z), & y(x_0) = y_0, \\ z'(x) = f(x, y, z), & z(x_0) = z_0. \end{cases}$$

Its Euler formula is

$$\begin{aligned} y_{n+1} &= y_n + hf(x_n, y_n, z_n), \quad y(x_0) = y_0, \\ z_{n+1} &= z_n + hg(x_n, y_n, z_n), \quad z(x_0) = z_0. \end{aligned}$$

6.4 Difference methods

The difference method is a fundamental method for solving differential equations. In this method, the derivatives in a differential equation are replaced by difference quotients.

6.4.1 The difference method of ordinary differential equations

Consider an ordinary differential equation with the first boundary condition

$$\begin{cases} y'' - p(x)y = q(x), & q(x) > 0 \quad (a \le x \le b), \\ y(a) = \alpha, \quad y(b) = \beta. \end{cases} \tag{6.4.1}$$

Take a partition $x_k = a + kh$ $(k = 0, \ldots, n)$, where $h = \frac{b-a}{n}$. Let $y = y(x)$ be the solution of the problem (6.4.1). We find the approximation value y_i of $y(x_i)$ $(i = 0, \ldots, n)$. For $i = 1, \ldots, n-1$, the second-order derivative $y''(x_i)$ is expressed approximately by the second-order central difference quotient, i.e.,

$$y''(x_i) \doteq \frac{y_{i+1} - 2y_i + y_{i-1}}{h^2}.$$

From this and (6.4.1), the difference equation is given by

$$\frac{y_{i+1} - 2y_i + y_{i-1}}{h^2} - p_i y_i \doteq q_i \quad (i = 1, \ldots, n-1),$$

where $p_i = p(x_i)$ and $q_i = q(x_i)$. Combining this with the boundary conditions $y_0 = \alpha$ and $y_n = \beta$, the problem (6.4.1) is reduced to the following system of linear equations

$$\begin{cases} \dfrac{y_{i+1} - 2y_i + y_{i-1}}{h^2} - p_i y_i \doteq q_i \quad (i = 1, \ldots, n-1), \\ y_0 = \alpha, \quad y_n = \beta. \end{cases}$$

Solving the system of linear equations, the numerical solution y_i $(i = 0, \ldots, n)$ of the problem (6.4.1) is obtained.

Consider an ordinary differential equation with the second boundary condition

$$\begin{cases} y'' - p(x)y = q(x), & q(x) > 0 \quad (a \le x \le b), \\ y'(a) = \alpha, \quad y'(b) = \beta. \end{cases}$$

Using the three-point form of the numerical differentiation formula (see Section 6.2)

$$y_0' \doteq \frac{-y_2 + 4y_1 - 3y_0}{2h},$$
$$y_n' \doteq \frac{3y_n - 4y_{n-1} + y_{n-2}}{2h},$$

the corresponding system of linear equations is given by

$$\begin{cases} y_{i+1} - 2y_i + y_{i-1} - p_i y_i \doteq q_i \quad (i = 1, \dots, n-1), \\ \dfrac{-y_2 + 4y_1 - 3y_0}{2h} \doteq \alpha, \quad \dfrac{3y_n - 4y_{n-1} + y_{n-2}}{2h} \doteq \beta. \end{cases}$$

Solving the system of linear equations, the numerical solution is obtained.

Consider an ordinary differential equation with the third boundary condition

$$\begin{cases} y'' - p(x)y = q(x), \quad q(x) > 0 \quad (a \le x \le b), \\ y'(a) - \alpha_0 y(a) = \alpha_1, \quad y'(b) + \beta_0 y(b) = \beta_1, \end{cases}$$

where $\alpha_0 \ge 0$, $\beta_0 \ge 0$, and $\alpha_0 + \beta_0 > 0$. The corresponding system of linear equations is given by

$$\begin{cases} \dfrac{y_{i+1} - 2y_i + y_{i-1}}{h^2} - p_i y_i \doteq q_i \quad (i = 1, \dots, n-1), \\ \dfrac{-y_2 + 4y_1 - 3y_0}{2h} - \alpha_0 y_0 \doteq \alpha_1, \quad \dfrac{3y_n - 4y_{n-1} + y_{n-2}}{2h} + \beta_0 y_n \doteq \beta_1. \end{cases}$$

Solving the system of linear equations, the numerical solution is obtained.

6.4.2 The difference method of elliptic equations

Consider an elliptic equation with the first boundary condition

$$\begin{cases} \Delta u := \dfrac{\partial^2 u}{\partial x^2} + \dfrac{\partial^2 u}{\partial y^2} = 0 & ((x,y) \in G), \\ u(x,y) = \varphi(x,y) & ((x,y) \in \Gamma), \end{cases} \tag{6.4.2}$$

where G is a bounded domain and Γ is the boundary of G.

Let $x_i = ih$, $y_j = j\tau$ $(i, j \in \mathbb{Z})$, where $h > 0$, $\tau > 0$. The point $(ih, j\tau)$ is called a *grid*, denoted by (i, j). If $(i, j) \in G$ and all four neighborhood points $(i-1, j)$, $(i+1, j)$, $(i, j-1)$, and $(i, j+1)$ belong to $G + \Gamma$, then (i, j) is called an *interior point*. If $(i, j) \in G$

and there is a neighborhood point that does not belong to G, then (i, j) is called a *boundary point.*

Replacing the second-order partial derivative in (6.4.2) by the second-order central difference quotient, we obtain the corresponding difference equation on each interior point

$$\frac{u_{i+1,j} - 2u_{i,j} + u_{i-1,j}}{h^2} + \frac{u_{i,j+1} - 2u_{i,j} + u_{i,j-1}}{\tau^2} \doteq 0,$$

where $u_{i,j} = u(x_i, y_j)$. This equation is called a *five-point scheme.*

Consider the simplest case that $h = \tau$ and G is a rectangle

$$G: \{0 \le x \le L,\ 0 \le y \le M\}.$$

and all boundary nodes lie on the boundary. Let $l = [L/h]$ and $m = [M/h]$. The boundary condition becomes

$$u_{i,j} = \varphi(ih, jh) \quad ((i = 0, l;\ j = 0, \dots, m) \text{ or } (i = 0, \dots, l;\ j = 0, m)).$$

So the problem (6.4.2) is reduced to the following system of $m + 1$ equations with $l + 1$ unknown

$$\begin{cases} u_{i,j} = \frac{1}{4}(u_{i+1,j} + u_{i-1,j} + u_{i,j+1} + u_{i,j-1}) & (i = 1, \dots, l-1;\ j = 1, \dots, m-1), \\ u_{i,j} = \varphi(ih, jh) & ((i = 0, l,\ j = 0, \dots, m) \text{ or } (i = 0, \dots, l;\ j = 0, m)). \end{cases}$$

It can be proved that this system of difference equations has a unique solution by the extremum principle. One may use the direct method and the iteration method to solve this system of linear equations.

6.4.3 The difference method of parabolic equations

Consider a parabolic equation with an initial condition and boundary conditions

$$\begin{cases} Lu := \frac{\partial u}{\partial t} - \frac{\partial^2 u}{\partial x^2} = 0 & (0 < x < 1, 0 < t \le T), \\ u(0, x) = \varphi(x) & (0 \le x \le 1), \\ u(t, 0) = \mu_1(t), \quad u(t, 1) = \mu_2(t) & (0 \le t \le T), \end{cases} \tag{6.4.3}$$

where $\varphi(0) = \mu_1(0)$ and $\varphi(1) = \mu_2(0)$.

Establish the following grids (x_k, t_j), where

$$x_k = kh \quad (k = 1, \dots, N), \quad Nh = 1,$$
$$t_j = j\tau \quad (j = 0, \dots, m_0), \quad m_0 = [T/\tau].$$

Using the numerical differential formula, we obtain

$$\frac{\partial u}{\partial t}(x_k, t_j) \doteq \frac{u(x_k, t_{j+1}) - u(x_k, t_j)}{\tau},$$
$$\frac{\partial^2 u}{\partial x^2}(x_k, t_j) \doteq \frac{u(x_{k+1}, t_j) - 2u(x_k, t_j) + u(x_{k-1}, t_j)}{h^2}.$$

By (6.4.3), it follows that

$$\frac{u_{k,j+1}-u_{k,j}}{\tau}-\frac{u_{k+1,j}-2u_{k,j}+u_{k-1,j}}{h^2}\doteq 0.$$

Denote $r=\tau/h^2$, this equality becomes

$$u_{k,j+1}\doteq(1-2r)u_{k,j}+r(u_{k+1,j}+u_{k-1,j})\quad(k=1,\ldots,N-1;\ j=0,\ldots,m_0-1).$$

The corresponding initial condition and boundary condition are

$$\begin{aligned}&u_{k,0}\doteq\varphi(kh) && (k=0,\ldots,N),\\ &u_{0,j}\doteq\mu_1(j\tau),\quad \mu_{N,j}\doteq\mu_2(j\tau) && (j=0,\ldots,m_0).\end{aligned}$$

From this, the problem (6.4.3) is reduced to the following system of linear equations:

$$\begin{cases}u_{k,j+1}\doteq(1-2r)u_{k,j}+r(u_{k+1,j}+u_{k-1,j}) & (k=1,\ldots,N-1;\ j=0,\ldots,m_0-1),\\ u_{k,0}\doteq\varphi(kh) & (k=0,\ldots,N),\\ u_{0,j}\doteq\mu_1(j\tau),\quad \mu_{N,j}\doteq\mu_2(j\tau) & (j=0,\ldots,m_0).\end{cases}$$

6.4.4 The difference method of hyperbolic equations

Consider a hyperbolic equation problem with boundary conditions

$$\begin{cases}\frac{\partial^2 u}{\partial x^2}-\frac{\partial^2 u}{\partial t^2}=f(x,t) & (0<x<1,0<t\le T),\\ u(0,x)=\varphi(x),\quad \frac{\partial u}{\partial t}(0,x)=\psi(x) & (0\le x\le 1),\\ u(t,0)=\Phi_0(t),\quad u(t,1)=\Phi_1(t) & (0\le t\le T).\end{cases}\tag{6.4.4}$$

Take grids (x_k,t_j), where

$$\begin{aligned}x_k&=kh\quad(k=0,\ldots,N),\\ t_j&=j\tau\quad(j=0,\ldots,m_0),\end{aligned}$$

where $h=\frac{1}{N}$ and $m_0=[T/\tau]$. Using the numerical differential formula, we get

$$\begin{aligned}\frac{\partial^2 u}{\partial x^2}(x_k,t_j)&\doteq\frac{u(x_{k+1},t_j)-2u(x_k,t_j)+u(x_{k-1},t_j)}{h^2}.\\ \frac{\partial^2 u}{\partial t^2}(x_k,t_j)&\doteq\frac{u(x_k,t_{j+1})-2u(x_k,t_j)+u(x_k,t_{j-1})}{\tau^2}.\end{aligned}$$

By (6.4.4), it follows that

$$\frac{u_{k+1,j}-2u_{k,j}+u_{k-1,j}}{h^2}-\frac{u_{k,j+1}-2u_{k,j}+u_{k,j-1}}{\tau^2}\doteq f_{k,j}$$

where $u(x_k, y_j) = u_{k,j}$ and $f(x_k, y_j) = f_{k,j}$. Denote $s = \tau/h$. This equation becomes

$$u_{k,j+1} \doteq s^2(u_{k+1,j} + u_{k-1,j}) + 2(1-s^2)u_{k,j} - u_{k,j-1} - s^2h^2 f_{k,j}$$
$$(k = 1, \dots, N-1;\ j = 1, \dots, m_0 - 1).$$

So the problem (6.4.4) is reduced to the following system of linear equations

$$\begin{cases} u_{k,j+1} \doteq s^2(u_{k+1,j} + u_{k-1,j}) + 2(1-s^2)u_{i,j} & \\ \qquad - u_{k,j-1} - s^2h^2 f_{k,j} & (k = 1, \dots, N-1;\ j = 1, \dots, m_0 - 1), \\ u_{k,0} \doteq \varphi_k, \quad u_{k,1} \doteq \varphi_k + \tau\psi_k & (k = 0, \dots, N), \\ u_{0,j} \doteq \Phi_0(j\tau) = \Phi_{0j}, \quad u_{N,j} \doteq \Phi_1(j\tau) = \Phi_{1j} & (j = 0, \dots, m_0). \end{cases}$$

This is an explicit difference scheme. Solving the system of linear equations, the numerical solution is obtained.

6.5 Finite element methods

The finite element method is another numerical method for solving differential equations. Its basic idea is to transfer a differential equation into an integral equation, partition the integral domain into finite subdomains (e.g. triangulation), then construct an interpolation polynomial of the generalized solution with nodes (e.g. vertices of triangle), and finally yield a system of linear equations in each subdomain.

6.5.1 The one-dimensional finite element method

Consider the boundary value problem of a second-order ordinary differential equation

$$\begin{cases} -(p(x)u'(x))' + q(x)u(x) = f(x) & (0 \le x \le b), \\ u(0) = 0, \quad p(b)u'(b) + \alpha u(b) = g, & \end{cases} \tag{6.5.1}$$

where $p(x)$, $q(x)$ are both continuous functions on $[0, b]$ and $p(x) \ge p_0 > 0$, $q(x) > 0\ (0 \le x \le b)$, $\alpha > 0$, and g is a constant. Its solution in the space $C^1([0, b]) \bigcap C^2((0, b))$ is called the *classical solution*.

(a) Generalized solution

We want to find the generalized solution of the problem (6.5.1) in the space $C^1([0, b])$.

Let $\varphi(x)$ be a continuously differentiable function on $[0, b]$ and satisfy the boundary condition $\varphi(0) = 0$. Multiplying both sides of the second-order ordinary equation in (6.5.1) by $\varphi(x)$, and then integrating both sides from 0 to b, we get

$$\int_0^b (-(p(x)u'(x))' + q(x)u(x))\varphi(x)\,dx = \int_0^b f(x)\varphi(x)\,dx. \tag{6.5.2}$$

Note that $\varphi(0) = 0$ and the boundary condition $p(b)u'(b) + \alpha u(b) = g$. Using the integration by parts gives

$$\int_0^b (p(x)u'(x))'\varphi(x)\,dx = p(b)u'(b)\varphi(b) - \int_0^b p(x)u'(x)\varphi'(x)\,dx$$
$$= (g - \alpha u(b))\varphi(b) - \int_0^b p(x)u'(x)\varphi'(x)\,dx.$$

From this and (6.5.2), it follows that

$$\int_0^b (p(x)\varphi'(x)u'(x) + q(x)\varphi(x)u(x))\,dx + \alpha u(b)\varphi(b) = \int_0^b f(x)\varphi(x)\,dx + g\varphi(b). \quad (6.5.3)$$

If $u \in C^1([0,b])$ satisfying $u(0) = 0$ and $\int_0^b ((u'(x))^2 + (u(x))^2)\,dx < \infty$, and (6.5.3) holds for any $\varphi \in C^1([0,b])$ satisfying $\varphi(0) = 0$, the function $u(x)$ is called a *generalized solution* of the problem (6.5.1).

(b) Discretization

Take a partition of the interval $[0,b]$: $0 \le x_0 < x_1 < \cdots < x_n = b$. For $i = 0,\dots,n-1$, denote by η_i each subinterval $[x_i, x_{i+1}]$ and denote by u_i the value of $u(x)$ at the node x_i. From $[0,b] = \bigcup_{i=0}^{n-1}[x_i, x_{i+1}] = \bigcup_{i=0}^{n-1}\eta_i$, it follows by (6.5.3) that

$$\sum_0^{n-1} U_i + \alpha u(b)\varphi(b) = \sum_0^{n-1} V_i + g\varphi(b), \quad (6.5.4)$$

where

$$U_i = \int_{\eta_i} (p(x)\varphi'(x)u'(x) + q(x)\varphi(x)u(x))\,dx,$$
$$V_i = \int_{\eta_i} f(x)\varphi(x)\,dx.$$

(c) Interpolation function

Introduce the linear interpolations of $u(x)$ and $\varphi(x)$ on η_i $(i = 0,\dots,n-1)$ as follows:

$$u(x) \doteq A_i(x)u_i + B_i(x)u_{i+1},$$
$$\varphi(x) \doteq A_i(x)\varphi_i + B_i(x)\varphi_{i+1} \quad (x \in \eta_i),$$

where $A_i(x) = \frac{x_{i+1}-x}{M_i}$ and $B_i(x) = \frac{x-x_i}{M_i}$ and $M_i = x_{i+1} - x_i$, and

$$u'(x) \doteq A_i'(x)u_i + B_i'(x)u_{i+1},$$
$$\varphi'(x) \doteq A_i'(x)\varphi_i + B_i'(x)\varphi_{i+1} \quad (x \in \eta_i),$$

where $A_i'(x) = -\frac{1}{M_i}$, $B_i'(x) = \frac{1}{M_i}$.

(d) Unit stiffness matrix and unit carrier vector

Introduce two 2-dimensional vectors $\boldsymbol{J}_i, \boldsymbol{J}_i^*$ and two 1×2 matrices $K_i(x), H_i(x)$ as follows:

$$\begin{gathered} \boldsymbol{J}_i = (u_i, u_{i+1})^{\mathrm{T}}, \qquad \boldsymbol{J}_i^* = (\varphi_i, \varphi_{i+1})^{\mathrm{T}}, \\ K_i(x) = (A_i(x), B_i(x)), \quad H_i(x) = (A_i'(x), B_i'(x)) \qquad (i = 0, \dots, n). \end{gathered} \tag{6.5.5}$$

So

$$\begin{aligned} u(x) &= K_i\boldsymbol{J}_i, \quad \varphi(x) = K_i\boldsymbol{J}_i^*, \\ u'(x) &= H_i\boldsymbol{J}_i, \quad \varphi'(x) = H_i\boldsymbol{J}_i^*. \end{aligned}$$

Note that

$$\begin{aligned} H_i\boldsymbol{J}_i^* &= (H_i\boldsymbol{J}_i^*)^{\mathrm{T}} = (\boldsymbol{J}_i^*)^{\mathrm{T}}H_i^{\mathrm{T}}, \\ K_i\boldsymbol{J}_i^* &= (K_i\boldsymbol{J}_i^*)^{\mathrm{T}} = (\boldsymbol{J}_i^*)^{\mathrm{T}}K_i^{\mathrm{T}}. \end{aligned}$$

So

$$\begin{aligned} \varphi'(x)u'(x) &= (\boldsymbol{J}_i^*)^{\mathrm{T}}H_i^{\mathrm{T}}H_i\boldsymbol{J}_i, \\ \varphi(x)u(x) &= (\boldsymbol{J}_i^*)^{\mathrm{T}}K_i^{\mathrm{T}}K_i\boldsymbol{J}_i \end{aligned}$$

and

$$\begin{aligned} U_i &= \int_{\eta_i} \left(p(x)\varphi'(x)u'(x) + q(x)\varphi(x)u(x)\right) \mathrm{d}x \\ &\doteq \int_{\eta_i} p(x)(\boldsymbol{J}_i^*)^{\mathrm{T}}H_i^{\mathrm{T}}H_i\boldsymbol{J}_i \,\mathrm{d}x + \int_{\eta_i} q(x)(\boldsymbol{J}_i^*)^{\mathrm{T}}K_i^{\mathrm{T}}K_i\boldsymbol{J}_i \,\mathrm{d}x. \end{aligned}$$

Since $\boldsymbol{J}_i$ and $\boldsymbol{J}_i^*$ are independent of x,

$$U_i \doteq (\boldsymbol{J}_i^*)^{\mathrm{T}}\left(\int_{\eta_i} p(x)H_i^{\mathrm{T}}H_i \,\mathrm{d}x + \int_{\eta_i} q(x)K_i^{\mathrm{T}}K_i \,\mathrm{d}x\right)\boldsymbol{J}_i.$$

Let $G^{\eta_i} = \int_{\eta_i}(p(x)H_i^{\mathrm{T}}H_i + q(x)K_i^{\mathrm{T}}K_i)\,\mathrm{d}x$ $(i = 0, \dots, n-1)$. Then

$$U_i \doteq (\boldsymbol{J}_i^*)^{\mathrm{T}}G^{\eta_i}\boldsymbol{J}_i \quad (i = 0, \dots, n-1). \tag{6.5.6}$$

Note that

$$\begin{aligned} H_i^{\mathrm{T}}H_i &= \begin{pmatrix} \frac{1}{M_i^2} & -\frac{1}{M_i^2} \\ -\frac{1}{M_i^2} & \frac{1}{M_i^2} \end{pmatrix}, \\ K_i^{\mathrm{T}}(x)K_i(x) &= \begin{pmatrix} A_i^2(x) & A_i(x)B_i(x) \\ A_i(x)B_i(x) & B_i^2(x) \end{pmatrix}. \end{aligned}$$

For $i = 0, \dots, n-1$,

$$\begin{aligned} G^{\eta_i} &= \int_{\eta_i} \left(p(x)H_i^{\mathrm{T}}H_i + q(x)K_i^{\mathrm{T}}(x)K_i(x)\right) \mathrm{d}x \\ &\doteq \begin{pmatrix} \frac{1}{M_i^2} & -\frac{1}{M_i^2} \\ -\frac{1}{M_i^2} & \frac{1}{M_i^2} \end{pmatrix} p_i + \int_{\eta_i} q(x) \begin{pmatrix} A_i^2(x) & A_i(x)B_i(x) \\ A_i(x)B_i(x) & B_i^2(x) \end{pmatrix} \mathrm{d}x \\ &= \begin{pmatrix} G_{i,i}^{\eta_i} & G_{i,i+1}^{\eta_i} \\ G_{i+1,i}^{\eta_i} & G_{i+1,i+1}^{\eta_i} \end{pmatrix} \end{aligned}$$

where $p_i \doteq \int_{\eta_i} p(x)\,\mathrm{d}x$ and

$$\begin{aligned} G^{\eta_i}_{i,i} &= \frac{p_i}{M_i^2} + \int_{\eta_i} q(x)A_i^2(x)\,\mathrm{d}x, \\ G^{\eta_i}_{i,i+1} &= G^{\eta_i}_{i+1,i} = -\frac{p_i}{M_i^2} + \int_{\eta_i} q(x)A_i(x)B_i(x)\,\mathrm{d}x, \\ G^{\eta_i}_{i+1,i+1} &= \frac{p_i}{M_i^2} + \int_{\eta_i} q(x)B_i^2(x)\,\mathrm{d}x \quad (i = 0, \dots, n-1). \end{aligned}$$

The matrix

$$\begin{pmatrix} G^{\eta_i}_{i,i} & G^{\eta_i}_{i,i+1} \\ G^{\eta_i}_{i+1,i} & G^{\eta_i}_{i+1,i+1} \end{pmatrix}$$

is called the *unit stiffness matrix*. It is a positive definite and symmetric matrix.

Substituting (6.5.6) into (6.5.4),

$$\sum_0^{n-1} (\boldsymbol{J}_i^*)^{\mathrm{T}} G^{\eta_i} \boldsymbol{J}_i + \alpha u(b)\varphi(b) = \sum_0^{n-1} V_i + g\varphi(b). \tag{6.5.7}$$

By (6.5.5),

$$\begin{aligned} (\boldsymbol{J}_{n-1}^*)^{\mathrm{T}} \begin{pmatrix} 0 & 0 \\ 0 & \alpha \end{pmatrix} \boldsymbol{J}_{n-1} &= (\varphi_{n-1}, \varphi_n) \begin{pmatrix} 0 & 0 \\ 0 & \alpha \end{pmatrix} \begin{pmatrix} u_{n-1} \\ u_n \end{pmatrix} = \alpha\varphi_n u_n = \alpha u(b)\varphi(b), \\ (\boldsymbol{J}_{n-1}^*)^{\mathrm{T}} \begin{pmatrix} 0 \\ g \end{pmatrix} &= (\varphi_{n-1}, \varphi_n) \begin{pmatrix} 0 \\ g \end{pmatrix} = g\varphi_n = g\varphi(b). \end{aligned}$$

From this and (6.5.7), it follows that

$$\sum_0^{n-1} (\boldsymbol{J}_i^*)^{\mathrm{T}} G^{\eta_i} \boldsymbol{J}_i + (\boldsymbol{J}_{n-1}^*)^{\mathrm{T}} \begin{pmatrix} 0 & 0 \\ 0 & \alpha \end{pmatrix} \boldsymbol{J}_{n-1} = \sum_0^{n-1} V_i + (\boldsymbol{J}_{n-1}^*)^{\mathrm{T}} \begin{pmatrix} 0 \\ g \end{pmatrix}. \tag{6.5.8}$$

Since $\boldsymbol{J}_i^*$ is independent of x, by (6.5.5), we get

$$V_i = \int_{\eta_i} f(x)\varphi(x)\,\mathrm{d}x \doteq \int_{\eta_i} (\boldsymbol{J}_i^*)^{\mathrm{T}} K_i^{\mathrm{T}}(x) f(x)\,\mathrm{d}x = (\boldsymbol{J}_i^*)^{\mathrm{T}} \int_{\eta_i} \begin{pmatrix} A_i(x) \\ B_i(x) \end{pmatrix} f(x)\,\mathrm{d}x = (\boldsymbol{J}_i^*)^{\mathrm{T}} \boldsymbol{F}^{e_i},$$

where

$$\boldsymbol{F}^{e_i} = \left(\int_{\eta_i} A_i(x) f(x)\,\mathrm{d}x, \int_{\eta_i} B_i(x) f(x)\,\mathrm{d}x \right)^{\mathrm{T}}.$$

The $\boldsymbol{F}^{e_i}$ is called a *unit carrier vector*. By (6.5.8), it follows that

$$\sum_0^{n-1} (\boldsymbol{J}_i^*)^{\mathrm{T}} G^{\eta_i} \boldsymbol{J}_i + (\boldsymbol{J}_{n-1}^*)^{\mathrm{T}} \begin{pmatrix} 0 & 0 \\ 0 & \alpha \end{pmatrix} \boldsymbol{J}_{n-1} = \sum_0^{n-1} (\boldsymbol{J}_i^*)^{\mathrm{T}} \boldsymbol{F}^{e_i} + (\boldsymbol{J}_{n-1}^*)^{\mathrm{T}} \begin{pmatrix} 0 \\ g \end{pmatrix}. \tag{6.5.9}$$

Let

$$\begin{aligned} \tilde{G}^{\eta_i} &= G^{\eta_i}, & \tilde{\boldsymbol{F}}^{e_i} &= \boldsymbol{F}^{e_i} \quad (i = 0, 1, \dots, n-2), \\ \tilde{G}^{\eta_{n-1}} &= G^{\eta_{n-1}} + \begin{pmatrix} 0 & 0 \\ 0 & \alpha \end{pmatrix}, & \tilde{\boldsymbol{F}}^{e_{n-1}} &= \boldsymbol{F}^{e_{n-1}} + \begin{pmatrix} 0 \\ g \end{pmatrix}. \end{aligned}$$

Then (6.5.9) becomes

$$\sum_0^{n-1} (J_i^*)^{\mathrm T} \tilde G^{\eta_i} J_i = \sum_0^{n-1} (J_i^*)^{\mathrm T} \tilde{\boldsymbol F}^{e_i}. \tag{6.5.10}$$

(e) Global stiffness matrix and global carrier vector

Let $\boldsymbol J = (u_0, u_1, \dots, u_n)^{\mathrm T}$ and $\boldsymbol J^* = (\varphi_0, \varphi_1, \dots, \varphi_n)^{\mathrm T}$, and let $\tilde{\boldsymbol F}^{e_i} = (\alpha_0, \alpha_1, \dots, \alpha_n)^{\mathrm T}$, where

$$\alpha_i = \int_{\eta_i} A_i(x) f(x)\,\mathrm dx, \quad \alpha_{i+1} = \int_{\eta_i} B_i(x) f(x)\,\mathrm dx, \quad \alpha_\nu = 0 \qquad (\nu \neq i, i+1),$$

and let $\hat G^{\eta_i} = (\beta_{\mu,\nu})_{\mu,\nu=0,\dots,n}$ $(i = 0, \dots, n-1)$, where

$$\begin{cases} \beta_{i,i} = \tilde G^{\eta_i}_{i,i}, \\ \beta_{i,i+1} = \beta_{i+1,i} = \tilde G^{\eta_i}_{i,i+1}, \\ \beta_{i+1,i+1} = \tilde G^{\eta_i}_{i+1,i+1}, \\ \beta_{n,n} = \tilde G^{\eta_{n-1}}_{n,n} + \alpha, \\ \beta_{\mu,\nu=0} \quad \text{(otherwise)}. \end{cases}$$

By (6.5.10),

$$(\boldsymbol J^*)^{\mathrm T} \left(\sum_0^{n-1} \hat G^{\eta_i} \right) \boldsymbol J_i = (\boldsymbol J^*)^{\mathrm T} \left(\sum_0^{n-1} \boldsymbol F^{e_i} \right).$$

Let $G = \sum_{i=0}^{n-1} \hat G^{\eta_i}$ and $\boldsymbol F = \sum_{i=0}^{n-1} \tilde{\boldsymbol F}^{e_i}$. Then

$$(\boldsymbol J^*)^{\mathrm T} (G \boldsymbol J_i - \boldsymbol F) = 0, \tag{6.5.11}$$

where G is called a *global stiffness matrix* and $\boldsymbol F$ is called a *global carrier vector*. The matrix G is a positive definite and symmetric matrix.

(f) Constraint condition

Define $\tilde{\boldsymbol F} = (0, \boldsymbol F^{e_1}, \dots, \boldsymbol F^{e_{n-1}})$, $\tilde{\boldsymbol J} = (0, u_1, \dots, u_n)^{\mathrm T}$, and $\tilde{\boldsymbol J}^* = (0, \varphi_1, \dots, \varphi_n)^{\mathrm T}$, and

$$\tilde G = \begin{pmatrix} 1 & 0 & \cdots & 0 \\ 0 & \beta_{11} & \cdots & \beta_{1n} \\ \vdots & \vdots & \ddots & \vdots \\ 0 & \beta_{n1} & \cdots & \beta_{nn} \end{pmatrix}.$$

From (6.5.11) and $\varphi(0) = 0$, and $u(0) = 0$, it follows that

$$\tilde{\boldsymbol J}^* (\tilde G \tilde{\boldsymbol J} - \tilde{\boldsymbol F}) = 0.$$

This equality holds for any function φ satisfying $\varphi \in C^1([0,b])$ and $\varphi(0) = 0$. So, deleting $\tilde{\boldsymbol{J}}^*$, we get

$$\tilde{G}\tilde{\boldsymbol{J}} = \tilde{\boldsymbol{F}}. \tag{6.5.12}$$

Note that $\tilde{G}$ is a positive definite and symmetric matrix, the solution $(0, u_1, \ldots, u_n)$ of the equation (6.5.12) has a unique solution. The $u_1, \ldots, u_n$ are approximate values of the solution of the problem (6.5.1) at nodes $x_1, \ldots, x_n$.

6.5.2 The two-dimensional finite element method

Consider the two-dimensional boundary value problem of the Poisson equation

$$\begin{cases} -\Delta u = f(x,y) & ((x,y) \in \Omega), \\ u|_{\partial\Omega} = u_0(x,y), \end{cases} \tag{6.5.13}$$

where $\Delta = \frac{\partial^2}{\partial x^2} + \frac{\partial^2}{\partial y^2}$ is the Laplace operator and Ω is a bounded domain in the plane.

(a) Generalized solution

Take $\varphi(x,y) \in C^2(\Omega) \bigcap C^1(\bar{\Omega})$ and $\varphi(x,y) = 0$ on $\partial\Omega$. Multiplying both sides of the Poisson equation in (6.5.13) by $\varphi(x,y)$, and then integrating both sides over Ω, we get

$$-\iint_\Omega (\Delta u)\varphi \,\mathrm{d}x\,\mathrm{d}y = \iint_\Omega f\varphi \,\mathrm{d}x\,\mathrm{d}y.$$

By the Green formula and $\varphi(x,y) = 0$ on $\partial\Omega$, the left-hand side is

$$\begin{aligned} -\iint_\Omega (\Delta u)\varphi \,\mathrm{d}x\,\mathrm{d}y &= \iint_\Omega \left(\frac{\partial u}{\partial x}\frac{\partial \varphi}{\partial x} + \frac{\partial u}{\partial y}\frac{\partial \varphi}{\partial y}\right)\mathrm{d}x\,\mathrm{d}y - \oint_{\partial\Omega} \varphi \frac{\partial u}{\partial \boldsymbol{n}}\,\mathrm{d}S \\ &= \iint_\Omega \left(\frac{\partial u}{\partial x}\frac{\partial \varphi}{\partial x} + \frac{\partial u}{\partial y}\frac{\partial \varphi}{\partial y}\right)\mathrm{d}x\,\mathrm{d}y. \end{aligned}$$

So

$$\iint_\Omega \left(\frac{\partial u}{\partial x}\frac{\partial \varphi}{\partial x} + \frac{\partial u}{\partial y}\frac{\partial \varphi}{\partial y}\right)\mathrm{d}x\,\mathrm{d}y = \iint_\Omega f\varphi \,\mathrm{d}x\,\mathrm{d}y. \tag{6.5.14}$$

If $u(x,y)$ is such that $\iint_\Omega (u^2 + \Delta u)\,\mathrm{d}x\,\mathrm{d}y < \infty$ and (6.5.14) holds for any $\varphi \in C^1(\bar{\Omega})$ satisfying $\varphi(x,y) = 0$ on $\partial\Omega$, then $u(x,y)$ is called a *generalized solution* of the problem (6.5.13).

(b) Discretization

Partition the integral domain Ω into a combination of triangle units e_k $(k = 1, \ldots, N_e)$. The vertices $p_i(x_i, y_i)$ $(i = 1, \ldots, N_e)$ of triangles are called *nodes*. Each node cannot be an interior point of sides of other triangles.

Let $u(x_i, y_i) = u_i$ $(i = 1, \ldots, N_e)$. Take a triangle unit e whose three vertices are P_i, P_j, P_m (anticlockwise order). Choose three constants a, b, c such that the function

$$\tilde{u}(x, y) = ax + by + c \tag{6.5.15}$$

attains values u_i, u_j, and u_m, respectively, i.e., a, b, and c satisfy the following equations:

$$\begin{cases} ax_i + by_i + c = u_i, \\ ax_j + by_j + c = u_j, \\ ax_m + by_m + c = u_m. \end{cases}$$

Its solution is

$$\begin{aligned} a &= \frac{1}{2\Delta_e}((y_j - y_m)u_i + (y_m - y_i)u_j + (y_i - y_j)u_m), \\ b &= -\frac{1}{2\Delta_e}((x_i - x_m)u_i + (x_m - x_i)u_j + (x_i - x_j)u_m), \\ c &= \frac{1}{2\Delta_e}((x_j y_m - x_m y_j)u_i + (x_m y_i - x_i y_m)u_j + (x_i y_j - x_j y_i)u_m), \end{aligned}$$

where

$$\Delta_e = \frac{1}{2}\begin{vmatrix} x_i & y_i & 1 \\ x_j & y_j & 1 \\ x_m & y_m & 1 \end{vmatrix}.$$

Since the order of P_i, P_j, P_m is anticlockwise, Δ_e is the area of the triangle unit $e = \Delta P_i P_j P_m$. From this and (6.5.15), the interpolation function is

$$\tilde{u}(x, y) = N_i(x, y)u_i + N_j(x, y)u_j + N_m(x, y)u_m,$$

where

$$\begin{aligned} N_i(x, y) &= \frac{1}{2\Delta_e}\left((y_j - y_m)x - (x_j - x_m)y + (x_j y_m - x_m y_j)\right), \\ N_j(x, y) &= \frac{1}{2\Delta_e}\left((y_m - y_i)x - (x_m - x_i)y + (x_m y_i - x_i y_m)\right), \\ N_m(x, y) &= \frac{1}{2\Delta_e}\left((y_i - y_j)x - (x_i - x_j)y + (x_i y_j - x_j y_i)\right). \end{aligned}$$

Here $N_i(x, y)$, $N_j(x, y)$, and $N_m(x, y)$ are all polynomials of degree 1 and satisfy $N_s(x_t, y_t) = \delta_{s,t}$ $(s, t = i, j, m)$, where $\delta_{s,t}$ is the Kronecker delta. These three functions are called *primary functions* of linear interpolation on the triangle unit e.

Denote

$$\begin{aligned} \boldsymbol{J}_e &= (u_i, u_j, u_m)^{\mathrm{T}}, \\ \boldsymbol{N} &= (N_i, N_j, N_m)^{\mathrm{T}}. \end{aligned}$$

So $\tilde{u} = \boldsymbol{N}^{\mathrm{T}} \boldsymbol{J}_e$.

The gradient vector of $\tilde{u}$ is

$$\nabla\tilde{u} = \begin{pmatrix} \frac{\partial\tilde{u}}{\partial x} \\ \frac{\partial\tilde{u}}{\partial y} \end{pmatrix} = \begin{pmatrix} \frac{\partial N_i}{\partial x} & \frac{\partial N_j}{\partial x} & \frac{\partial N_m}{\partial x} \\ \frac{\partial N_i}{\partial y} & \frac{\partial N_j}{\partial y} & \frac{\partial N_m}{\partial y} \end{pmatrix} \begin{pmatrix} u_i \\ u_j \\ u_m \end{pmatrix}.$$

Note that

$$\begin{aligned} \frac{\partial N_i}{\partial x} &= \frac{1}{2\Delta_e}(y_j - y_m), & \frac{\partial N_i}{\partial y} &= -\frac{1}{2\Delta_e}(x_j - x_m), \\ \frac{\partial N_j}{\partial x} &= \frac{1}{2\Delta_e}(y_m - y_i), & \frac{\partial N_j}{\partial y} &= -\frac{1}{2\Delta_e}(x_m - x_i), \\ \frac{\partial N_m}{\partial x} &= \frac{1}{2\Delta_e}(y_i - y_j), & \frac{\partial N_m}{\partial y} &= -\frac{1}{2\Delta_e}(x_i - x_j). \end{aligned}$$

Then $\nabla\tilde{u} = B\boldsymbol{J}_e$, where

$$B = \frac{1}{2\Delta_e}\begin{pmatrix} y_j - y_m & y_m - y_i & y_i - y_j \\ x_m - x_j & x_i - x_m & x_j - x_i \end{pmatrix}. \tag{6.5.16}$$

(c) Unit stiffness matrix and unit carrier vector

Let $\tilde{\varphi}(x, y)$ be the interpolation function of φ on the triangle unit e. Then

$$\tilde{\varphi}(x, y) = N_i(x, y)\varphi_i + N_j(x, y)\varphi_j + N_m(x, y)\varphi_m = \boldsymbol{N}^{\mathrm{T}}\boldsymbol{J}_e^*,$$

where

$$\begin{aligned} &\boldsymbol{J}_e^* = (\varphi_i, \varphi_j, \varphi_m)^{\mathrm{T}}, \\ &\boldsymbol{N} = (N_i, N_j, N_m)^{\mathrm{T}}\boldsymbol{N} = (N_i, N_j, N_m)^{\mathrm{T}}, \\ &\varphi_i = \varphi(x_i, y_i), \quad \varphi_j = \varphi(x_j, y_j), \quad \varphi_m = \varphi(x_m, y_m). \end{aligned}$$

Similar to the argument of $\nabla\tilde{u}$, we get

$$\nabla\tilde{\varphi} = B\boldsymbol{J}_e^*,$$

where B is stated in (6.5.16).

Replacing u and φ by the interpolation functions $\tilde{u}$ and $\tilde{\varphi}$ in (6.5.14), respectively, noticing that $\sum_{n=1}^{N_e} e_n = \Omega$, we get

$$\sum_{n=1}^{N_e} \iint_{e_n} \left(\frac{\partial\tilde{u}}{\partial x}\frac{\partial\tilde{\varphi}}{\partial x} + \frac{\partial\tilde{u}}{\partial y}\frac{\partial\tilde{\varphi}}{\partial y} \right) \mathrm{d}x\,\mathrm{d}y = \sum_{n=1}^{N_e} \iint_{e_n} f\varphi\,\mathrm{d}x\,\mathrm{d}y. \tag{6.5.17}$$

By $\nabla\tilde{u} = B\boldsymbol{J}_e$ and $\nabla\tilde{\varphi} = B\boldsymbol{J}_e^*$, each term on the left-hand side of (6.5.17) is

$$\begin{aligned} \iint_e \left(\frac{\partial\tilde{u}}{\partial x}\frac{\partial\tilde{\varphi}}{\partial x} + \frac{\partial\tilde{u}}{\partial y}\frac{\partial\tilde{\varphi}}{\partial y} \right) \mathrm{d}x\,\mathrm{d}y &= \iint_e (\nabla\tilde{\varphi})^{\mathrm{T}}(\nabla\tilde{u})\,\mathrm{d}x\,\mathrm{d}y \\ &= \iint_e (\boldsymbol{J}_e^*)^{\mathrm{T}}B^{\mathrm{T}}B\boldsymbol{J}_e\,\mathrm{d}x\,\mathrm{d}y = (\boldsymbol{J}_e^*)^{\mathrm{T}}K_e\boldsymbol{J}_e, \end{aligned} \tag{6.5.18}$$

where $K_e = \iint_e B^{\mathrm{T}} B \, \mathrm{d}x \, \mathrm{d}y$ is called a *unit stiffness matrix*.

Since B is a constant matrix and the area of e is Δ_e,

$$K_e = \Delta_e B^{\mathrm{T}} B = \begin{pmatrix} K_{ii}^e & K_{ij}^e & K_{im}^e \\ K_{ji}^e & K_{jj}^e & K_{jm}^e \\ K_{mi}^e & K_{mj}^e & K_{mm}^e \end{pmatrix},$$

where

$$K_{st}^e = \Delta_e \left(\frac{\partial N_s}{\partial x} \frac{\partial N_t}{\partial x} + \frac{\partial N_s}{\partial y} \frac{\partial N_t}{\partial y} \right) = \frac{1}{4\Delta_e} (a_s a_t + b_s b_t) \quad (s, t = i, j, m)$$

and

$$a_i = y_j - y_m, \quad a_j = y_m - y_i, \quad a_m = y_i - y_j,$$
$$b_i = x_m - x_j, \quad b_j = x_i - x_m, \quad b_m = x_j - x_i.$$

Note that $f\varphi = \varphi^{\mathrm{T}} f = (\boldsymbol{J}_e^*)^{\mathrm{T}} \boldsymbol{N} f$. Since $\boldsymbol{J}_e^*$ is independent of x and y, each term on the right-hand side of (6.5.17) is

$$\iint_e f\varphi \, \mathrm{d}x \, \mathrm{d}y = \iint_e (\boldsymbol{J}_e^*)^{\mathrm{T}} \boldsymbol{N} f \, \mathrm{d}x \, \mathrm{d}y = (\boldsymbol{J}_e^*)^{\mathrm{T}} \boldsymbol{F}_e,$$

where

$$\boldsymbol{F}_e = \iint_{\boldsymbol{e}} \boldsymbol{N} f \, \mathrm{d}x \, \mathrm{d}y = (F_i^e, F_j^e, F_m^e)^{\mathrm{T}}.$$

The vector $\boldsymbol{F}_e$ is called a *unit carrier vector* and $F_s^e = \iint_e N_s f \, \mathrm{d}x \, \mathrm{d}y$ $(s = i, j, m)$.

(d) Global stiffness matrix and global carrier vector

Substituting representations of unit stiffness matrix and unit carrier vector into (6.5.17). For the convenience superposition, each K_e is extended to an $N_e \times N_e$ matrix $\tilde{K}_e$, each $\boldsymbol{F}_e$ is extended to N_e-dimensional vectors $\tilde{\boldsymbol{F}}_e$, and $\boldsymbol{J}_e^*$ and $\boldsymbol{J}_e$ are extended to $\boldsymbol{J}^*$ and $\boldsymbol{J}$, respectively. So

$$(\boldsymbol{J}^*)^{\mathrm{T}} \left(\sum_1^{N_e} \tilde{K}_{e_n} \right) \boldsymbol{J} = (\boldsymbol{J}^*)^{\mathrm{T}} \left(\sum_1^{N_e} \tilde{\boldsymbol{F}}_{e_n} \right).$$

This equality holds for arbitrary $\boldsymbol{J}^*$. Deleting $(\boldsymbol{J}^*)^{\mathrm{T}}$, we get

$$\left(\sum_1^{N_e} \tilde{K}_{e_n} \right) \boldsymbol{J} = \left(\sum_1^{N_e} \tilde{\boldsymbol{F}}_{e_n} \right).$$

Let $K = \sum_{n=1}^{N_e} K_{e_n}$ and $\boldsymbol{F} = \sum_{n=1}^{N_e} \boldsymbol{F}_{e_n}$. Then $K\boldsymbol{J} = \boldsymbol{F}$. This is a system of linear equations and the matrices K and F are called a *global stiffness matrix* and a *global carrier vector*, respectively. Rewriting this system of linear equations using constraint conditions, the desired solution can be obtained.

6.6 Wavelet methods

Starting from the classical Galerkin method and Daubechies wavelets, we introduce the wavelet-Galerkin method for solving differential and integral equations.

6.6.1 Classical Galerkin method

Consider a problem of linear differential equations

$$Lu = f, \tag{6.6.1}$$

where L is a linear differential operator, f is a known function, and u is a undetermined function. Choosing a set of test functions $u_1, \ldots, u_n$, we try to approximate to u by a linear combination $u = \sum_{k=1}^{n} c_k u_k$ of these test functions. Since L is a linear operator,

$$Lu = \sum_{1}^{n} c_k L u_k .$$

Choose coefficients c_k $(k = 1, \ldots, n)$ such that some norm

$$\|Lu - f\| = \left\| \sum_{1}^{n} c_k L u_k - f \right\|$$

attains the minimal value. The function u is an approximation solution of (6.6.1). The norm may be the norm of $C([a, b])$ or the norm of $L^2([a, b])$, i.e.,

$$\|f\| = \max_{a \le x \le b} |f(x)| \quad \text{or} \quad \|f\| = \left(\int_a^b |f(x)|^2 \, dx \right)^{1/2} .$$

The interpolation form of the Galerkin method is

$$\sum_{k=1}^{n} c_k (Lu_k)(x_i) = f(x_i) \quad (i = 1, \ldots, n),$$

i.e., choose $c_1, \ldots, c_n$ such that Lu is an interpolation function. The function

$$u(x) = \sum_{1}^{n} c_k u_k(x)$$

is an approximation solution of (6.6.1) in the interpolation sense.

The weak form of the Galerkin method is often used. Choose c_k $(k = 1, \ldots, n)$ such that

$$(u_l, Lu)_{L^2} = (u_l, f)_{L^2} \quad (l = 1, \ldots, n), \tag{6.6.2}$$

i.e.,

$$\sum_{1}^{n} c_k(u_l, Lu_k)_{L^2} = (u_l, f)_{L^2} \quad (l = 1, \dots, n).$$

Denote $(f_l, \dots, f_n)^{\mathrm{T}}$ and $g_{lk} = (u_l, Lu_k)_{L^2}$. From the system of linear equations

$$\sum_{k=1}^{n} g_{lk} c_k = f_l,$$

we solve out c_k $(k = 1, \dots, n)$. The function $u = \sum_{k=1}^{n} c_k u_k$ satisfies (6.6.2).

For example, consider the Dirichlet problem

$$\begin{cases} \Delta u = 0 & \text{in } \Omega, \\ u(x, y) = g & \text{on } \partial\Omega. \end{cases}$$

or write as

$$Lu = \boldsymbol{f}, \quad \text{where } Lu = (\Delta u, u|_{\partial\Omega})^{\mathrm{T}} \text{ and } \boldsymbol{f} = (0, g)^{\mathrm{T}}.$$

Let u_k be the real part of z^k. Then u_k $(k = 1, \dots, n)$ are harmonic functions and their linear combination $u = \sum_{k=1}^{n} c_k u_k$ is also a harmonic function. So $\Delta u = 0$. Take m points (x_l, y_l) $(l = 1, \dots, m)$ distributed on $\partial\Omega$ uniformly, where $m > n$, and then use the Remez algorithm (see Section 3.3) to find c_k $(k = 1, \dots, n)$ such that

$$\max_{1 \le l \le m} \left| \sum_{k=1}^{n} c_k u_k(x_l, y_l) - g(x_l, y_l) \right|$$

attains the minimal value. So this linear combination u is an approximation solution of the Dirichlet problem.

In Galerkin method, one often takes a subset of a normal orthogonal basis as test functions such that their linear combination approximates to the solution. Moreover, when the cardinal numbers of subsets increase to ∞, the approximate solutions tend to the exact solution.

Wavelet bases have many nice properties such as compact support, smoothness, and vanishing moment. Therefore, the wavelet-Galerkin method has been developing recently. To give the wavelet-Galerkin method, we first state the construction of compactly supported wavelets.

6.6.2 Compactly supported wavelets

If ψ is an orthogonal wavelet with p vanishing moments, then it has a support of size larger than or equal to $2p - 1$. Daubechies wavelet has a minimum size support in these wavelets.

Given a trigonometric polynomial $H(\omega)$ satisfying

$$|H(\omega)|^2 = 2 \left| \cos \frac{\omega}{2} \right|^{2p} P\left(\sin^2 \frac{\omega}{2}\right),$$

where

$$P(y) = \sum_{k=0}^{p-1} C_{p-1+k}^{k} y^k, \quad C_{p-1+k}^{k} = \frac{(p-1+k)!}{k!\,(p-1)!},$$

one uses the Riesz lemma to construct a minimum degree polynomial $R(z)$ by factorization method such that

$$|R(\mathrm{e}^{-\mathrm{i}\omega})|^2 = P\left(\sin^2 \frac{\omega}{2}\right),$$

and let

$$H(\omega) = \sqrt{2}\,\mathrm{e}^{-\mathrm{i}(p-1)\omega} \left(\frac{1+\mathrm{e}^{-\mathrm{i}\omega}}{2}\right)^p R(\mathrm{e}^{-\mathrm{i}\omega}) = \sum_{0}^{2p-1} c_k \mathrm{e}^{-\mathrm{i}k\omega}.$$

The bi-scale equation based on the coefficient c_k

$$\frac{1}{2}\varphi(t) = \sum_{0}^{2p-1} c_k \varphi(2t-k) \tag{6.6.3}$$

can be solved by the ordinary iteration method as follows.

Define an operator T as

$$T(\varphi(t)) = 2\sum_{0}^{2p-1} c_k \varphi(2t-k).$$

Take a compactly supported function $\varphi_0(t)$ such as

$$\varphi_0(t) = N_2(t) = \begin{cases} t & (0 \le t < 1), \\ 2-t & (1 \le t < 2), \\ 0 & \text{otherwise}, \end{cases}$$

where $N_2(t)$ is a two-order B spline (see Section 4.4). Let

$$\varphi_n(t) = T(\varphi_{n-1}(t)) \quad (n \in \mathbb{Z}_+).$$

Then $\varphi_k(t)$ converges as $k \to \infty$, the limit is $\varphi^*(t)$, and $\varphi^*(t)$ is a continuous solution of (6.6.3) with compact support $[0, 2p-1]$. Under the additional condition

$$\sum_{k=0}^{2p-1} \varphi^*(k) = 1,$$

the values of the scaling function $\varphi^*(t)$ at dyadic decimal $k/2^m$ are given. The corresponding wavelet function ψ^* is given by

$$\psi^*(t) = -2 \sum_{2-2p}^{1} (-1)^k c_{1-k} \varphi(2t-k)$$

(see Section 3.5). The function ψ^* is called a *Daubechies wavelet*. It has p vanishing moments and support $[-p+1, p]$. The larger p is, the smoother the Daubechies wavelet is.

6.6.3 Wavelet-Galerkin method

Consider the problem of one-dimensional differential equation with Dirichlet boundary condition

$$\begin{cases} Lu(x) = f(x) & (0 \le x \le 1), \\ u(0) = u(1) = 0, \end{cases} \tag{6.6.4}$$

where f is a real-valued and continuous function on $[0, 1]$ and L is a uniformly elliptic differential operator.

Let $\psi_{mn}(x) = 2^{\frac{m}{2}}\psi(2^m x - n)$ be a wavelet basis for $L^2([0, 1])$ with boundary condition

$$\psi_{mn}(0) = \psi_{mn}(1) = 0.$$

Let Λ be a set of indices (m, n). For each $(m, n) \in \Lambda$, ψ_{mn} is a two-order continuously differentiable function and

$$\psi_{mn}(x) = 0 \quad \text{if} \quad \left| x - \frac{n}{2^m} \right| > \frac{1}{2^m}. \tag{6.6.5}$$

Let the approximation solution of (6.6.4) be

$$u_p = \sum_{(m,n)\in\Lambda} c_{mn}\psi_{mn},$$

where c_{mn} $((m, n) \in \Lambda)$ are undetermined coefficients. Select c_{mn} $((m, n) \in \Lambda)$ such that

$$(Lu_p, \psi_{jk}) = (f, \psi_{jk}) \quad ((j, k) \in \Lambda).$$

This implies that

$$\sum_{(m,n)\in\Lambda} (L\psi_{mn}, \psi_{jk})\, c_{mn} = (f, \psi_{jk}) \quad ((j, k) \in \Lambda). \tag{6.6.6}$$

Denote

$$\alpha_{mnjk} = (L\psi_{mn}, \psi_{jk}), \qquad y_{jk} = (f, \psi_{jk}),$$
$$A = (\alpha_{mnjk})_{(j,k),(m,n)\in\Lambda}, \quad C = (c_{mn})_{(m,n)\in\Lambda}, \quad Y = (Y_{jk})_{(j,k)\in\Lambda},$$

where (m, n) and (j, k) represent the rows and columns of A, respectively. So (6.6.6) is reduced to the following system of linear equations with unknown c_{mn}

$$\sum_{(m,n)\in\Lambda} \alpha_{mnjk}c_{mn} = y_{jk} \quad ((j, k) \in \Lambda)$$

which is equivalent to $AC = Y$. From (6.6.5), it is seen that A is a sparse matrix. If the condition number of A is high, choose a matrix B such that the system $AC = Y$ is replaced by the equivalent system $BAC = BY$, where BA has low condition number. Denote

$$BA = M, \quad BY = V.$$

The system of the linear equation $MC = V$ is stable. From this, the undetermined coefficients c_{mn} are solved. So we get the approximation solution u_p.

6.6.4 Numerical solution of integral equations

Numerical calculation of integral operators plays a key role in solving integral equations. Consider a linear integral operator

$$Tf(x) = \int_{\mathbb{R}} K(x, y)f(y)\,dy$$

and $K(x, y) = 0$ for $(x, y) \in \mathbb{R}^2 \setminus [0, 1]^2$. Take N points (x_i, y_i) $(i = 1, \dots, N)$, where

$$0 = x_1 < x_2 < \cdots < x_N = 1,$$
$$0 = y_1 < y_2 < \cdots < y_N = 1.$$

Then

$$Tf(x_i) = \sum_1^N K(x_i, y_i)f(y_i)\Delta y_i \quad (\Delta y_i = y_i - y_{i-1}).$$

From this, the calculation of $Tf(x)$ is reduced to a product of an $N \times N$ matrix and an N-dimensional vector. So the number of multiply operations is $O(N^2)$. This number is very large.

When T is a convolution operator, i.e.,

$$(Tf)(x) = \int_{\mathbb{R}} K(x - y)f(y)\,dy,$$

taking Fourier transform,

$$(\hat{Tf})(\omega) = \hat{K}(\omega)\hat{f}(\omega).$$

Its discrete form is

$$(\hat{Tf})(\omega_i) = \hat{K}(\omega_i)\hat{f}(\omega_i) \quad (i = 1, \dots, N).$$

The corresponding number of operations decreases from $O(N^2)$ to $O(N \log N)$.

When the linear operator is not in a convolution form, the Fourier transform approach does not work. Since Daubechies wavelet expansion enables the discretization matrix corresponding to the integral operator T to be reduced to a sparse matrix, the number of operations is only $O(N \log N)$ for wavelet-based numerical computation of integral operators.

Take a slightly modified Daubechies wavelet with $M - 1$ vanishing moments, its scaling function φ and wavelet ψ satisfy

$$\varphi(x) = 2 \sum_0^{2M-1} h_{k+1}\varphi(2x - k),$$
$$\psi(x) = 2 \sum_0^{2M-1} g_{k+1}\varphi(2x - k),$$

where $g_k = (-1)^{k-1}h_{2M-k+1}$ $(k = 1, \dots, 2M)$.

(a) Wavelet expansion of $K(x, y)$

The bivariate Daubechies wavelets are $\varphi(x)\psi(y)$, $\psi(x)\varphi(y)$, $\psi(x)\psi(y)$, where $\varphi(x)$ and $\psi(x)$ are the Daubechies scaling function and the corresponding wavelet. The bivariate Daubechies wavelet basis is

$$\{\varphi_{jm}(x)\psi_{jn}(y), \quad \psi_{jm}(x)\varphi_{jn}(y), \quad \psi_{jm}(x)\psi_{jn}(y)\}_{j,m,n\in\mathbb{Z}}.$$

Expand the kernel function $K(x, y)$ into a series with respect to this basis

$$K(x,y) = \sum_{j,k,l} \alpha_{jkl}\varphi_{jk}(x)\psi_{jl}(y) + \sum_{j,k,l} \beta_{jkl}\psi_{jk}(x)\varphi_{jl}(y) + \sum_{j,k,l} \gamma_{jkl}\psi_{jk}(x)\psi_{jl}(y), \tag{6.6.7}$$

where the wavelet coefficients are

$$\alpha_{jkl} = \iint_{\mathbb{R}^2} K(x,y)\varphi_{jk}(x)\psi_{jl}(y)\,\mathrm{d}x\,\mathrm{d}y,$$

$$\beta_{jkl} = \iint_{\mathbb{R}^2} K(x,y)\psi_{jk}(x)\varphi_{jl}(y)\,\mathrm{d}x\,\mathrm{d}y,$$

$$\gamma_{jkl} = \iint_{\mathbb{R}^2} K(x,y)\psi_{jk}(x)\psi_{jl}(y)\,\mathrm{d}x\,\mathrm{d}y.$$

(b) Wavelet expansion of $Tf(x)$

By $Tf(x) = \int_{\mathbb{R}} K(x,y)f(y)\,\mathrm{d}y$ and (6.6.7), it follows that

$$Tf(x) = \sum_{j,k,l} \alpha_{jkl}d_{jl}\varphi_{jk}(x) + \sum_{j,k,l} \beta_{jkl}c_{jl}\psi_{jk}(x) + \sum_{j,k,l} \gamma_{jkl}d_{jl}\psi_{jk}(x), \tag{6.6.8}$$

where

$$d_{jl} = \int_{\mathbb{R}} f(y)\psi_{jl}(y)\,\mathrm{d}y,$$

$$c_{jl} = \int_{\mathbb{R}} f(y)\varphi_{jl}(y)\,\mathrm{d}y.$$

Define three matrices α^j, β^j, γ^j and two vectors $\tilde{\boldsymbol{d}}^j$, $\tilde{\boldsymbol{c}}^j$ as

$$\begin{aligned} \alpha^j = (\alpha_{jkl})_{k,l}, \quad \beta^j = (\beta_{jkl})_{k,l}, \quad \gamma^j = (\gamma_{jkl})_{k,l}, \\ \tilde{\boldsymbol{d}}^j = \gamma^j\boldsymbol{d}^j + \beta^j\boldsymbol{c}^j, \qquad \tilde{\boldsymbol{c}}^j = \alpha^j\boldsymbol{d}^j, \end{aligned} \tag{6.6.9}$$

where $\boldsymbol{d}^j = (d_{jl})_l$ and $\boldsymbol{c}^j = (c_{jl})_l$ are vectors. From this and (6.6.8),

$$Tf(x) = \sum_j \sum_k \left(\tilde{d}^j_k\psi_{jk}(x) + \tilde{c}^j_k\varphi_{jk}(x)\right).$$

(c) Numerical calculation of $Tf(x)$

Let $N = 2^n$. Take c_k^0 $(k = 1, \ldots, N)$ as N samples which are regarded as the average of f on dyadic intervals with length 2^{-n} and $\{c_k^0\}$ is a 2^n-periodic sequence. So we get a $(2N-2)$-dimensional vector

$$\tilde{\boldsymbol{f}} = (d_{11}, \ldots, d_{1\frac{N}{2}}, c_{11}, \ldots, c_{1\frac{N}{2}}, d_{21}, \ldots, d_{2,\frac{N}{4}}, c_{21}, \ldots, c_{2\frac{N}{4}}, \ldots, d_{n1} c_{n1})^{\mathrm{T}}.$$

Take

$$T^N f(x) = \sum_{j=1}^{n} \sum_{k=1}^{2^{n-j}} \left(\tilde{d}_k^j \psi_{jk}(x) + \tilde{c}_k^j \varphi_{jk}(x) \right). \tag{6.6.10}$$

Then $Tf(x) \doteq T^N f(x)$. We only need compute $T^N f(x)$.

Formula (6.6.9) can be represented by a product of the matrix A and the vector $\tilde{f}$, where

$$A = \begin{pmatrix} A_1 & & \\ & \ddots & \\ & & A_n \end{pmatrix}, \quad A_j = \begin{pmatrix} \gamma^j & \beta^j \\ \alpha^j & 0 \end{pmatrix}, \tag{6.6.11}$$

and

$$\begin{aligned} \alpha^j &= (\alpha_{jkl})_{k,l=1,\ldots,2^{n-j},\, j=1,\ldots,n}, \\ \beta^j &= (\beta_{jkl})_{k,l=1,\ldots,2^{n-j},\, j=1,\ldots,n}, \\ \gamma^j &= (\gamma_{jkl})_{k,l=1,\ldots,2^{n-j},\, j=1,\ldots,n}. \end{aligned}$$

(d) Calderon–Zygmund operators

Calderon–Zygmund operators are most important integral operators which are widely applied in integral equations. Let T be a Calderon–Zygmund operator whose kernel function satisfies the following three conditions:

$$\begin{aligned} |K(x, y)| &\le C(|x - y|^{-1}), \\ \left| \frac{\partial^M}{\partial x} K(x, y) \right| + \left| \frac{\partial^M}{\partial y} K(x, y) \right| &\le C(|x - y|^{-1-M}), \\ \left| \int_{I \times I} K(x, y) \, \mathrm{d}x \, \mathrm{d}y \right| &\le C|I| \quad \text{for any dyadic interval } I, \end{aligned}$$

where C is a constant and $M \ge 1$. Then the wavelet coefficients of $K(x, y)$ satisfy

$$|\alpha_{jkl}| + |\beta_{jkl}| + |\gamma_{jkl}| = O\left((1 + |k - l|)^{-1-M} \right).$$

(e) Modified integral operators

Given $\varepsilon > 0$, take

$$B \ge \left(\frac{C}{\varepsilon} \log_2 N \right)^{\frac{1}{M}}.$$

Let $A = (A_{\mu\nu})$ be stated in (6.6.11). Define $A^{(B)} = (A^{(B)}_{\mu\nu})$, where

$$A^{(B)}_{\mu\nu} = \begin{cases} A_{\mu\nu}, & |\mu - \nu| \leq B, \\ 0, & |\mu - \nu| > B. \end{cases}$$

Let $\alpha^{jB}, \beta^{jB}, \gamma^{jB}$ be obtained from $\alpha^j, \beta^j, \gamma^j$, respectively, by replacing A by $A^{(B)}$ and $\tilde{\boldsymbol{d}}^{jB}$, $\tilde{\boldsymbol{c}}^{jB}$ be obtained by replacing $\alpha^j, \beta^j, \gamma^j$ by $\alpha^{jB}, \beta^{jB}, \gamma^{jB}$ in (6.6.9), respectively. Denote

$$\boldsymbol{d}^{jB} = (d_1^{jB}, \ldots, d_{2^{n-j}}^{jB})^{\mathrm{T}},$$
$$\boldsymbol{c}^{jB} = (c_1^{jB}, \ldots, c_{2^{n-j}}^{jB})^{\mathrm{T}}.$$

The *modified integral operator* is defined as

$$T^{NB}f(x) = \sum_{j=1}^{n} \sum_{k=1}^{2^{n-j}} \left(\widetilde{d_k^{jB}} \psi_{jk}(x) + \widetilde{c_k^{jB}} \varphi_{jk}(x) \right).$$

Then

$$\| T^{NB}f - T^N(f) \|_{L^2} \leq \frac{D}{B^M} \log_2 N \, \|f\|_{L^2} \leq \varepsilon,$$

where T_N is stated in (6.6.10). So we only need compute $(T^{NB}f)(x)$. The corresponding total number of operations is just $O(N \log N)$. From this, we can see that the numerical solution of integral equations $Tf(x) = g(x)$ via wavelet methods is fast.

Further reading

[1] Goyal K, Mehra M. Fast diffusion wavelet method for partial differential equations. *Applied Mathematical Modelling*. 2016(40):5000–5025.

[2] Nguyen K, Dabdub D. Two-level time-marching scheme using splines for solving the advection equation. *Atmospheric Environment*. 2001(35):1627–1637.

[3] Garcin M, Guegan D. Wavelet shrinkage of a noisy dynamical system with non-linear noise impact. *Physica D: Nonlinear Phenomena*. 2016(325):126–145.

[4] Ravansalar M, Rajaee T, Zounemat-Kermani M. A wavelet-linear genetic programming model for sodium (Na+) concentration forecasting in rivers. *Journal of Hydrology*. 2016(537):398–407.

[5] Lateb M, Meroney RM, Yataghene M, Fellouah H, Saleh F, Boufadel MC. On the use of numerical modelling for near-field pollutant dispersion in urban environments? A review. *Environmental Pollution*. 2016(208):271–283.

[6] Tan Z, Dong J, Xiao Y, Tu J. Numerical simulation of diurnally varying thermal environment in a street canyon under haze-fog conditions. *Atmospheric Environment*. 2015(119):95–106.

[7] Taghinia J, Rahman MM, Siikonen T. Numerical simulation of airflow and temperature fields around an occupant in indoor environment. *Energy and Buildings*. 2015(104):199–207.

[8] Zhang S, Xia Z, Wang T. A real-time interactive simulation framework for watershed decision making using numerical models and virtual environment. *Journal of Hydrology*. 2013(493): 95–104.

[9] Yu W, Chen XJ, Wu GH, Liu J, Hearn GE. A fast numerical method for trimaran wave resistance prediction. *Ocean Engineering*. 2015(107):70–84.

[10] Rai AC, Lin CH, Chen Q. Numerical modeling of particle generation from ozone reactions with human-worn clothing in indoor environments. *Atmospheric Environment*. 2015(102):145–155.

[11] Lefebvre A, Paarlberg AJ, Ernstsen VB, Winter C. Flow separation and roughness lengths over large bedforms in a tidal environment: A numerical investigation. *Continental Shelf Research*. 2014(91):57–69.

[12] Felde VA, Bjune AE, Grytnes JA, Birks HJB. A comparison of novel and traditional numerical methods for the analysis of modern pollen assemblages from major vegetation-landform types. *Review of Palaeobotany and Palynology*. 2014(210):22–36.

[13] Sarbu I, Pacurar C. Experimental and numerical research to assess indoor environment quality and schoolwork performance in university classrooms. *Building and Environment*. 2015(93): 141–154.

[14] Chen H, Janbakhsh S, Larsson U, Moshfegh B. Numerical investigation of ventilation performance of different air supply devices in an office environment. *Building and Environment*. 2015(90):37–50.

[15] Pennell KG, Scammell MK, McClean MD, Suuberg EM, Roghani AMM, Ames J, Friguglietti L, Indeglia PA, Shen R, Yao Y, Heiger-Bernays WJ. Field data and numerical modeling: A multiple lines of evidence approach for assessing vapor intrusion exposure risks. *Science of the Total Environment*. 2016(556):291–301.

7 Optimization

Optimization problems arise in almost all areas of environmental science. Optimization can be used to reduce overall product cost, minimize negative environment impacts and maximize the probability of making a correct decision. Many numerical methods can solve these optimization problems quickly with/without some constraints, especially, linear and nonlinear optimization problems. In this chapter we will introduce main optimization techniques, including Newton's method, steepest descent method, Newton–Raphson method, variational method, simplex method, Fermat method, KKT optimality conditions and primal/dual pairs.

7.1 Newton's method and steepest descent method

Newton's algorithm is one of the known methods for solving unconstrained optimization problems. We start from the one-dimensional version.

Let $f: \mathbb{R} \to \mathbb{R}$ be an objective function and we will minimize f, i.e., solve the equation $f'(x) = 0$. At first, $x_0 \in \mathbb{R}$ is taken as an initial point. If $f'(x_0) = 0$, the algorithm stops. Otherwise, $f'(x_0) \neq 0$. Ignoring all nonlinear terms in the Taylor series of $f'(x)$ at x_0

$$f'(x) = f'(x_0) + (x - x_0)f''(x_0) + (x - x_0)^2 \frac{f'''(x_0)}{2!} + (x - x_0)^3 \frac{f^{(4)}(x_0)}{3!} + \cdots,$$

the nonlinear equation $f'(x) = 0$ can be approximated by the linear equation $f'(x_0) + (x - x_0)f''(x_0) = 0$. Since $f''(x_0) > 0$, the equation yields the first iterate

$$x_1 = x_0 - \frac{f'(x_0)}{f''(x_0)} \quad (f'(x_0) \neq 0).$$

The process is repeated at x_1, leading to a new iterate x_2, and repeatedly to a new iterate x_n. The generic recurrence equation is given by

$$x_{k+1} = x_k - \frac{f'(x_k)}{f''(x_k)} \quad (f'(x_k) \neq 0,\ f''(x_k) > 0).$$

The one-dimensional classical Newton's algorithm has a local convergence property as follows.

Let $f: \mathbb{R} \to \mathbb{R}$ be twice smooth with $f'(x^*) = 0$ and $f''(x^*) > 0$, where $x^* \in \mathbb{R}$. Denote $M^* = f''(x^*)$. If there are $r > 0$ and $M > 0$ such that for any $x', x'' \in (x^* - r, x^* + r)$,

$$|f''(x') - f''(x'')| \leq M\,|x' - x''|, \tag{7.1.1}$$

then the iterates $\{x_k\}_{k=0,1,\ldots} \in (x^* - r, x^* + r)$ converge to x^* quadratically, i.e., $|x_{k+1} - x^*| \leq C|x_k - x^*|^2$, where $C = \frac{M}{M^* - Mr}$ $(r < \frac{M^*}{M})$.

DOI 10.1515/9783110424904-008

In fact, by (7.1.1) and $Mr < M^*$, it follows that for $x \in (x^* - r, x^* + r)$,

$$|f''(x)| = |f''(x^*) + f''(x) - f''(x^*)| \geq |f''(x^*)| - |f''(x) - f''(x^*)| \geq M^* - Mr. \quad (7.1.2)$$

Since $f(x)$ is twice smooth, the Lagrange form of the Taylor expansion is

$$f'(x_k) = f'(x^*) + (x_k - x^*)f''(\bar{x}),$$

where $\bar{x} \in (x_k, x^*)$ or $\bar{x} \in (x^*, x_k)$. From this and $\frac{f'(x^*)}{f''(x_k)} = 0$ (since $f'(x^*) = 0$), it follows that

$$\begin{aligned} x_{k+1} - x^* &= \left(x_k - \frac{f'(x_k)}{f''(x_k)} \right) - x^* + \frac{f'(x^*)}{f''(x_k)} \\ &= \frac{(x_k - x^*)f''(x_k) - (f'(x_k) - f'(x^*))}{f''(x_k)} \\ &= \frac{(x_k - x^*)(f''(x_k) - f''(\bar{x}))}{f''(x_k)}. \end{aligned}$$

Therefore, by (7.1.1) and (7.1.2), for $x_k \in (x^* - r, x^* + r)$,

$$|x_{k+1} - x^*| = \frac{|x_k - x^*|\,|f''(\bar{x}) - f''(x_k)|}{|f''(x_k)|} \leq C|x_k - x^*|\,|\bar{x} - x_k|,$$

where $C = \frac{M}{M^* - Mr}$. Note that $\bar{x} \in (x_k, x^*)$ or $\bar{x} \in (x^*, x_k)$. Clearly, $|\bar{x} - x_k| \leq |x_k - x^*|$. So

$$|x_{k+1} - x^*| \leq C|x_k - x^*|^2,$$

i.e., the iterates $\{x_k\}_{k=0,1,\dots}$ converge to x^* quadratically.

Classical Newton's method

Let $f: \mathbb{R}^n \to \mathbb{R}$ be a twice smooth function and its Hessian matrix

$$\nabla^2 f(\boldsymbol{x}) = \begin{pmatrix} \frac{\partial^2 f(\boldsymbol{x})}{\partial x_1^2} & \frac{\partial^2 f(\boldsymbol{x})}{\partial x_1 \partial x_2} & \cdots & \frac{\partial^2 f(\boldsymbol{x})}{\partial x_1 \partial x_n} \\ \frac{\partial^2 f(\boldsymbol{x})}{\partial x_2 \partial x_1} & \frac{\partial^2 f(\boldsymbol{x})}{\partial x_2^2} & \cdots & \frac{\partial^2 f(\boldsymbol{x})}{\partial x_2 \partial x_n} \\ \vdots & \vdots & \ddots & \vdots \\ \frac{\partial^2 f(\boldsymbol{x})}{\partial x_n \partial x_1} & \frac{\partial^2 f(\boldsymbol{x})}{\partial x_n \partial x_2} & \cdots & \frac{\partial^2 f(\boldsymbol{x})}{\partial x_n^2} \end{pmatrix}$$

be positive definite, where $\boldsymbol{x} = (x_1, \dots, x_n)$. The classical Newton's algorithm involves the following three steps:

Step 1. Start from an initial point $\boldsymbol{x}_0 \in \mathbb{R}^n$.

Step 2. Stop if $\nabla f(\boldsymbol{x}_k) = 0$, where $\boldsymbol{x}_k \in \mathbb{R}^n$. Otherwise, the Newton direction

$$\boldsymbol{d}_k = -\nabla f(\boldsymbol{x}_k)(\nabla^2 f(\boldsymbol{x}_k))^{-1} \quad (\boldsymbol{d}_k \in \mathbb{R}^n),$$

where $(\nabla^2 f(\boldsymbol{x}))^{-1}$ is the inverse matrix of the Hessian matrix.

Step 3. The next iterate $\boldsymbol{x}_{k+1} = \boldsymbol{x}_k + \boldsymbol{d}_k$. Return to Step 2, the process is repeated at $\boldsymbol{x}_{k+1}$.

The classic Newton method is related closely to the concept of the descent direction.

Definition 7.1.1. Let $f: \mathbb{R}^n \to \mathbb{R}$ be a smooth function. The derivative of f at $\boldsymbol{x}$ in direction $\boldsymbol{d}$ is defined as

$$\frac{\partial f(\boldsymbol{x})}{\partial \boldsymbol{d}} = \lim_{\lambda \to 0} \frac{f(\boldsymbol{x} + \lambda \boldsymbol{d}) - f(\boldsymbol{x})}{\lambda},$$

where $\boldsymbol{x}, \boldsymbol{d} \in \mathbb{R}^n$ and $\lambda \in \mathbb{R}$. If

$$\frac{\partial f(\boldsymbol{x})}{\partial \boldsymbol{d}} < 0,$$

then $\boldsymbol{d}$ is said to be a *descent direction* for f at $\boldsymbol{x}$.

Note that

$$\begin{aligned}
&\frac{f(\boldsymbol{x} + \lambda \boldsymbol{d}) - f(\boldsymbol{x})}{\lambda} \\
&= \frac{f(x_1 + \lambda d_1, x_2 + \lambda d_2, x_3 + \lambda d_3, \ldots, x_n + \lambda d_n) - f(x_1, x_2, x_3, \ldots, x_n)}{\lambda} \\
&= \frac{f(x_1 + \lambda d_1, x_2 + \lambda d_2, x_3 + \lambda d_3, \ldots, x_n + \lambda d_n) - f(x_1, x_2 + \lambda d_2, x_3 + \lambda d_3, \ldots, x_n + \lambda d_n)}{\lambda} \\
&\quad + \frac{f(x_1, x_2 + \lambda d_2, x_3 + \lambda d_3, \ldots, x_n + \lambda d_n) - f(x_1, x_2, x_3 + \lambda d_3, \ldots, x_n + \lambda d_n)}{\lambda} + \cdots \\
&\quad + \frac{f(x_1, x_2, \ldots, x_{n-1}, x_n + \lambda d_n) - f(x_1, x_2, \ldots, x_{n-1}, x_n)}{\lambda}.
\end{aligned}$$

Let $\lambda \to 0$. Then

$$\frac{\partial f(\boldsymbol{x})}{\partial \boldsymbol{d}} = d_1 \frac{\partial f(\boldsymbol{x})}{\partial x_1} + d_2 \frac{\partial f(\boldsymbol{x})}{\partial x_2} + \cdots + d_n \frac{\partial f(\boldsymbol{x})}{\partial x_n} = \langle \nabla f(\boldsymbol{x}), \boldsymbol{d} \rangle,$$

where

$$\nabla f(\boldsymbol{x}) = \left(\frac{\partial f(\boldsymbol{x})}{\partial x_1}, \ldots, \frac{\partial f(\boldsymbol{x})}{\partial x_n} \right)$$

is the gradient of f at $\boldsymbol{x}$. From this and Definition 7.1.1, it is seen that if f is smooth, then every direction $\boldsymbol{d}$ satisfying $\langle \nabla f(\boldsymbol{x}), \boldsymbol{d} \rangle < 0$ is a descent direction at $\boldsymbol{x}$.

Since the Hessian matrix $\nabla^2 f(\boldsymbol{x})$ is positive definite, its inverse matrix $(\nabla^2 f(\boldsymbol{x}))^{-1}$ exists and is also positive definite, and for all $\nabla f(\boldsymbol{x}_k) \neq 0$,

$$\langle \nabla f(\boldsymbol{x}_k), \nabla f(\boldsymbol{x}_k)(\nabla^2 f(\boldsymbol{x}_k))^{-1} \rangle > 0$$

or $\langle \nabla f(\boldsymbol{x}_k), \boldsymbol{d}_k \rangle < 0$. By Definition 7.1.1, the Newton direction $\boldsymbol{d}_k$ is a descent direction.

The classical Newton's algorithm has a local convergence property as follows.

Let $f: \mathbb{R}^n \to \mathbb{R}$ be twice smooth with $\nabla f(\boldsymbol{x}^*) = 0$ and $\nabla^2 f(\boldsymbol{x}^*)$ be positive definite. If there are $r > 0$ and $M > 0$ such that the Hessian matrix satisfies the locally Lipschitz condition, i.e., for any $\boldsymbol{x}', \boldsymbol{x}'' \in B(\boldsymbol{x}^*, r)$,

$$\|\nabla^2 f(\boldsymbol{x}') - \nabla^2 f(\boldsymbol{x}'')\| \leq M |\boldsymbol{x}' - \boldsymbol{x}''|, \tag{7.1.3}$$

where $B(\boldsymbol{x}^*, r)$ is a ball with the center $\boldsymbol{x}^*$ and radius r, then the iterates $\boldsymbol{x}_k \in B(\boldsymbol{x}^*, r)$ $(k = 0, 1, \dots)$ converge to $\boldsymbol{x}^*$ quadratically, i.e., $|\boldsymbol{x}_{k+1} - \boldsymbol{x}^*| \le C|\boldsymbol{x}_k - \boldsymbol{x}^*|^2$, where $C = \frac{M\|I\|}{M^* - Mr}$ $(r < \frac{M^*}{M})$ and $M^* = \|\nabla^2 f(\boldsymbol{x}^*)\|$.

In fact, by (7.1.3) and $Mr < M^*$, it follows that for $\boldsymbol{x} \in B(\boldsymbol{x}^*, r)$,

$$\begin{aligned}\|\nabla^2 f(\boldsymbol{x})\| &= \|\nabla^2 f(\boldsymbol{x}^*) + \nabla^2 f(\boldsymbol{x}) - \nabla^2 f(\boldsymbol{x}^*)\| \\ &\ge \|\nabla^2 f(\boldsymbol{x}^*)\| - \|\nabla^2 f(\boldsymbol{x}) - \nabla^2 f(\boldsymbol{x}^*)\| \\ &\ge M^* - Mr.\end{aligned}$$

Since $\nabla^2 f(\boldsymbol{x})(\nabla^2 f(\boldsymbol{x}))^{-1} = I$, where I is the unit matrix and $(\nabla^2 f(\boldsymbol{x}))^{-1}$ is the inverse matrix of $\nabla^2 f(\boldsymbol{x})$, for $\boldsymbol{x} \in B(\boldsymbol{x}^*, r)$,

$$\|(\nabla^2 f(\boldsymbol{x}))^{-1}\| = \frac{\|I\|}{\|\nabla^2 f(\boldsymbol{x})\|} \le \frac{\|I\|}{M^* - Mr}. \tag{7.1.4}$$

Since $\nabla f(\boldsymbol{x})$ is a continuously differentiable function, the Lagrange form of its Taylor expansion is

$$\nabla f(\boldsymbol{x}_k) = \nabla f(\boldsymbol{x}^*) + (\boldsymbol{x}_k - \boldsymbol{x}^*)\nabla^2 f(\bar{\boldsymbol{x}}),$$

where $\bar{\boldsymbol{x}} \in (\boldsymbol{x}_k, \boldsymbol{x}^*)$ or $\bar{\boldsymbol{x}} \in (\boldsymbol{x}^*, \boldsymbol{x})$. From this and $\nabla f(\boldsymbol{x}^*)(\nabla^2 f(\boldsymbol{x}_k))^{-1} = 0$ (since $\nabla f(\boldsymbol{x}^*) = 0$), it follows that

$$\begin{aligned}\boldsymbol{x}_{k+1} - \boldsymbol{x}^* &= \left(\boldsymbol{x}_k - \nabla f(\boldsymbol{x}_k)(\nabla^2 f(\boldsymbol{x}_k))^{-1}\right) - \boldsymbol{x}^* + \nabla f(\boldsymbol{x}^*)(\nabla^2 f(\boldsymbol{x}_k))^{-1} \\ &= (\boldsymbol{x}_k - \boldsymbol{x}^*)\nabla^2 f(\boldsymbol{x}_k)(\nabla^2 f(\boldsymbol{x}_k))^{-1} - (\nabla f(\boldsymbol{x}_k) - \nabla f(\boldsymbol{x}^*))(\nabla^2 f(\boldsymbol{x}_k))^{-1} \\ &= (\boldsymbol{x}_k - \boldsymbol{x}^*)(\nabla^2 f(\boldsymbol{x}_k) - \nabla^2 f(\bar{\boldsymbol{x}}))(\nabla^2 f(\boldsymbol{x}_k))^{-1}.\end{aligned}$$

By (7.1.3) and (7.1.4), it follows that for $\boldsymbol{x}_k \in B(\boldsymbol{x}^*, r)$,

$$\begin{aligned}|\boldsymbol{x}_{k+1} - \boldsymbol{x}^*| &= |\boldsymbol{x}_k - \boldsymbol{x}^*|\,\|\nabla^2 f(\boldsymbol{x}_k) - \nabla^2 f(\bar{\boldsymbol{x}})\|\,\|(\nabla^2 f(\boldsymbol{x}_k))^{-1}\| \\ &\le C|\boldsymbol{x}_k - \boldsymbol{x}^*|\,|\boldsymbol{x}_k - \bar{\boldsymbol{x}}|,\end{aligned}$$

where $C = \frac{M\|I\|}{M^* - Mr}$. Note that $\bar{\boldsymbol{x}} \in (\boldsymbol{x}_k, \boldsymbol{x}^*)$ or $\bar{x} \in (\boldsymbol{x}^*, \boldsymbol{x}_k)$. Clearly, $|\boldsymbol{x}_k - \bar{\boldsymbol{x}}| \le |\boldsymbol{x}_k - \boldsymbol{x}^*|$. So

$$|\boldsymbol{x}_{k+1} - \boldsymbol{x}^*| \le C|\boldsymbol{x}_k - \boldsymbol{x}^*|^2,$$

i.e., the iterates $\{\boldsymbol{x}_k\}_{k=0,1,\dots} \in B(\boldsymbol{x}^*, r)$ converge to $\boldsymbol{x}^*$ quadratically.

The steepest descent method was introduced by Cauchy. Different to the classical Newton's method, steepest descent method chooses a descent direction such that f decreases most quickly and includes a line minimization step.

Steepest descent method

Let $f : \mathbb{R}^n \to \mathbb{R}$ be a smooth function. The steepest descent algorithm involves the following three steps:

Step 1. Start at an initial point $\boldsymbol{x}_0 \in \mathbb{R}^n$.
Step 2. Stop if $\nabla f(\boldsymbol{x}_k) = 0$, where $\boldsymbol{x}_k \in \mathbb{R}^n$. Otherwise, the descent direction is

$$\boldsymbol{d}_k = -\nabla f(\boldsymbol{x}_k) \quad (\boldsymbol{d}_k \in \mathbb{R}^n).$$

Step 3. The next iterate $\boldsymbol{x}_{k+1} = \boldsymbol{x}_k + \mu_k \boldsymbol{d}_k$, where the step size $\mu_k > 0$ minimizes the function $f(\boldsymbol{x}_k + \mu \boldsymbol{d}_k)$ with respect to μ on $[0, \infty)$. Return to Step 2, the process is repeated at $\boldsymbol{x}_{k+1}$.

The steepest descent algorithm possesses the following convergence property:

The steepest descent algorithm stops after a finite number of steps at a point $\boldsymbol{x}_k$ where $\nabla f(\boldsymbol{x}_k) = 0$. Otherwise, the algorithm generates an infinite sequence of points $\{\boldsymbol{x}_k\}_{k=0,1,\dots}$. According to Bolzano–Weierstrass theorem, the infinite sequence $\{\boldsymbol{x}_k\}_{k=0,1,\dots}$ must have at least one convergent subsequence. Without loss of generality, we still denote this convergent subsequence by $\{\boldsymbol{x}_k\}_{k=0,1,\dots}$. Let $\boldsymbol{x}^*$ be its limit point. Now we prove $\nabla f(\boldsymbol{x}^*) = 0$.

If $\nabla f(\boldsymbol{x}^*) \neq 0$, then $\boldsymbol{d}^* = -\nabla f(\boldsymbol{x}^*)$ is a descent direction, i.e., there is a $\mu^* > 0$ so that $f(\boldsymbol{x}^* + \mu \boldsymbol{d}^*) < f(\boldsymbol{x}^*)$ for any $\mu \in (0, \mu^*)$. Note that

$$\begin{aligned} \boldsymbol{d}_k &= -\nabla f(\boldsymbol{x}_k), \\ \boldsymbol{d}^* &= -\nabla f(\boldsymbol{x}^*) \neq 0. \end{aligned}$$

Since $f(\boldsymbol{x})$ is a smooth function,

$$\boldsymbol{d}_k \to \boldsymbol{d}^* \neq 0.$$

Combining this with $\boldsymbol{x}_k \to \boldsymbol{x}^*$, it follows from $\boldsymbol{x}_{k+1} = \boldsymbol{x}_k + \mu_k \boldsymbol{d}_k$ that $\mu_k \to 0$. This implies $\mu_k \in (0, \mu^*)$.

By the assumption that μ_k minimizes the function $f(\boldsymbol{x}_k + \mu \boldsymbol{d}_k)$ with respect to μ on $[0, \infty)$, since $f(\boldsymbol{x}_k)$ is independent of μ, it is clear that μ_k also minimizes the function $f(\boldsymbol{x}_k + \mu \boldsymbol{d}_k) - f(\boldsymbol{x}_k)$ with respect to μ on $[0, \infty)$. So, for any $\mu_k \in (0, \mu^*)$ and $\mu \in (0, \mu^*)$,

$$f(\boldsymbol{x}_k + \mu_k \boldsymbol{d}_k) - f(\boldsymbol{x}_k) \le f(\boldsymbol{x}_k + \mu \boldsymbol{d}_k) - f(\boldsymbol{x}_k). \tag{7.1.5}$$

Since $f(\boldsymbol{x})$ is smooth and $\boldsymbol{x}_k \to \boldsymbol{x}^*$, and $\mu_k \to 0$, the left-hand side of (7.1.5) is

$$f(\boldsymbol{x}_k + \mu_k \boldsymbol{d}_k) - f(\boldsymbol{x}_k) \to 0.$$

Since $f(\boldsymbol{x})$ is a smooth function and $\boldsymbol{d}_k = -\nabla f(\boldsymbol{x}_k)$, the right-hand side of (7.1.5) is

$$f(\boldsymbol{x}_k + \mu \boldsymbol{d}_k) - f(\boldsymbol{x}_k) = \langle \mu \boldsymbol{d}_k, \nabla f(\boldsymbol{x}_k) \rangle + o(\mu) = -\mu \langle \nabla f(\boldsymbol{x}_k), \nabla f(\boldsymbol{x}_k) \rangle + o(\mu),$$

where $\frac{o(\mu)}{|\mu|} \to 0$ as $\mu \to 0$. Let $k \to \infty$ in (7.1.5) and choose μ close enough to 0. Note that $\mu > 0$ and the assumption $\langle \nabla f(\boldsymbol{x}^*), \nabla f(\boldsymbol{x}^*) \rangle > 0$. Then

$$0 \le -\mu \langle \nabla f(\boldsymbol{x}^*), \nabla f(\boldsymbol{x}^*) \rangle < 0.$$

This is a contradiction. Thus $\nabla f(\boldsymbol{x}^*) = 0$.

Newton's method includes a minimization step and a step size. It is a development of the classical Newton's method and is a variant of the steepest descent method.

Newton's method

Let $f\colon \mathbb{R}^n \to \mathbb{R}$ be a twice smooth function and its Hessian matrix

$$\nabla^2 f(\boldsymbol{x}) = \begin{pmatrix} \frac{\partial^2 f(\boldsymbol{x})}{\partial x_1^2} & \frac{\partial^2 f(\boldsymbol{x})}{\partial x_1 \partial x_2} & \cdots & \frac{\partial^2 f(\boldsymbol{x})}{\partial x_1 \partial x_n} \\ \frac{\partial^2 f(\boldsymbol{x})}{\partial x_2 \partial x_1} & \frac{\partial^2 f(\boldsymbol{x})}{\partial x_2^2} & \cdots & \frac{\partial^2 f(\boldsymbol{x})}{\partial x_2 \partial x_n} \\ \vdots & \vdots & \ddots & \cdots \\ \frac{\partial^2 f(\boldsymbol{x})}{\partial x_n \partial x_1} & \frac{\partial^2 f(\boldsymbol{x})}{\partial x_n \partial x_2} & \cdots & \frac{\partial^2 f(\boldsymbol{x})}{\partial x_n^2} \end{pmatrix}$$

be positive definite, where $\boldsymbol{x} = (x_1, \dots, x_n)$. Newton's algorithm involves the following three steps:

Step 1. Start from an initial point $\boldsymbol{x}_0 \in \mathbb{R}^n$.

Step 2. Stop if $\nabla f(\boldsymbol{x}_k) = 0$, where $\boldsymbol{x}_k \in \mathbb{R}^n$. Otherwise, the Newton direction is

$$\boldsymbol{d}_k = -\nabla f(\boldsymbol{x}_k)(\nabla^2 f(\boldsymbol{x}_k))^{-1} \quad (\boldsymbol{d}_k \in \mathbb{R}^n),$$

where $(\nabla^2 f(\boldsymbol{x}))^{-1}$ is the inverse matrix of the Hessian matrix.

Step 3. The next iterate $\boldsymbol{x}_{k+1} = \boldsymbol{x}_k + \mu_k \boldsymbol{d}_k$, where the step size $\mu_k > 0$ minimizes the function $f(\boldsymbol{x}_k + \mu \boldsymbol{d}_k)$ with respect to μ on $[0, \infty)$. Return to Step 2, the process is repeated at $\boldsymbol{x}_{k+1}$.

Newton's algorithm has the same convergence as the steepest descent algorithm.

It stops after a finite number of steps at a point $\boldsymbol{x}_k$, where $\nabla f(\boldsymbol{x}_k) = 0$. Otherwise, the algorithm generates an infinite sequence of points $\{\boldsymbol{x}_k\}_{k=0,1,\dots}$. By the Bolzano–Weierstrass theorem, the infinite sequence $\{\boldsymbol{x}_k\}_{k=0,1,\dots}$ must have at least one convergent subsequence. Without loss of generality, we still denote this convergent subsequence by $\{\boldsymbol{x}_k\}_{k=0,1,\dots}$. Let $\boldsymbol{x}^*$ be its limit point. Now we prove $\nabla f(\boldsymbol{x}^*) = 0$.

If $\nabla f(\boldsymbol{x}^*) \neq 0$, since the Hessian matrix is positive definite, it is invertible and its inverse matrix is also positive definite. By the definition of positive definite, the inverse matrix satisfies that

$$\langle \nabla f(\boldsymbol{x}^*), \nabla f(\boldsymbol{x}^*)(\nabla^2 f(\boldsymbol{x}^*))^{-1} \rangle > 0, \tag{7.1.6}$$

or $\langle \nabla f(\boldsymbol{x}^*), \boldsymbol{d}_k \rangle < 0$. By Definition 7.1.1, $\boldsymbol{d}^* = -\nabla f(\boldsymbol{x}^*)(\nabla^2 f(\boldsymbol{x}^*))^{-1}$ is also a descent direction. So, there is a $\mu^* > 0$ so that $f(\boldsymbol{x}^* + \mu \boldsymbol{d}^*) < f(\boldsymbol{x}^*)$ for any $\mu \in (0, \mu^*)$. Note that

$$\begin{aligned} \boldsymbol{d}_k &= -\nabla f(\boldsymbol{x}_k)(\nabla^2 f(\boldsymbol{x}_k))^{-1}, \\ \boldsymbol{d}^* &= -\nabla f(\boldsymbol{x}^*)(\nabla^2 f(\boldsymbol{x}^*))^{-1} \neq 0. \end{aligned}$$

Since $f\colon \mathbb{R}^n \to \mathbb{R}$ is a twice smooth function and $\boldsymbol{x}_k \to \boldsymbol{x}^*$,

$$\boldsymbol{d}_k \to \boldsymbol{d}^* \neq 0.$$

Combining this with $\boldsymbol{x}_k \to \boldsymbol{x}^*$, it follows from $\boldsymbol{x}_{k+1} = \boldsymbol{x}_k + \mu_k \boldsymbol{d}_k$ that $\mu_k \to 0$. This implies $\mu_k \in (0, \mu^*)$.

By the assumption that μ_k minimizes the function $f(\boldsymbol{x}_k + \mu \boldsymbol{d}_k)$ with respect to μ on $[0, \infty)$, but $f(\boldsymbol{x}_k)$ is independent of μ, so μ_k also minimizes the function $f(\boldsymbol{x}_k + \mu \boldsymbol{d}_k) - f(\boldsymbol{x}_k)$ with respect to μ on $[0, \infty)$. This implies that for any $\mu_k \in (0, \mu^*)$ and $\mu \in (0, \mu^*)$,

$$f(\boldsymbol{x}_k + \mu_k \boldsymbol{d}_k) - f(\boldsymbol{x}_k) \le f(\boldsymbol{x}_k + \mu \boldsymbol{d}_k) - f(\boldsymbol{x}_k). \tag{7.1.7}$$

Since $f(x)$ is smooth and $\boldsymbol{x}_k \to \boldsymbol{x}^*$, and $\mu_k \to 0$, the left-hand side of (7.1.7) is

$$f(\boldsymbol{x}_k + \mu_k \boldsymbol{d}_k) - f(\boldsymbol{x}_k) \to 0.$$

Since $f(\boldsymbol{x})$ is smooth and $\boldsymbol{d}_k = -\nabla f(\boldsymbol{x}_k)(\nabla^2 f(\boldsymbol{x}_k))^{-1}$, the right-hand side of (7.1.7) is

$$f(\boldsymbol{x}_k + \mu \boldsymbol{d}_k) - f(\boldsymbol{x}_k) = \langle \nabla f(\boldsymbol{x}_k), \mu \boldsymbol{d}_k \rangle + o(\mu) = -\mu \langle \nabla f(\boldsymbol{x}_k), \nabla f(\boldsymbol{x}_k)(\nabla^2 f(\boldsymbol{x}_k))^{-1} \rangle + o(\mu),$$

where $\frac{o(\mu)}{|\mu|} \to 0$ as $\mu \to 0$. Let $k \to \infty$ in (7.1.7) and choose μ close enough to 0. Note that $\mu > 0$ and (7.1.6). It is clear that

$$0 \le -\mu \langle \nabla f(\boldsymbol{x}^*), \nabla f(\boldsymbol{x}^*)(\nabla^2 f(\boldsymbol{x}^*))^{-1} \rangle < 0.$$

This is a contradiction. Thus $\nabla f(\boldsymbol{x}^*) = 0$.

Denote $G(\boldsymbol{x}) = \nabla f(\boldsymbol{x})$ and $G_k(\boldsymbol{x}) = \frac{\partial f(x)}{\partial x_k}$. The Newton method is viewed as systems of n linear/nonlinear equations in n unknown variables

$$G(\boldsymbol{x}) = \begin{pmatrix} G_1(\boldsymbol{x}) \\ \vdots \\ G_n(\boldsymbol{x}) \end{pmatrix} = \begin{pmatrix} 0 \\ \vdots \\ 0 \end{pmatrix}$$

with the Jacobian matrix

$$\nabla G(\boldsymbol{x}) = \begin{pmatrix} \frac{\partial G_1(\boldsymbol{x})}{\partial x_1} & \cdots & \frac{\partial G_1(\boldsymbol{x})}{\partial x_n}, \\ \vdots & \ddots & \vdots \\ \frac{\partial G_n(\boldsymbol{x})}{\partial x_1} & \cdots & \frac{\partial G_n(\boldsymbol{x})}{\partial x_n} \end{pmatrix}$$

and Newton's algorithm involves a line minimization. Inspired by it, we have the following method.

Newton–Raphson method

Assume that $\nabla G(\boldsymbol{x})$ is invertible. The algorithm involves three steps:

Step 1. Start from an initial point $\boldsymbol{x}_0 \in \mathbb{R}^n$.

Step 2. Stop if $G(\boldsymbol{x}_k) = 0$, where $\boldsymbol{x}_k \in \mathbb{R}^n$. Otherwise, the *Newton–Raphson direction*

$$\boldsymbol{d}_k = -G(\boldsymbol{x}_k)(\nabla G(\boldsymbol{x}_k))^{-1} \quad (\boldsymbol{d}_k \in \mathbb{R}^n),$$

where $(\nabla G(\boldsymbol{x}))^{-1}$ is the inverse matrix of the Jacobian matrix $\nabla G(\boldsymbol{x})$.

Step 3. The next iterate $\boldsymbol{x}_{k+1} = \boldsymbol{x}_k + \boldsymbol{d}_k$. Return to Step 2, the process is repeated at $\boldsymbol{x}_{k+1}$.

The Newton–Raphson method has the same local convergence property as the classical Newton method

Let $G\colon \mathbb{R}^n \to \mathbb{R}^n$ be a twice smooth function with $G(\boldsymbol{x}^*) = 0$ and $\nabla G(\boldsymbol{x}^*)$ invertible. If there are $r > 0$ and $M \geq 0$ such that for any $\boldsymbol{x}', \boldsymbol{x}'' \in B(\boldsymbol{x}^*, r)$,

$$\|\nabla G(\boldsymbol{x}') - \nabla G(\boldsymbol{x}'')\| \leq M\,|\boldsymbol{x}' - \boldsymbol{x}''|,$$

then the iterates $\{\boldsymbol{x}_k\}_{k=0,1,\dots} \in B(\boldsymbol{x}^*, r)$ converge to $\boldsymbol{x}^*$ quadratically, i.e., $|\boldsymbol{x}_{k+1} - \boldsymbol{x}^*| \leq C|\boldsymbol{x}_k - \boldsymbol{x}^*|^2$, where $C = \frac{M\|I\|}{M^* - Mr}$ $(r < \frac{M^*}{M})$ and $M^* = \|\nabla G(\boldsymbol{x}^*)\|$.

In fact, similar to the argument of (7.1.4), for $\boldsymbol{x} \in B(\boldsymbol{x}^*, r)$,

$$\|(\nabla G(\boldsymbol{x}))^{-1}\| = \frac{\|I\|}{\|\nabla G(\boldsymbol{x})\|} \leq \frac{\|I\|}{M^* - Mr}.$$

Since $G(\boldsymbol{x})$ is continuously differentiable, the Lagrange form of the Taylor expansion shows that for $\boldsymbol{x}_k \in B(\boldsymbol{x}^*, r)$,

$$G(\boldsymbol{x}_k) = G(\boldsymbol{x}^*) + (\boldsymbol{x}_k - \boldsymbol{x}^*)\nabla G(\bar{\boldsymbol{x}}),$$

where $\bar{\boldsymbol{x}} \in (\boldsymbol{x}_k, \boldsymbol{x}^*)$ or $\bar{\boldsymbol{x}} \in (\boldsymbol{x}^*, \boldsymbol{x}_k)$. From this and $G(\boldsymbol{x}^*)(\nabla G(\boldsymbol{x}_k))^{-1} = 0$ (since $G(\boldsymbol{x}^*) = 0$), it follows that

$$\begin{aligned}\boldsymbol{x}_{k+1} - \boldsymbol{x}^* &= \boldsymbol{x}_k + \boldsymbol{d}_k - \boldsymbol{x}^* + G(\boldsymbol{x}^*)(\nabla G(\boldsymbol{x}_k))^{-1}\\ &= \boldsymbol{x}_k - G(\boldsymbol{x}_k)(\nabla G(\boldsymbol{x}_k))^{-1} - \boldsymbol{x}^* + G(\boldsymbol{x}^*)(\nabla G(\boldsymbol{x}_k))^{-1}\\ &= (\boldsymbol{x}_k - \boldsymbol{x}^*)(\nabla G(\boldsymbol{x}_k) - \nabla G(\bar{\boldsymbol{x}}))(\nabla G(\boldsymbol{x}_k))^{-1},\end{aligned}$$

and so

$$\begin{aligned}|\boldsymbol{x}_{k+1} - \boldsymbol{x}^*| &= |\boldsymbol{x}_k - \boldsymbol{x}^*|\,\|\nabla G(\boldsymbol{x}_k) - \nabla G(\bar{\boldsymbol{x}})\|\,\|(\nabla G(\boldsymbol{x}_k))^{-1}\|\\ &\leq C|\boldsymbol{x}_k - \boldsymbol{x}^*|^2,\end{aligned}$$

where $C = \frac{M\|I\|}{M^* - Mr}$, i.e., the iterates $\{\boldsymbol{x}_k\}_{k=0,1,\dots} \in B(\boldsymbol{x}^*, r)$ converge to $\boldsymbol{x}^*$ quadratically.

7.2 The variational method

The variational method is a useful method for solving the optimization problem of functionals, i.e., for finding a function f such that the functional $\upsilon(f)$ with boundary conditions attains the minimal value

$$\min \upsilon(f) \quad \text{subject to}$$

$$\begin{cases} \upsilon(f) = \int_a^b F(x, f(x), f'(x))\,\mathrm{d}x,\\ f(a) = y_0, \quad f(b) = y_1 \quad \text{(boundary conditions)}, \end{cases} \tag{7.2.1}$$

where F is a second-order differentiable function and f is a second-order continuously differentiable function. For the optimization problem (7.2.1), if the functional $\upsilon(f)$ attains the minimal value, then f must satisfy the *Euler equation*

$$F_f - \frac{\mathrm{d}}{\mathrm{d}x} F_{f'} = 0.$$

In fact, for a small parameter α, let

$$I(\alpha) = \int_a^b F\left(x, f(x) + \alpha\eta(x), f'(x) + \alpha\eta'(x)\right) \mathrm{d}x,$$

where η is any differentiable function and $\eta(a) = \eta(b) = 0$. Clearly, $I(0) = \upsilon(f)$. Differentiating both sides,

$$I'(\alpha) = \int_a^b \left(F_{f+\alpha\eta}\eta(x) + F_{f'+\alpha\eta'}\eta'(x)\right) \mathrm{d}x.$$

Let $\alpha = 0$. Then

$$I'(0) = \int_a^b \left(F_f\eta(x) + F_{f'}\eta'(x)\right) \mathrm{d}x = \int_a^b F_f\eta(x)\,\mathrm{d}x + \int_a^b F_{f'}\eta'(x)\,\mathrm{d}x.$$

By integration by parts, the second integral on the right-hand side is

$$\int_a^b F_{f'}\eta'(x)\,\mathrm{d}x = F_{f'}\eta(x)\big|_a^b - \int_a^b \left(\frac{\mathrm{d}}{\mathrm{d}x} F_{f'}\right)\eta(x)\,\mathrm{d}x = -\int_a^b \left(\frac{\mathrm{d}}{\mathrm{d}x} F_{f'}\right)\eta(x)\,\mathrm{d}x.$$

Therefore,

$$I'(0) = \int_a^b F_f\eta(x)\,\mathrm{d}x - \int_a^b \left(\frac{\mathrm{d}}{\mathrm{d}x} F_{f'}\right)\eta(x)\,\mathrm{d}x = \int_a^b \left(F_f - \frac{\mathrm{d}}{\mathrm{d}x} F_{f'}\right)\eta(x)\,\mathrm{d}x.$$

If $\upsilon(f)$ attains the minimal value, then $I'(0) = 0$, i.e.,

$$\int_a^b \eta(x)\left(F_f - \frac{\mathrm{d}}{\mathrm{d}x} F_{f'}\right) \mathrm{d}x = 0.$$

Since $\eta(x)$ is arbitrary, this implies that f must satisfy $F_f - \frac{\mathrm{d}}{\mathrm{d}x} F_{f'} = 0$.

The generalization of the optimization problem (7.2.1) is as follows:

$$\min \upsilon(f) \quad \text{subject to}$$

$$\begin{cases} \upsilon(f) = \int_a^b F\,(x, f_1(x), \ldots, f_n(x), f_1'(x), \ldots, f_n'(x))\,\mathrm{d}x, \\ f_i(a) = y_{i0}, \quad f_i(b) = y_{i1} \quad (i = 1, \ldots, n) \quad \text{(boundary conditions)}, \end{cases}$$

where F is a second-order differentiable function and f_i $(i = 1, \dots, n)$ are second-order continuously differentiable functions. Similarly, for the generalized optimization problem, if the functional $\upsilon(f)$ attains the minimal value, then f_i must satisfy the Euler equation

$$F_{f_i} - \frac{\mathrm{d}}{\mathrm{d}x} F_{f_i'} = 0 \quad (i = 1, \dots, n).$$

The conditional optimization problem of functionals is to find f, g such that the functional $\upsilon(f, g)$ with an additional condition and boundary conditions attains the minimal value, i.e., to solve the following conditional optimization problem:

$$\min \upsilon(f, g) \quad \text{subject to}$$
$$\begin{cases} \upsilon(f, g) = \int_a^b F(x, f, f', g, g')\,\mathrm{d}x, \\ G(x, f, g) = 0, \\ f(a) = y_0,\ f(b) = y_1, \quad g(a) = z_0,\ g(b) = z_1, \end{cases}$$

where $G(x, f, g) = 0$ is the additional condition and $f(a) = y_0$, $f(b) = y_1$, $g(a) = z_0$ and $g(b) = z_1$ are boundary conditions. If the functional $\upsilon(f, g)$ attains the minimal value, then f and g must satisfy

$$\frac{\mathrm{d}}{\mathrm{d}x} F^*_{f'} - F^*_f = 0,$$
$$\frac{\mathrm{d}}{\mathrm{d}x} F^*_{g'} - F^*_g = 0,$$

where $F^* = F + \lambda(x) G$ and

$$\lambda(x) = \frac{\frac{\mathrm{d}}{\mathrm{d}x} F_{f'} - F_f}{G_f} = \frac{\frac{\mathrm{d}}{\mathrm{d}x} F_{g'} - F_g}{G_g}.$$

In fact, the additional condition $G(x, f, g) = 0$ determines that g is a function of x and f, say, $g = \varphi(x, f)$. So

$$F(x, f, f', g, g') = F(x, f, f', \varphi, \varphi_x + \varphi_f f') =: \tilde{F}(x, f, f'),$$
$$\upsilon(f, g) = \upsilon(f, \varphi(x, f)) =: \tilde{\upsilon}(f),$$
$$\varphi(a, y_0) = z_0, \quad \varphi(b, y_1) = z_1.$$

Then the conditional optimization problem of functionals is reduced to the optimization problem (7.2.1), i.e.,

$$\min \tilde{\upsilon}(f) \quad \text{subject to}$$
$$\begin{cases} \tilde{\upsilon}(f) = \int_a^b \tilde{F}(x, f, f')\,\mathrm{d}x, \\ f(a) = y_0, \quad f(b) = y_1. \end{cases}$$

If the functional $\upsilon(f, g)$ attains the minimal value, i.e., $\tilde{\upsilon}(f)$ attains the minimal value, then the Euler equation

$$\frac{\mathrm{d}}{\mathrm{d}x} \tilde{F}_{f'} - \tilde{F}_f = 0$$

holds. Note that

$$\begin{aligned}\tilde{F}_f &= F_f + F_g\varphi_f + F_{g'}(\varphi_{xf} + \varphi_{ff}f'),\\ \tilde{F}_{f'} &= F_{f'} + F_{g'}\varphi_f,\\ \frac{\mathrm{d}}{\mathrm{d}x}\tilde{F}_{f'} &= \frac{\mathrm{d}}{\mathrm{d}x}F_{f'} + \varphi_f\frac{\mathrm{d}}{\mathrm{d}x}F_{g'} + F_{g'}(\varphi_{fx} + \varphi_{ff}f').\end{aligned}$$

The Euler equation becomes that

$$\frac{\mathrm{d}}{\mathrm{d}x}F_{f'} - F_f + \varphi_f\left(\frac{\mathrm{d}}{\mathrm{d}x}F_{g'} - F_g\right) = 0.$$

On the other hand, differentiating the equation $G(x, f, g) = 0$ with respect to f,

$$G_f + G_g\varphi_f = 0,$$

or $\varphi_f = -G_f/G_g$. So the Euler equation further becomes that

$$\frac{\mathrm{d}}{\mathrm{d}x}F_{f'} - F_f - \frac{G_f}{G_g}\left(\frac{\mathrm{d}}{\mathrm{d}x}F_{g'} - F_g\right) = 0,$$

i.e.,

$$\frac{\frac{\mathrm{d}}{\mathrm{d}x}F_{f'} - F_f}{G_f} = \frac{\frac{\mathrm{d}}{\mathrm{d}x}F_{g'} - F_g}{G_g} = \lambda(x)$$

which is equivalent to

$$\begin{aligned}&\frac{\mathrm{d}}{\mathrm{d}x}F_{f'} - (F_f + \lambda(x)G_f) = 0,\\ &\frac{\mathrm{d}}{\mathrm{d}x}F_{g'} - (F_g + \lambda(x)G_g) = 0.\end{aligned} \tag{7.2.2}$$

Let $F^* = F + \lambda(x)G$. Note that

$$\begin{aligned}\frac{\mathrm{d}}{\mathrm{d}x}F^*_{f'} &= \frac{\mathrm{d}}{\mathrm{d}x}F_{f'}, \quad F^*_f = F_f + \lambda(x)G_f,\\ \frac{\mathrm{d}}{\mathrm{d}x}F^*_{g'} &= \frac{\mathrm{d}}{\mathrm{d}x}F_{g'} \quad F^*_g = F_g + \lambda(x)G_g.\end{aligned}$$

Then, the conditions (7.2.2) are further equivalent to the following conditions:

$$\begin{aligned}\frac{\mathrm{d}}{\mathrm{d}x}F^*_{f'} - F^*_f &= 0,\\ \frac{\mathrm{d}}{\mathrm{d}x}F^*_{g'} - F^*_g &= 0.\end{aligned}$$

The variational method is frequently used in data compression by combining with dyadic wavelet transform. Denote by $\theta(t)$ the Gauss function, i.e.,

$$\theta(t) = \frac{1}{\sqrt{2\pi\alpha}}\mathrm{e}^{-\frac{t^2}{2\alpha}} \quad (\alpha > 0).$$

It is clear that

$$\int_{\mathbb{R}} \theta(t)\,\mathrm{d}t = \frac{1}{\sqrt{\pi}} \int_{-\infty}^{\infty} \frac{1}{\sqrt{2\alpha}} \mathrm{e}^{-\frac{t^2}{2\alpha}} = 1.$$

Let $\theta_\lambda(t) = \frac{1}{\lambda}\theta(\frac{t}{\lambda})$ $(\lambda > 0)$. Using change of variables $t = \lambda u$ gives

$$\int_{\mathbb{R}} \theta_\lambda(t)\,\mathrm{d}t = \int_{-\infty}^{\infty} \frac{1}{\lambda}\theta\left(\frac{t}{\lambda}\right)\mathrm{d}t = \int_{-\infty}^{\infty} \theta(u)\,\mathrm{d}u = 1.$$

Let f be a one-dimensional signal. The convolution of f and θ_λ is a mean with weight θ_λ,

$$(f * \theta_\lambda)(t) = \int_{\mathbb{R}} f(\tau)\theta_\lambda(t-\tau)\,\mathrm{d}\tau,$$

and is an infinitely differentiable function. Note that $\theta_{2^{-m}}(t) = 2^m\theta(2^m t)$. Then

$$\frac{\mathrm{d}}{\mathrm{d}t}(f * \theta_{2^{-m}})(t) = 2^m \frac{\mathrm{d}}{\mathrm{d}t}\left(\int_{\mathbb{R}} f(\tau)\theta(2^m(t-\tau))\,\mathrm{d}\tau\right) = 2^{2m}\int_{\mathbb{R}} f(\tau)\theta'(2^m(t-\tau))\,\mathrm{d}\tau.$$

Let $\psi(t) = -\theta'(t)$. Its Fourier transform satisfies $A \le \sum_{m\in\mathbb{Z}}|\hat{\psi}(\frac{\omega}{2^m})|^2 \le B$ (stability condition), where A and B are positive constants. Since $\theta(t)$ is an even function, ψ is an odd function and

$$\frac{\mathrm{d}}{\mathrm{d}t}(f * \theta_{2^{-m}})(t) = 2^m(W_\psi^m f)(t) \quad (m \in \mathbb{Z}),$$

where $W_\psi^m f$ is the dyadic wavelet transform.

Assume that $|(W_\psi^m f)(t)|$ attains the maximal values on points $\{t_n^m\}_{n,m\in\mathbb{Z}}$. Denote by $\{X_n^m\}_{m,n\in\mathbb{Z}}$ these maximal values on points $\{t_n^m\}_{n,m\in\mathbb{Z}}$. Once $W_\psi^m f$ $(m \in \mathbb{Z})$ are found by data $\{t_n^m, X_n^m\}_{m,n\in\mathbb{Z}}$, the signal f is obtained immediately using the inversion formula of dyadic wavelet transform. Thus, it remains to reconstruct $W_\psi^m f$ by data $\{t_n^m, X_n^m\}_{n,m\in\mathbb{Z}}$. This is reduced to finding $h(t)$ satisfying the following:

$$\begin{gathered}(W_\psi^m h)(t_n^m) = X_n^m \quad (m, n \in \mathbb{Z}),\\ \int_{\mathbb{R}} 2^m h(t)\psi\left(2^m(t - t_n^m)\right)\mathrm{d}t = X_n^m.\end{gathered} \tag{7.2.3}$$

Since the values of $|W_n^m h|$ at t_n^m are known, the first equation of (7.2.3) can be replaced approximately by finding h such that the integral

$$\|W_\psi^m h\|_2^2 = \int_{\mathbb{R}} |(W_\psi^m h)(t)|^2\,\mathrm{d}t \quad (m \in \mathbb{Z})$$

is as small as possible, and the integral

$$\left\|\frac{\mathrm{d}}{\mathrm{d}t}W_\psi^m h\right\|_2^2 = \int_{\mathbb{R}} \left|\frac{\mathrm{d}}{\mathrm{d}t}(W_\psi^m h)(t)\right|^2\mathrm{d}t$$

is also required to be as small as possible to avoid that new extreme points occur. Both require that

$$\|h\|_*^2 = \sum_n \left(\left\| W_\psi^m h \right\|_2^2 + 2^{-2m} \left\| \frac{\mathrm{d}}{\mathrm{d}t} W_\psi^m h \right\|_2^2 \right)$$

attains the minimal value. It is easy to prove that $\|h\|_* < \infty$ for $f \in L^2(\mathbb{R})$.

Let the space K consist of all differentiable function sequences $F = \{F_m\}_{m \in \mathbb{Z}}$ satisfying

$$\sum_m \left(\|F_m\|_2^2 + 2^{-2m} \|F_m'\|_2^2 \right) < \infty.$$

For $\{F_m\}_{m\in\mathbb{Z}} \in K$, define a boundary linear operator W_ψ^{-1} from K to $L^2(\mathbb{R})$ by

$$W_\psi^{-1}(F) = \frac{1}{2\pi} \sum_m F_m * H_m^*,$$

where $H_m^* = 2^m \psi^*(2^m t)$ and ψ^* is the dyadic dual wavelet of ψ. Let Γ be the closed subset of $F = \{F_m\}_{m\in\mathbb{Z}} \in K$ satisfying the condition $F_m(t_n^m) = X_n^m$ $(m, n \in \mathbb{Z})$, and let V denote the space of dyadic wavelet transform $W_\psi f$, where $f \in L^2(\mathbb{R})$. Then it is desired to find $h \in V \bigcap \Gamma$ such that $\|h\|_K$ attains the minimal value.

Define a linear bounded operator from K to V as $P_V = W_\psi \circ W_\psi^{-1}$. When $F \in V$, $P_V F = F$. Since ψ is a real-valued odd function, it can be proved that P_V is an orthogonal project operator from K to V.

Now we find the project of $F \in K$ to the closed subset Γ.

Let $P_\Gamma(F) = h$. Then $\|F - h^*\|_K^2$ $(h^* \in \Gamma)$ attains the minimal value at $h^* = h = \{h_m^*\}_{m\in\mathbb{Z}}$. Let

$$\varepsilon_m(t) = F_m(t) - h_m^*(t). \tag{7.2.4}$$

Then

$$\|F - h^*\|_K^2 = \sum_m \left(\|\varepsilon_m\|_2^2 + 2^{-2m} \|\varepsilon_m'\|_2^2 \right).$$

When $h_m^* = h_m$, this formula attains the minimal value. Note that

$$\begin{aligned} \varepsilon_m(t_n^m) &= F_m(t_n^m) - X_n^m, \\ \varepsilon_m(t_{n+1}^m) &= F_m(t_{n+1}^m) - X_{n+1}^m. \end{aligned} \tag{7.2.5}$$

When $h_m^* = h_m$, the integral

$$\int_{t_n^m}^{t_{n+1}^m} \left(|\varepsilon_m(t)|^2 + 2^{-2m} |\varepsilon_m'(t)|^2 \right) \mathrm{d}t \tag{7.2.6}$$

attains the minimal value. Let $y = \varepsilon_m(t)$. Take $H = y^2 + 2^{-2m}(y')^2$. Then

$$\begin{aligned} H_y &= 2\varepsilon_m(t), \\ \frac{\mathrm{d}}{\mathrm{d}t} H_{y'} &= \frac{\mathrm{d}}{\mathrm{d}t}(2^{-2m+1} y') = 2^{-2m+1} \varepsilon_m''(t), \end{aligned}$$

and

$$\int_{t_n^m}^{t_{n+1}^m} \left(|\varepsilon_n(t)|^2 + 2^{-2m}|\varepsilon_m'(t)|^2\right) \mathrm{d}t = \int_{t_n^m}^{t_{n+1}^m} H(y(t), y'(t))\,\mathrm{d}t.$$

The variational method says that if y satisfies the Euler equation $H_y - \frac{\mathrm{d}}{\mathrm{d}t}H_{y'} = 0$, then the integral

$$\int_{t_n^m}^{t_{n+1}^m} H(y(t), y'(t))\,\mathrm{d}t$$

attains the minimal value. Thus, if $\varepsilon_m(t)$ satisfies the Euler equation

$$\varepsilon_m(t) - 2^{-2m}\varepsilon_m''(t) = 0 \quad (t_n^m \le t \le t_{n+1}^m) \tag{7.2.7}$$

and (7.2.5), then the integral (7.2.6) attains the minimal value. The solution of (7.2.7) is

$$\varepsilon_m(t) = \alpha_{m,n}\mathrm{e}^{2^m t} + \beta_{m,n}\mathrm{e}^{-2^m t} \quad (t_n^m \le t \le t_{n+1}^m),$$

where $\alpha_{m,n}$ and $\beta_{m,n}$ are determined by (7.2.5). This implies from $P_\Gamma(F) = h$ and (7.2.4) that

$$P_\Gamma(\{F_m\}) = h_m(t) = F_m(t) - (\alpha_{m,n}e^{2^m t} + \beta_{m,n}e^{-2^m t}) \quad (t_n^m \le t \le t_{n+1}^m,\ n \in \mathbb{Z}).$$

Finally, denote $P = P_V \circ P_\Gamma$. Let F be a sequence of null functions. Then $W_\psi^m f$ $(m \in \mathbb{Z})$ are reconstructed by data $\{\, t_n^m, X_n^m \,\}_{n,m\in\mathbb{Z}}$ as follows:

$$\{(W_\psi^m f)(t)\} \approx \lim_{v\to\infty} P(\psi) \circ P(\psi) \circ \cdots \circ P(\psi), \quad \text{where } v \text{ is the number of } P(\psi).$$

Finally, the signal f is reconstructed immediately by using the inversion formula of dyadic wavelet transform.

7.3 The simplex method

The simplex method is applied for solving linear optimization models. Linear optimization models may have an objective function that is either to be maximized or to be minimized; they may have variables that are either nonnegative or free; and they may have constraints that are either equalities or inequalities.

7.3.1 Pivot operations

Pivot operations are used for solving systems of linear equations or for solving linear optimization models. Here consider a system of m linear equations with n unknown variables

$$\begin{cases} a_{11}x_1 + a_{12}x_2 + \cdots + a_{1n}x_n = b_1, \\ a_{21}x_1 + a_{22}x_2 + \cdots + a_{2n}x_n = b_2, \\ \vdots \\ a_{m1}x_1 + a_{m2}x_2 + \cdots + a_{mn}x_n = b_m. \end{cases}$$

Any *entry* (i, j) consists of the row index i and the column index j. The chosen entry (i, j) in order to obtain a new system of linear equations from the current system is called a *pivot entry*. The resulting calculation of the new system from the current system is called a *pivot operation*. An indexing function ϕ of the row indices into the column indices records that for each row k ($k = 1, \dots, m$), the $\phi(k)$-th column has coefficient 1 in the entry $(k, \phi(k))$ and coefficient 0 in the entries $(i, \phi(k))$ ($i = 1, \dots, m$, $i \neq k$).

Consider the k-th equation of the current system

$$a_{k1}x_1 + a_{k2}x_2 + \cdots + a_{kn}x_n = b_k.$$

There are three cases.

Case 1. $a_{kj} = 0$ for all $j = 1, \dots, n$ and $b_k = 0$. Then any values $x_1, \dots, x_n$ satisfy the k-th equation. The k-th equation is vacuous. In this case, define the indexing function as $\phi(k) = 0$.

Case 2. $a_{kj} = 0$ for all $j = 1, \dots, n$ but $b_k \neq 0$. In this case, the k-th equation has no solution.

Case 3. There are several nonzero coefficients a_{kj}. The gradient rule advises to choose the column index l so that the absolute value of a_{kl} ($a_{kl} \neq 0$) is as large as possible. In this case, the (k, l) entry is chosen as a pivot entry (i.e., $\phi(k) = l$), and then a pivot operation is performed by Gauss–Jordan elimination. This produces a new linear system from the current system. The process is as follows.

Divide the k-th equation by a_{kl} to produce the new k-th equation, where the coefficient of x_l is 1,

$$a^{(1)}_{k1}x_1 + \cdots + a^{(1)}_{k,l-1}x_{l-1} + x_l + a^{(1)}_{k,l+1}x_{l+1} + \cdots + a^{(1)}_{kn}x_n = b^{(1)}_k, \quad \phi(k) = l,$$

where

$$a^{(1)}_{kj} = \frac{a_{kj}}{a_{kl}} \quad (j = 1, \dots, n),$$
$$b^{(1)}_k = \frac{b_k}{a_{kl}}.$$

Multiply the new k-th equation by a_{il} ($i = 1, \dots, m$; $i \neq k$) so that the coefficient of x_l becomes a_{il}, and then subtract this equation from the i-th equation of the current system. This produces the new i-th equation, where the coefficient of x_l is 0,

$$a^{(1)}_{i1}x_1 + \cdots + a^{(1)}_{i,l-1}x_{l-1} + 0x_l + a^{(1)}_{k,l+1}x_{l+1} + \cdots + a^{(1)}_{in}x_n = b^{(1)}_i,$$

where for $i = 1, \ldots, m,\ i \neq k$,

$$a_{ij}^{(1)} = a_{ij} - \frac{a_{kj}}{a_{kl}} a_{il} \quad (j = 1, \ldots, n),$$
$$b_i^{(1)} = b_i - \frac{b_k}{a_{kl}} a_{il}.$$

Therefore, the resulting new system from the current system by a pivot operation is

$$\begin{cases} a_{11}^{(1)} x_1 + \cdots + a_{1,l-1}^{(1)} x_{l-1} + 0 + a_{1,l+1}^{(1)} x_{l+1} + \cdots + a_{1n}^{(1)} x_n = b_1^{(1)}, \\ \vdots \\ a_{k1}^{(1)} x_1 + \cdots + a_{k,l-1}^{(1)} x_{l-1} + x_l + a_{k,l+1}^{(1)} x_{l+1} + \cdots + a_{kn}^{(1)} x_n = b_k^{(1)}, \qquad \phi(k) = l, \\ \vdots \\ a_{m1}^{(1)} x_1 + \cdots + a_{m,l-1}^{(1)} x_{l-1} + 0 + a_{m,l+1}^{(1)} x_{l+1} + \cdots + a_{mn}^{(1)} x_n = b_m^{(1)}, \end{cases}$$

where

$$a_{kj}^{(1)} = \frac{a_{kj}}{a_{kl}} \quad (j = 1, \ldots, n),$$
$$b_k^{(1)} = \frac{b_k}{a_{kl}},$$

and for $i = 1, \ldots, m,\ i \neq l$,

$$a_{ij}^{(1)} = a_{ij} - \frac{a_{kj}}{a_{kl}} a_{il} \quad (j = 1, \ldots, n),$$
$$b_i^{(1)} = b_i - \frac{b_k}{a_{kl}} a_{il},$$

which are called *pivoting formulas*. These formulas are extremely important for solving linear system and linear optimization models. It can be verified that the new system and the current system are equivalent.

7.3.2 Basic variables and nonbasic variables

The standard linear optimization model with equality constraints and nonnegative variables is the form

$$(L):\ \max y \quad \text{subject to}$$
$$\begin{cases} y + c_1 x_1 + c_2 x_2 + \cdots + c_n x_n = d, \\ a_{11} x_1 + a_{12} x_2 + \cdots + a_{1n} x_n = b_1, \\ \vdots \\ a_{m1} x_1 + a_{m2} x_2 + \cdots + a_{mn} x_n = b_m, \\ x_1 \geq 0,\ x_2 \geq 0,\ \ldots,\ x_n \geq 0, \end{cases}$$

where b_i, c_j have any values. Assume that for each k ($k = 1, \dots, m$), the coefficients of the $\phi(k)$-th column satisfy

$$\begin{aligned} a_{k\phi(k)} &= 1, \\ a_{i\phi(k)} &= 0, \quad (i = 1, \dots, m,\ i \neq k), \\ c_{\phi(k)} &= 0, \end{aligned}$$

where ϕ is an indexing function of row indices into column indices. Then the corresponding variables $x_{\phi(k)}$ are called *basic variables* and the remaining variables are called *nonbasic variables*. The solution

$$\begin{cases} x_{\phi(k)} = b_k & (k = 1, \dots, m), \\ x_j = 0 & \text{otherwise} \end{cases}$$

is called the *basic solution* of the model (L). The objective function of the model (L) is rewritten as

$$y = -c_1x_1 - \dots - c_nx_n + d.$$

Under the above assumption condition, the coefficients of basic variables are equal to zero, i.e., $c_{\phi(k)} = 0$. Since nonbasic variables themselves in the basic solution of the model (L) are equal to zero, i.e., $x_j = 0$ $(j \neq \phi(k))$, the value of the objective function is reduced to $y = d$.

Definition 7.3.1. Under the above assumption condition,
- if $b_i \geq 0$ for all $i = 1, \dots, m$ and all c_j have any values, the model (L) is said to be in *basic form*;
- if $c_j \geq 0$ for all $j = 1, \dots, n$ and all b_i have any values, the model (L) is said to be in *dual basic form*.

7.3.3 The simplex algorithm

The discovery of the simplex method is due to George Dantzig. This method is applied for solving the standard linear optimization models with equality constraints and nonnegative variables in basic form.

Given a standard linear optimization model with equality constraints and nonnegative variables in basic form as follows:

$$(L^b) \quad \max y \quad \text{subject to}$$

$$\begin{cases} y + c_1x_1 + c_2x_2 + \dots + c_nx_n = d, \\ a_{11}x_1 + a_{12}x_2 + \dots + a_{1n}x_n = b_1, \\ \vdots \\ a_{m1}x_1 + a_{m2}x_2 + \dots + a_{mn}x_n = b_m, \\ x_1 \geq 0,\ x_2 \geq 0,\ \dots,\ x_n \geq 0, \end{cases}$$

where $b_i \geq 0$ for all $i = 1, \ldots, m$ and all c_j have any values, its matrix form is that

$$(L^b): \max y \quad \text{subject to}$$
$$\begin{cases} y + \langle \mathbf{c}, \boldsymbol{x} \rangle = d, \\ A\boldsymbol{x}^{\mathrm{T}} = \boldsymbol{b}^{\mathrm{T}}, \\ \boldsymbol{x} \in \mathbb{R}_+^n, \end{cases}$$

where $\boldsymbol{b} = (b_1, \ldots, b_m) \geq \mathbf{0}$ and $\mathbf{c} = (c_1, \ldots, c_n)$ have any values, and $\mathbb{R}_+^n = [0, \infty)^n$, and

$$A = \begin{pmatrix} a_{11} & \cdots & a_{1n} \\ \vdots & \ddots & \vdots \\ a_{m1} & \cdots & a_{mn} \end{pmatrix}.$$

The simplex algorithm consists of three cases.

Case 1. $c_j \geq 0$ for all $j = 1, \ldots, n$. Then the basic solution of the model (L^b) is optimal and the corresponding value of the objective function is d. The algorithm stops. In fact, if $x_1^*, x_2^*, \ldots, x_n^*$ is the basic solution of the model (L^b), then

$$y(x_1^*, x_2^*, \ldots, x_n^*) = d.$$

On the other hand, for any solution $x_1, x_2, \ldots, x_n$, since $c_j, x_j \geq 0$ $(j = 1, \ldots, n)$, the objective function

$$y(x_1, \ldots, x_n) = -c_1 x_1 - \cdots - c_n x_n + d \leq d.$$

Thus, the basic solution is optimal.

Case 2. There is a column l so that $c_l < 0$ and $a_{il} \leq 0$ for all $i = 1, \ldots, m$. Then the model (L^b) has no feasible solution. The algorithm stops.
In fact, since $c_l < 0$, x_l must be a nonbasic variable. It is known that for basic variables,

$$\begin{aligned} a_{k\phi(k)} &= 1, \\ a_{i\phi(k)} &= 0 \quad (i \neq k), \\ c_{\phi(k)} &= 0, \end{aligned}$$

and for nonbasic variables, $x_j = 0$ $(j \neq l)$. So the k-th equation of the model (L^b) reduces to

$$x_{\phi(k)} + a_{kl} x_l = b_k \quad \text{or} \quad x_{\phi(k)} = b_k - a_{kl} x_l,$$

where $b_k \geq 0$ and $a_{kl} \leq 0$, and the objective function of the model (L^b) reduces to

$$y = -c_l x_l + d,$$

where $c_l < 0$. Thus, y can be made arbitrarily large by choosing x_l sufficiently large.

Case 3. There is a column l so that $c_l < 0$ and $a_{il} > 0$ for some $i = 1, \dots, m$. If there are several columns l so that $c_l < 0$, then the gradient rule advises to choose the column l so that c_l is the most negative, and then choose the row index k so that

$$\begin{cases} a_{kl} > 0, \\ \frac{b_k}{a_{kl}} = \min_{1 \le i \le m} \left\{ \frac{b_i}{a_{il}} : a_{il} > 0 \right\}. \end{cases}$$

Pivoting on the (k, l) entry, this produces a new model (L_1^b) from the model (L^b) by a pivot operation

$$(L_1^b) \quad \max y \quad \text{subject to}$$

$$\begin{cases} y + c_1^{(1)} x_1 + \cdots + c_{l-1}^{(1)} x_{l-1} + 0 + c_{l+1}^{(1)} x_{l+1} + \cdots + c_n^{(1)} x_n = d^{(1)}, \\ a_{11}^{(1)} x_1 + \cdots + a_{1,l-1}^{(1)} x_{l-1} + 0 + a_{1,l+1}^{(1)} x_{l+1} + \cdots + a_{1n}^{(1)} x_n = b_1^{(1)}, \\ \vdots \\ a_{k1}^{(1)} x_1 + \cdots + a_{k,l-1}^{(1)} x_{l-1} + x_l + a_{k,l+1}^{(1)} x_{l+1} + \cdots + a_{kn}^{(1)} x_n = b_k^{(1)}, \quad \phi(k) = l, \\ \vdots \\ a_{m1}^{(1)} x_1 + \cdots + a_{m,l-1}^{(1)} x_{l-1} + 0 + a_{m,l+1}^{(1)} x_l + \cdots + a_{mn}^{(1)} x_n = b_m^{(1)}, \\ x_1 \ge 0, \ x_2 \ge 0, \ \dots, \ x_n \ge 0, \end{cases}$$

where the new coefficients are given by the pivoting formulas

$$a_{kj}^{(1)} = \frac{a_{kj}}{a_{kl}} \quad (j = 1, \dots, n),$$
$$b_k^{(1)} = \frac{b_k}{a_{kl}},$$

and the pivoting formulas for $i = 1, \dots, m$, $i \neq k$,

$$a_{ij}^{(1)} = a_{ij} - \frac{a_{kj}}{a_{kl}} a_{il} \quad (j = 1, \dots, n), \qquad b_i^{(1)} = b_i - \frac{b_k}{a_{kl}} a_{il},$$
$$c_j^{(1)} = c_j - \frac{a_{kj}}{a_{kl}} c_l \quad (j = 1, \dots, n), \qquad d^{(1)} = d - \frac{b_k}{a_{kl}} c_l.$$

Repeat the above process with (L_1^b) until a newer model is in Case 1 or Case 2.

7.3.4 The dual simplex method

The dual simplex method is used for solving the standard linear optimization models with equality constraints and nonnegative variables in dual basic form. Given a standard linear optimization model with equality constraints and nonnegative variables in dual basic form as follows:

$$(L^d): \max y \quad \text{subject to}$$

$$\begin{cases} y + c_1x_1 + \cdots + c_nx_n = d, \\ a_{11}x_1 + a_{12}x_2 + \cdots + a_{1n}x_n = b_1, \\ \vdots \\ a_{m1}x_1 + a_{m2}x_2 + \cdots + a_{mn}x_n = b_m, \\ x_1 \geq 0, \quad x_2 \geq 0, \quad \ldots, \quad x_n \geq 0, \end{cases}$$

where $c_j \geq 0$ for all $j = 1, \ldots, n$ and all b_i have any values. Its matrix form is that

$$(L^d): \max y \quad \text{subject to}$$

$$\begin{cases} y + \langle \mathbf{c}, \mathbf{x} \rangle = d, \\ A\mathbf{x}^{\mathrm{T}} = \mathbf{b}^{\mathrm{T}}, \\ \mathbf{x} \in \mathbb{R}_+^n, \end{cases}$$

where $\mathbf{c} = (c_1, \ldots, c_n) \geq \mathbf{0}$ and $\mathbf{b} = (b_1, \ldots, b_m)$ have any values, and $\mathbf{x} = (x_1, \ldots, x_n)$, and

$$A = \begin{pmatrix} a_{11} & \cdots & a_{1n} \\ \vdots & \ddots & \vdots \\ a_{m1} & \cdots & a_{mn} \end{pmatrix}.$$

The dual simplex algorithm consists of three cases.

Case 1. $b_i \geq 0$ for all $i = 1, \ldots, m$. This is just Case 1 of the above simplex algorithm. Thus, the basic solution of the model (L^d) is optimal and the corresponding value of the objective function is d. The algorithm stops.

Case 2. There is a row k so that $b_k < 0$ and $a_{kl} \geq 0$ for all $l = 1, \ldots, n$. Then the model (L^d) has no feasible solution. The algorithm stops.
In fact, for any $x_1 \geq 0, \ldots, x_n \geq 0$, since $a_{kl} \geq 0$ for all $l = 1, \ldots, n$,

$$a_{k1}x_1 + \cdots + a_{kn}x_n \geq 0.$$

But $b_k < 0$. So $x_1, \ldots, x_n$ cannot satisfy the k-th equation, and so the model (L^d) has no feasible solution.

Case 3. There is a row k so that $b_k < 0$ and $a_{kl} < 0$ for some $l = 1, \ldots, n$. If there are several rows k so that $b_k < 0$, then the dual gradient rule advises to choose the row index k so that b_k is the most negative, and then choose the column index l so that

$$\begin{cases} a_{kl} < 0, \\ \frac{c_l}{a_{kl}} = \max_{1 \leq j \leq n} \left\{ \frac{c_j}{a_{kj}} : a_{kj} < 0 \right\}. \end{cases}$$

Pivoting on the (k, l) entry, this produces a new model (L_1^d) from the current model (L^d) by a pivot operation. Repeat the above process with (L_1^d) until a newer model is in Case 1 or Case 2.

7.3.5 Slack variables

Any linear optimization model with inequality constraints and nonnegative variables can be converted into an equivalent standard linear optimization model with equality constraints and nonnegative variables by introducing new variables. The introduced new variables are called *slack variables*. This is a major creative idea in the development of the simplex method. The types often used are as follows.

Type 1
The model is a linear optimization model with inequality constraints and nonnegative variables as follows:

$$(L) \quad \max y \quad \text{subject to}$$
$$\begin{cases} y + c_1x_1 + c_2x_2 + \cdots + c_nx_n = d, \\ a_{11}x_1 + a_{12}x_2 + \cdots + a_{1n}x_n \le b_1, \\ \quad \vdots \\ a_{m1}x_1 + a_{m2}x_2 + \cdots + a_{mn}x_n \le b_m, \\ x_1 \ge 0, \ x_2 \ge 0, \ \dots, \ x_n \ge 0, \end{cases}$$

where $b_i \ge 0$ for all $i = 1, \dots, m$ and all c_j have any values. Its matrix form is that

$$(L): \ \max y \quad \text{subject to}$$
$$\begin{cases} y + \langle \boldsymbol{c}, \boldsymbol{x} \rangle = d, \\ A\boldsymbol{x}^{\mathrm{T}} \le \boldsymbol{b}^{\mathrm{T}}, \\ \boldsymbol{x} \in \mathbb{R}_+^n, \end{cases}$$

where $\boldsymbol{b} = (b_1, \dots, b_m) \ge \boldsymbol{0}$ and $\boldsymbol{c} = (c_1, \dots, c_n)$ have any values, and $\boldsymbol{x} = (x_1, \dots, x_n)$, and

$$A = \begin{pmatrix} a_{11} & \cdots & a_{1n} \\ \vdots & \ddots & \vdots \\ a_{m1} & \cdots & a_{mn} \end{pmatrix}.$$

Introduce m new variables x_{n+i} $(i = 1, \dots, m)$ (i.e., m slack variables) to convert inequality constraints to equality constraints. This produces a standard linear optimization model with equality constraints and nonnegative variables in basic form as follows:

(L^b) max y subject to

$$\begin{cases} y + c_1x_1 + \cdots + c_nx_n + 0 + 0 + 0 + \cdots + 0 = d, & \\ a_{11}x_1 + a_{12}x_2 + \cdots + a_{1n}x_n + x_{n+1} + 0 + 0 + \cdots + 0 = b_1, & \phi(1) = n+1, \\ a_{21}x_1 + a_{22}x_2 + \cdots + a_{2n}x_n + 0 + x_{n+2} + 0 + \cdots + 0 = b_2, & \phi(2) = n+2, \\ \vdots & \\ a_{m1}x_1 + a_{m2}x_2 + \cdots + a_{mn}x_n + 0 + 0 + \cdots + 0 + x_{n+m} = b_m, & \phi(m) = n+m, \\ x_1 \ge 0,\ x_2 \ge 0,\ \ldots,\ x_{n+m} \ge 0, & \end{cases}$$

where $b_i \ge 0$ for all $i = 1, \ldots, m$ and all c_j have any values. Here ϕ is an indexing function of row indices into column indices and $x_{n+i} \ge 0$ $(i = 1, \ldots, m)$. It is seen that when $\phi(k) = n + k$ $(k = 1, \ldots, m)$,

$$\begin{aligned} a_{k\phi(k)} &= 1, \\ a_{i\phi(k)} &= 0 \quad (i = 1, \ldots, m,\ i \ne k), \\ c_{\phi(k)} &= 0. \end{aligned}$$

Thus, the slack variables x_{n+k} $(k = 1, \ldots, m)$ are basic variables and the original variables x_j $(j = 1, \ldots, n)$ are nonbasic variables. Let the nonbasic variables $x_j = 0$ $(j = 1, \ldots, n)$. Then $x_{n+i} = b_i$ $(i = 1, \ldots, m)$. So

$$\begin{aligned} x_j &= 0 \quad (j = 1, \ldots, n), \\ x_{n+i} &= b_i \quad (i = 1, \ldots, m) \end{aligned}$$

is the basic solution of the model (L^b).

Applying the simplex method to the model (L^b), we will obtain the optimal solution and the corresponding value of the objective function.

Example 7.3.2. Given a linear optimization model with inequality constraints and nonnegative variables,

$$(L):\ \max y \quad \text{subject to}$$

$$\begin{cases} y - 4x_1 - 5x_2 = 25, \\ x_1 + 2x_2 \le 12, \\ x_1 + x_2 \le 9, \\ 4x_1 + 5x_2 \le 40, \\ x_1 \le 0,\ x_2 \ge 0, \end{cases}$$

try to solve it by the simplex method.

Solution. It is clear that the set of feasible solutions of the model (L) is a polygonal region with vertices $(0,0)$, $(0,6)$, $(6,3)$, and $(9,0)$.

Introduce three slack variables x_3, x_4, x_5 to convert inequality to equality constraints. This produces a standard linear optimization model with equality constraints and nonnegative variables in basic form as follows:

$$(L^b): \max y \quad \text{subject to}$$

$$\begin{cases} y - 4x_1 - 5x_2 + 0 + 0 + 0 = 25, & \\ x_1 + 2x_2 + x_3 + 0 + 0 = 12, & \phi(1) = 3, \\ x_1 + x_2 + 0 + x_4 + 0 = 9, & \phi(2) = 4, \\ 4x_1 + 5x_2 + 0 + 0 + x_5 = 40, & \phi(3) = 5, \\ x_1 \le 0,\ x_2 \ge 0,\ x_3 \ge 0,\ x_4 \ge 0,\ x_5 \ge 0, & \end{cases}$$

where ϕ is an indexing function of row indices into column indices. It is seen that the basic variables of the model (L^b) are x_3, x_4, x_5 and the nonbasic variables of the model (L^b) are x_1, x_2. Let nonbasic variables $x_1 = x_2 = 0$. Then $x_3 = 12$, $x_4 = 9$, $x_5 = 40$. So the basic solution of the model (L^b) is

$$x_1 = 0, \quad x_2 = 0, \quad x_3 = 12, \quad x_4 = 9, \quad x_5 = 40$$

and the corresponding value of the objective function is $y(0,0) = 25$ (see Table 7.1).

Tab. 7.1: First simplex tableau.

Basic Variables	Equation Number	Coefficient of y	x_1	x_2	x_3	x_4	x_5	Right Side of Equation
y	0	1	−4	−5	0	0	0	25
x_3	1	0	1	2	1	0	0	12
x_4	2	0	1	1	0	1	0	9
x_5	3	0	4	5	0	0	1	40

Here the point $(0,0)$ is the first vertex of the set of feasible solutions of the model (L). Since $c_1 = -4$ and $c_2 = -5$, the model (L^b) is not in Case 1 of the simplex algorithm. So the basic solution is not optimal.

In the model (L^b), $c_1 = -4$, and $c_2 = -5$. The gradient rule advises to choose the column index $l = 2$ since $c_2 = -5$ is the most negative. In this column,

$$a_{12} = 2 > 0, \quad a_{22} = 1 > 0, \quad a_{32} = 5 > 0,$$

so the model (L^b) is in Case 3 of the simplex algorithm. Choose the row index $k = 1$ since $a_{12} > 0$ and

$$\frac{b_k}{a_{k2}} = \min\left\{\frac{b_1}{a_{12}}, \frac{b_2}{a_{22}}, \frac{b_3}{a_{32}}\right\} = \min\{6, 9, 8\} = 6 = \frac{b_1}{a_{12}}.$$

Pivoting on the $(1, 2)$ entry, this produces a new model (L_1^b) from the model (L^b) by the first pivot operation

$$(L_1^b):\ \max y \quad \text{subject to}$$

$$\begin{cases} y - \frac{3}{2}x_1 + 0 + \frac{5}{2}x_3 + 0 + 0 = 55, \\ \frac{1}{2}x_1 + x_2 + \frac{1}{2}x_3 + 0 + 0 = 6, & \phi(1) = 2, \\ \frac{1}{2}x_1 + 0 - \frac{1}{2}x_3 + x_4 + 0 = 3, & \phi(2) = 4, \\ \frac{3}{2}x_1 + 0 - \frac{5}{2}x_3 + 0 + x_5 = 10, & \phi(3) = 5, \\ x_1 \le 0,\ x_2 \ge 0,\ x_3 \ge 0,\ x_4 \ge 0,\ x_5 \ge 0. \end{cases}$$

It is seen that the basic variables of the model (L_1^b) are x_2, x_4, x_5 and the nonbasic variables of the model (L_1^b) are x_1, x_3. Let nonbasic variables $x_1 = x_3 = 0$. Then $x_2 = 6$, $x_4 = 3$, $x_5 = 10$. So the basic solution of the model (L_1^b) is

$$x_1 = 0, \quad x_2 = 6, \quad x_3 = 0, \quad x_4 = 3, \quad x_5 = 10$$

and the corresponding value of the objective function is $y(0, 6) = 55$ (see Table 7.2). Here the point $(0, 6)$ is the second vertex of the set of feasible solutions of the model (L). Since $c_1 = -\frac{3}{2}$, the model (L_1^b) is not in Case 1 of the simplex algorithm. So the basic solution is not optimal.

Tab. 7.2: Second simplex tableau.

Basic Variables	Equation Number	Coefficient of						Right Side of Equation
		y	x_1	x_2	x_3	x_4	x_5	
y	0	1	$-\frac{3}{2}$	0	$\frac{5}{2}$	0	0	55
x_2	1	0	$\frac{1}{2}$	1	$\frac{1}{2}$	0	0	6
x_4	2	0	$\frac{1}{2}$	0	$-\frac{1}{2}$	1	0	3
x_5	3	0	$\frac{3}{2}$	0	$-\frac{5}{2}$	0	1	10

In the model (L_1^b), $c_1 = -\frac{3}{2}$. The chosen column index is $l = 1$. In this column, $a_{11} = \frac{1}{2} > 0$, $a_{21} = \frac{1}{2} > 0$, and $a_{31} = \frac{3}{2} > 0$, so the model (L_1^b) is in Case 3 of the simplex algorithm. Choose the row index $k = 2$ since $a_{21} > 0$ and

$$\frac{b_k}{a_{k1}} = \min\left\{\frac{b_1}{a_{11}}, \frac{b_2}{a_{21}}, \frac{b_3}{a_{31}}\right\} = \min\left\{12, 6, \frac{20}{3}\right\} = 6 = \frac{b_2}{a_{21}}.$$

Pivoting on the $(2, 1)$ entry, this produces a new model (L_2^b) from the model (L_1^b) by the second pivot operation

$$(L_2^b):\ \max y \quad \text{subject to}$$

$$\begin{cases} y + 0 + 0 + x_3 + 3x_4 + 0 = 64, & \\ 0 + x_2 + x_3 - x_4 + 0 = 3, & \phi(1) = 2, \\ x_1 + 0 - x_3 + 2x_4 + 0 = 6, & \phi(2) = 1, \\ 0 + 0 - x_3 - 3x_4 + x_5 = 1, & \phi(3) = 5, \\ x_1 \le 0,\ x_2 \ge 0,\ x_3 \ge 0,\ x_4 \ge 0,\ x_5 \ge 0. & \end{cases}$$

It is seen that the basic variables of the model (L_2^b) are x_1, x_2, x_5 and the nonbasic variables of the model (L_2^b) are x_3, x_4. Let nonbasic variables $x_3 = x_4 = 0$. Then $x_1 = 6$, $x_2 = 3$, $x_5 = 1$. In the model (L_2^b), $c_j \ge 0$ for all $j = 1, 2, 3, 4, 5$. So the model (L_2^b) is in Case 1 of the simplex algorithm. The basic solution of the model (L_2^b)

$$x_1 = 6, \quad x_2 = 3, \quad x_3 = 0, \quad x_4 = 0, \quad x_5 = 1$$

is the optimal solution of the model (L) and the corresponding value of the objective function is $y(6, 3) = 64$. Here the point $(6, 3)$ is the third vertex of the set of feasible solutions of the model (L) (see Table 7.3). □

Tab. 7.3: Third simplex tableau.

Basic Variables	Equation Number	Coefficient of						Right Side of Equation
		y	x_1	x_2	x_3	x_4	x_5	
y	0	1	0	0	1	3	0	64
x_1	1	0	0	1	1	−1	0	3
x_2	2	0	1	0	−1	2	0	6
x_5	3	0	0	0	−1	−3	1	1

From Example 7.3.2, it is seen that the calculations applying the simplex method to solve the standard linear optimization model in basic form are to journey from one vertex of the set of feasible solutions of the given linear optimization model with inequality constraints and nonnegative variables to another such vertex by the pivot operations.

Type 2

The model is a linear optimization model with inequality constraints and nonnegative variables

$$(L) \quad \min z \quad \text{subject to}$$

$$\begin{cases} z = c_1x_1 + c_2x_2 + \cdots + c_nx_n + d, \\ a_{11}x_1 + a_{12}x_2 + \cdots + a_{1n}x_n \geq b_1, \\ a_{21}x_1 + a_{22}x_2 + \cdots + a_{2n}x_n \geq b_2, \\ \vdots \\ a_{m1}x_1 + a_{m2}x_2 + \cdots + a_{mn}x_n \geq b_m, \\ x_1 \geq 0, \ x_2 \geq 0, \ , \ldots, \ x_n \geq 0, \end{cases}$$

where $c_j \geq 0$ for all $j = 1, \ldots, n$ and all b_i have any values. Its matrix form is that

$$(L): \ \min z \quad \text{subject to}$$

$$\begin{cases} z = \langle \mathbf{c}, \mathbf{x} \rangle + d, \\ A\mathbf{x}^{\mathrm{T}} \geq \mathbf{b}^{\mathrm{T}}, \mathbf{x} \in \mathbb{R}_+^n, \end{cases}$$

where $\mathbf{c} = (c_1, \ldots, c_n) \geq 0$ and $\mathbf{b} = (b_1, \ldots, b_m)$ have any values, and $\mathbf{x} = (x_1, \ldots, x_n)$, and

$$A = \begin{pmatrix} a_{11} & \cdots & a_{1n} \\ \vdots & \ddots & \vdots \\ a_{m1} & \cdots & a_{mn} \end{pmatrix}.$$

Let $y = -z$. In the model (L), converting minimization of z to maximization of y, and then multiplying the inequalities by -1, produces a linear optimization model (L^*) with inequality constraints and nonnegative variables from the current model (L) as follows:

$$(L^*) \quad \max y \quad \text{subject to}$$

$$\begin{cases} -y = c_1x_1 + c_2x_2 + \cdots + c_nx_n + d, \\ -a_{11}x_1 - a_{12}x_2 - \cdots - a_{1n}x_n \leq -b_1, \\ \vdots \\ -a_{m1}x_1 - a_{m2}x_2 - \cdots - a_{mn}x_n \leq -b_m, \\ x_1 \geq 0, \ x_2 \geq 0, \ \ldots, \ x_n \geq 0, \end{cases}$$

where $c_j \geq 0$ for all $j = 1, \ldots, n$ and all b_i have any values.

Introducing m slack variables x_{n+i} $(i = 1, \ldots, m)$, the model (L^*) is further converted to a standard linear optimization model with equality constraints and nonnegative variables in dual basic form as follows:

(L^d): $\max y$ subject to

$$\begin{cases} y + c_1x_1 + c_2x_2 + \cdots + c_nx_n + 0 + 0 + \cdots + 0 = -d, & \\ -a_{11}x_1 - a_{12}x_2 - \cdots - a_{1n}x_n + x_{n+1} + 0 + \cdots + 0 = -b_1, & \phi(1) = n+1, \\ -a_{21}x_1 - a_{22}x_2 - \cdots - a_{2n}x_n + 0 + x_{n+2} + \cdots + 0 = -b_2, & \phi(2) = n+2, \\ \vdots & \\ -a_{m1}x_1 - a_{m2}x_2 - \cdots - a_{mn}x_n + 0 + 0 + \cdots + 0 + x_{n+m} = -b_m, & \phi(m) = n+m, \\ x_1 \geq 0, \ x_2 \geq 0, \ \ldots, \ x_{n+m} \geq 0, & \end{cases}$$

where $c_j \geq 0$ for all $j = 1, \ldots, n$ and all b_i have any values. Here $x_{n+i} \geq 0$ $(i = 1, \ldots, m)$.

Applying the dual simplex method to the model (L^d), we will obtain the optimal solution of the model (L^d) and the corresponding value of the objective function.

Example 7.3.3. Given a linear optimization model with inequality constraints and nonnegative variables

$$(L): \min z \quad \text{subject to}$$

$$\begin{cases} z = 3x_1 + 4x_2, \\ 4x_1 + 2x_2 \geq 18, \\ 2x_1 + 5x_2 \geq 25, \\ 2x_1 + 3x_2 \geq 19, \end{cases}$$

where $x_1 \geq 0$ and $x_2 \geq 0$, try to solve the model (L) by the dual simplex method.

Solution. Let $y = -z$. Converting minimization of z to maximization of y, multiplying the inequalities by -1, and then introducing three slack variables x_3, x_4, x_5, produces a standard linear optimization model with equality constraints and nonnegative variables in dual basic form as follows:

$$(L^d): \max y \quad \text{subject to}$$

$$\begin{cases} y + 3x_1 + 4x_2 + 0 + 0 + 0 = 0, & \\ -4x_1 - 2x_2 + x_3 + 0 + 0 = -18, & \phi(1) = 3, \\ -2x_1 - 5x_2 + 0 + x_4 + 0 = -25, & \phi(2) = 4, \\ -2x_1 - 3x_2 + 0 + 0 + x_5 = -19, & \phi(3) = 5, \\ x_1 \geq 0, \ x_2 \geq 0. & \end{cases}$$

where $c_1 = 3$ and $c_2 = 4$ are both nonnegative.

Since $b_1 = -18 < 0$, $b_2 = -25 < 0$, and $b_3 = -19 < 0$, the dual gradient rule advises to choose the second row $k = 2$ since $b_2 = -25$ is the most negative. In this row, $a_{21} = -2 < 0$ and $a_{22} = -5 < 0$. So the model (L^d) is in Case 3 of the dual simplex algorithm. Note that $c_1 = 3$ and $c_2 = 4$. Choose the column $l = 2$ since

$$\begin{cases} a_{22} < 0, \\ \frac{c_l}{a_{2l}} = \max\left\{\frac{c_1}{a_{21}}, \frac{c_2}{a_{22}}\right\} = \max\left\{-\frac{3}{2}, -\frac{4}{5}\right\} = -\frac{4}{5} = \frac{c_2}{a_{22}}. \end{cases}$$

Pivoting on the $(2, 2)$ entry, this produces a newer model (L_1^d) from the model (L^d) by the first pivot operation

$$(L_1^d):\ \max y \quad \text{subject to}$$

$$\begin{cases} y + \frac{7}{5}x_1 + 0 + 0 + \frac{4}{5}x_4 + 0 = -20, & \\ -\frac{16}{5}x_1 + 0 + x_3 - \frac{2}{5}x_4 + 0 = -8, & \phi(1) = 3, \\ \frac{2}{5}x_1 + x_2 + 0 - \frac{1}{5}x_4 + 0 = 5, & \phi(2) = 2, \\ -\frac{4}{5}x_1 + 0 + 0 - \frac{3}{5}x_4 + x_5 = -4, & \phi(3) = 5, \\ x_1 \geq 0,\ x_2 \geq 0, & \end{cases}$$

where $c_1 = \frac{7}{5}$ and $c_4 = \frac{4}{5}$ are both nonnegative. It is seen that basic variables of the model (L_1^d) are x_2, x_3, x_5 and nonbasic variables of the model (L_1^d) are x_1, x_4. Let nonbasic variables $x_1 = x_4 = 0$. Then $x_2 = 5$, $x_3 = -8$, $x_5 = -4$. So the basic solution of the model (L_1^d) is

$$x_1 = 0, \quad x_2 = 5, \quad x_3 = -8, \quad x_4 = 0, \quad x_5 = -4$$

and the corresponding value of the objective function is $y(0, 5) = -20$ (see Table 7.4). Since $b_1 = -8$ and $b_3 = -4$, the model (L_1^d) is not in Case 1 of the dual simplex algorithm. So the solution is not the optimal one.

Tab. 7.4: First dual simplex tableau.

Basic Variables	Equation Number	Coefficient of y	x_1	x_2	x_3	x_4	x_5	Right Side of Equation
y	0	1	$\frac{7}{5}$	0	0	$\frac{4}{5}$	0	−20
x_2	1	0	$-\frac{16}{5}$	0	1	$-\frac{2}{5}$	0	−8
x_3	2	0	$\frac{2}{5}$	1	0	$-\frac{2}{5}$	0	5
x_5	3	0	$-\frac{4}{5}$	0	0	$-\frac{3}{5}$	1	−4

In the model (L_1^d), since $b_1 = -8$ and $b_3 = -4$, the dual gradient rule advises to choose the row $k = 1$ since $b_1 = -8$ is the most negative. In this row, $b_1 = -8$, and $a_{11} = -\frac{16}{5} < 0$ and $a_{14} = -\frac{2}{5} < 0$, so the model (L_1^d) is in Case 3 of the dual simplex algorithm.

Note that $c_1 = \frac{7}{5}$ and $c_4 = \frac{4}{5}$. Choose the column $l = 1$ since

$$\begin{cases} a_{11} < 0, \\ \frac{c_l}{a_{1l}} = \max\left\{\frac{c_1}{a_{11}}, \frac{c_4}{a_{14}}\right\} = \max\left\{-\frac{7}{16}, -2\right\} = -\frac{7}{16} = \frac{c_1}{a_{11}}. \end{cases}$$

Pivoting on the $(1,1)$ entry, this produces a newer model (L_2^d) from the model (L_1^d) by the second pivot operation

$$(L_2^d): \ \max y \quad \text{subject to}$$

$$\begin{cases} y + 0 + 0 + \frac{7}{16}x_3 + \frac{5}{8}x_4 + 0 = -\frac{47}{2}, \\ x_1 + 0 - \frac{5}{16}x_3 + \frac{1}{8}x_4 + 0 = \frac{5}{2}, & \phi(1) = 1, \\ 0 + x_2 + \frac{1}{8}x_3 - \frac{1}{4}x_4 + 0 = 4, & \phi(2) = 2, \\ 0 + 0 - \frac{1}{4}x_3 - \frac{1}{2}x_4 + x_5 = -2, & \phi(3) = 5, \\ x_1 \geq 0, \ x_2 \geq 0, \end{cases}$$

where $c_3 = \frac{7}{16}$ and $c_4 = \frac{5}{8}$ are nonnegative. It is seen that basic variables of the model (L_2^d) are x_1, x_2, x_5 and nonbasic variables of the model (L_2^d) are x_3, x_4. Let nonbasic variables $x_3 = x_4 = 0$. Then $x_1 = \frac{5}{2}$, $x_2 = 4$, $x_5 = -2$. So the basic solution of the model (L_2^d) is

$$x_1 = \frac{5}{2}, \quad x_2 = 4, \quad x_3 = 0, \quad x_4 = 0, \quad x_5 = -2$$

and the corresponding value of the objective function is $y(\frac{5}{2}, 4) = -\frac{47}{2}$ (see Table 7.5). Since $b_3 = -2 < 0$, the model (L_2^d) is not in Case 1 of the dual simplex algorithm, and so the solution is not optimal.

In the model (L_2^d), choose the row $k = 3$ since $b_3 = -2$ is the only negative value. In this row, $a_{33} = -\frac{1}{4} < 0$ and $a_{34} = -\frac{1}{2} < 0$. So the model (L_1^d) is in Case 3 of the dual simplex algorithm. Note that $c_3 = \frac{7}{16}$ and $c_4 = \frac{5}{8}$. Choose the column $l = 4$ since

$$\begin{cases} a_{34} < 0, \\ \frac{c_l}{a_{3l}} = \max\left\{\frac{c_3}{a_{33}}, \frac{c_4}{a_{34}}\right\} = \max\left\{-\frac{7}{4}, -\frac{5}{4}\right\} = -\frac{5}{4} = \frac{c_4}{a_{34}}. \end{cases}$$

Tab. 7.5: Second dual simplex tableau.

Basic Variables	Equation Number	Coefficient of						Right Side of Equation
		y	x_1	x_2	x_3	x_4	x_5	
y	0	1	0	0	$\frac{7}{16}$	$\frac{5}{8}$	0	$-\frac{47}{2}$
x_1	1	0	1	0	$-\frac{5}{16}$	$\frac{1}{8}$	0	$\frac{5}{2}$
x_2	2	0	0	1	$\frac{1}{8}$	$-\frac{1}{4}$	0	4
x_5	3	0	0	0	$-\frac{1}{4}$	$-\frac{1}{2}$	1	-2

Pivoting on the $(3,4)$ entry, this produces a newer model (L_3^d) from the model (L_2^d) by the third pivot operation

$$(L_3^d):\ \max y \quad \text{subject to}$$

$$\begin{cases} y + 0 + 0 + \frac{1}{8}x_3 + 0 + \frac{5}{4}x_5 = -26, & \\ x_1 + 0 - \frac{3}{8}x_3 + 0 + \frac{1}{4}x_5 = 2, & \phi(1) = 1, \\ 0 + x_2 + \frac{1}{4}x_3 + 0 - \frac{1}{2}x_5 = 5, & \phi(2) = 2, \\ 0 + 0 + \frac{1}{2}x_3 + x_4 - 2x_5 = 4, & \phi(3) = 4, \\ x_1 \geq 0,\ x_2 \geq 0, & \end{cases}$$

where $c_3 = \frac{1}{8}$ and $c_4 = \frac{5}{4}$ are both nonnegative. It is seen that the basic variables of the model (L_3^d) are x_1, x_2, x_4 and nonbasic variables of the model (L_3^d) are x_3, x_5. Let $x_3 = x_5 = 0$. Then $x_1 = 2$, $x_2 = 5$, $x_4 = 4$. Note that $b_i > 0$ for all $i = 1, 2, 3$ in the model (L_3^d). So the model (L_3^d) is in Case 1 of the dual simplex algorithm. So the basic solution of the model (L_3^d)

$$x_1 = 2, \quad x_2 = 5, \quad x_3 = 0, \quad x_4 = 4, \quad x_5 = 0.$$

is the optimal solution of the model (L) and the corresponding value of the objective function is $y(2, 5) = -26$ (see Table 7.6). Here $(2, 5)$ is a vertex of the set of feasible solutions of the model (L).

Tab. 7.6: Third dual simplex tableau.

Basic Variables	Equation Number	Coefficient of						Right Side of Equation
		y	x_1	x_2	x_3	x_4	x_5	
y	0	1	0	0	$\frac{1}{8}$	0	$\frac{5}{4}$	-26
x_1	1	0	1	0	$-\frac{3}{8}$	0	$\frac{1}{4}$	2
x_2	2	0	0	1	$\frac{1}{4}$	0	$-\frac{1}{2}$	5
x_4	3	0	0	0	$\frac{1}{2}$	1	-2	4

Note that $\min z = -\max(-z) = -\max y$ and $\max y = -26$. Then $\min z = 26$. □

Consider the following standard linear optimization model, satisfying the assumption condition given in Section 7.3.2, with equality constraints and nonnegative variables:

$$(L):\ \min z \quad \text{subject to}$$

$$\begin{cases} z = c_1x_1 + c_2x_2 + \cdots + c_nx_n + d, \\ a_{11}x_1 + a_{12}x_2 + \cdots + a_{1n}x_n = b_1, \\ \quad \vdots \\ a_{m1}x_1 + a_{m2}x_2 + \cdots + a_{mn}x_n = b_m, \\ x_1 \geq 0,\ x_2 \geq 0,\ \ldots,\ x_n \geq 0, \end{cases}$$

where the first $h-1$ right-hand sides $b_i \geq 0$ $(i = 1, \ldots, h-1)$, b_i $(i = h, \ldots, m)$, and c_j $(j = 1, \ldots, n)$ have any values. The model (L) will be converted to a model in basic form.

For convenience, without loss of generality, assume that in the model (L), the first $m-1$ right-hand side $b_i \geq 0$ $(i = 1, \ldots, m-1)$ and b_m has any value. Then the model (L) is converted to a model in basic form below.

Focusing on the m-th row of the model (L), there are three cases.

Case 1. $b_m \geq 0$. Then the model (L) is the desired model. The algorithm stops.

Case 2. $b_m < 0$ and $a_{mj} \geq 0$ for all j. Then $a_{m1}x_1 + \cdots + a_{mn}x_n \geq 0$ for any solution $x_j \geq 0$ $(j = 1, \ldots, n)$ but $b_m < 0$. Thus the m-th equation has no solution. The algorithm stops.

Case 3. $b_m < 0$ and $a_{ml} < 0$ for some l. Then the chosen column index is l. Regard the first $(m-1)$ rows as the constraints and the m-th row as the objective row. There are two alternatives.

- If $a_{il} \leq 0$ for all $i = 1, \ldots, m-1$, then the chosen row index is m. So the (m, l) entry in the objective row is chosen as the pivot entry, this produces a new model (L_1) with coefficients $c_j^{(1)}, a_{ij}^{(1)}, b_i^{(1)}$, where

$$\begin{cases} b_m^{(1)} = \frac{b_m}{a_{ml}} > 0, \\ b_i^{(1)} = b_i - \frac{b_m}{a_{ml}} a_{il} \geq 0 \quad (i = 1, \ldots, m-1) \end{cases}$$

  since $b_i \geq 0$ $(i = 1, \ldots, m-1)$. Thus the model (L_1) is in basic form. The algorithm stops.
- If $a_{il} > 0$ for some i $(1 \leq i \leq m-1)$, then the model (L) is in Case 3 of the simplex algorithm. So the chosen row index k is such that $a_{kl} > 0$ and

$$\frac{b_k}{a_{kl}} = \min_{1 \leq j \leq m-1} \left\{ \frac{b_j}{a_{jl}} : a_{jl} > 0 \right\}.$$

Pivoting on the (k, l) entry, this produces a new model (L_1) from the model (L). If the new model (L_1) is in basic form, then the algorithm stops. Otherwise, repeat the above process with the model (L_1) until a newer model is in basic form.

Type 3

The model is a linear optimization model with nonnegative variables, μ inequality constraints and $m-\mu$ equality constraints as follows:

$$(L)\quad \min z \quad \text{subject to}$$

$$\begin{cases} z = c_1x_1 + c_2x_2 + \cdots + c_nx_n + d, \\ a_{11}x_1 + a_{12}x_2 + \cdots + a_{1n}x_n \geq b_1, \\ \quad\vdots \\ a_{\mu 1}x_1 + a_{\mu 2}x_2 + \cdots + a_{\mu n}x_n \geq b_\mu, \\ a_{\mu' 1}x_1 + a_{\mu' 2}x_2 + \cdots + a_{\mu' n}x_n = b_{\mu'}, \\ \quad\vdots \\ a_{m1}x_1 + a_{m2}x_2 + \cdots + a_{mn}x_n = b_m, \\ x_1 \geq 0,\ x_2 \geq 0,\ \ldots,\ x_n \geq 0, \end{cases}$$

where c_j and b_i have any values and $\mu' = \mu + 1$. Its matrix form is

$$(L):\ \min z \quad \text{subject to}$$

$$\begin{cases} z = \langle \mathbf{c}, \mathbf{x} \rangle + d, \\ A_1\mathbf{x}^{\mathrm{T}} \geq \mathbf{b}_1^{\mathrm{T}}, \\ A_2\mathbf{x}^{\mathrm{T}} = \mathbf{b}_2^{\mathrm{T}}, \\ \mathbf{x} \in \mathbb{R}_+^n, \end{cases}$$

where $\mathbf{b}_1 = (b_1, \ldots, b_\mu)$, $\mathbf{b}_2 = (b_{\mu'}, \ldots, b_m)$, $\mathbf{c} = (c_1, \ldots, c_n)$, $\mathbf{x} = (x_1, \ldots, x_n)$, and

$$A_1 = \begin{pmatrix} a_{11} & \cdots & a_{1n} \\ \vdots & \ddots & \vdots \\ a_{\mu 1} & \cdots & a_{\mu n} \end{pmatrix}, \quad A_2 = \begin{pmatrix} a_{\mu' 1} & \cdots & a_{\mu' n} \\ \vdots & \ddots & \vdots \\ a_{m1} & \cdots & a_{mn} \end{pmatrix}.$$

Let $y = -z$. Converting minimization of z to maximization of y and multiplying the inequalities by -1, and then introducing μ slack variables x_{n+i} $(i = 1, \ldots, \mu)$, produces a standard linear optimization model (L^*), satisfying the assumption given in Section 7.3.2, with equality constraints and $n + \mu$ nonnegative variables as follows:

$$(L^*) \quad \max y \quad \text{subject to}$$

$$\begin{cases} y + c_1x_1 + c_2x_2 + \cdots + c_nx_n + 0 + 0 + \cdots + 0 = -d, \\ -a_{11}x_1 - a_{12}x_2 - \cdots - a_{1n}x_n + x_{n+1} + 0 + 0 + \cdots + 0 = -b_1, \\ -a_{21}x_1 - a_{22}x_2 - \cdots - a_{2n}x_n + 0 + x_{n+2} + 0 + \cdots + 0 = -b_2, \\ \vdots \\ -a_{\mu 1}x_1 - a_{\mu 2}x_2 - \cdots - a_{\mu n}x_n + 0 + 0 + \cdots + 0 + x_{n+\mu} = -b_\mu, \\ a_{\mu' 1}x_1 + a_{\mu' 2}x_2 + \cdots + a_{\mu' n}x_n = b_{\mu'}, \\ \vdots \\ a_{m1}x_1 + a_{m2}x_2 + \cdots + a_{mn}x_n = b_m, \\ x_1 \geq 0,\ x_2 \geq 0,\ \ldots,\ x_n \geq 0,\ x_{n+1} \geq 0,\ \ldots,\ x_{n+\mu} \geq 0 \end{cases}$$

where c_j and b_i have any values. Applying the above method, the model (L^*) can be converted to a standard linear optimization model with equality constraints and nonnegative variables in basic form.

Example 7.3.4. Solve the following linear optimization model:

$$(L):\ \min z \quad \text{subject to}$$

$$\begin{cases} z + x_1 + 2x_2 - 2x_3 = 0, \\ x_1 - 2x_2 + 2x_3 \geq 2, \\ 2x_1 - x_2 + x_3 \geq 2, \\ -2x_1 - x_2 + x_3 \geq 4, \\ x_1 \geq 0,\ x_2 \geq 0,\ x_3 \geq 0. \end{cases}$$

Solution. Let $y = -z$. Converting minimization to maximization, multiplying the inequalities by -1, and then introducing slack variables x_4, x_5, x_6, produces a standard linear optimization model (L^*) with equality constraints and nonnegative variables as follows:

$$(L^*):\ \max y \quad \text{subject to}$$

$$\begin{cases} y - x_1 - 2x_2 + 2x_3 + 0 + 0 + 0 = 0, \\ -x_1 + 2x_2 - 2x_3 + x_4 + 0 + 0 = -2, \\ -2x_1 + x_2 - x_3 + 0 + x_5 + 0 = -2, \\ 2x_1 + x_2 - x_3 + 0 + 0 + x_6 = -4, \\ x_1 \geq 0,\ x_2 \geq 0,\ x_3 \geq 0,\ x_4 \geq 0,\ x_5 \geq 0,\ x_6 \geq 0. \end{cases}$$

Looking at the first row of (L^*), $b_1 = -2 < 0$ and $a_{11} = -1$, $a_{13} = -2$. Choose the column index $l = 3$ as the column index of the pivot entry. Pivoting on the $(1, 3)$ entry produces a model (L_1) from the model (L^*)

$$(L_1):\ \max y \quad \text{subject to}$$

$$\begin{cases} y - x_1 - 2x_2 + 2x_3 + 0 + 0 + 0 = 0, \\ \frac{1}{2}x_1 - x_2 + x_3 - \frac{1}{2}x_4 + 0 + 0 = 1, \\ -\frac{3}{2}x_1 + 0 + 0 - \frac{1}{2}x_4 + x_5 + 0 = -1, \\ \frac{5}{2}x_1 + 0 + 0 - \frac{1}{2}x_4 + 0 + x_6 = -3, \\ x_1 \geq 0,\ x_2 \geq 0,\ x_3 \geq 0,\ x_4 \geq 0,\ x_5 \geq 0,\ x_6 \geq 0. \end{cases}$$

Looking at the first row of the model (L_1), $b_1 = 1 > 0$. The model (L_1) is in Case 1. So we turn to look at its second row. Since $b_2 = -1 < 0$, $a_{21} = -\frac{3}{2} < 0$, and $a_{24} = -\frac{1}{2} < 0$, the model (L_1) is in Case 3. Choose the column index $l = 1$. Regard the first row as the only constraint and the second row as the objective row. Only $a_{11} = \frac{1}{2} > 0$, the chosen row index is $k = 1$.

Pivoting on the $(1, 1)$ entry, this produces a model (L_2) from (L_1)

$$(L_2):\ \max y \quad \text{subject to}$$

$$\begin{cases} y + 0 - 4x_2 + 4x_3 - x_4 + 0 + 0 = 2, \\ x_1 - 2x_2 + 2x_3 - x_4 + 0 + 0 = 2, \\ 0 - 3x_2 + 3x_3 - 2x_4 + x_5 + 0 = 2, \\ 0 + 5x_2 - 5x_3 + 2x_4 + 0 + x_6 = -8, \\ x_1 \geq 0,\ x_2 \geq 0,\ x_3 \geq 0,\ x_4 \geq 0,\ x_5 \geq 0,\ x_6 \geq 0. \end{cases}$$

Looking at the second row of the model (L_2), $b_2 = 2 > 0$, the model (L_2) is in Case 1. So we turn to look at its third row. Since $b_3 = -8 < 0$ and $a_{33} = -5 < 0$, the model (L_2) is in Case 3. The chosen column index is $l = 3$. Regard the first and second rows as the constraints and the third row as the objective row. Since $a_{13} = 2 > 0$ and $a_{23} = 3 > 0$, the model (L_2) is in the second alternative of Case 3. From $a_{13} = 2 > 0$ and $a_{23} = 3 > 0$, it follows that

$$\frac{b_k}{a_{kl}} = \min\left\{\frac{b_1}{a_{13}}, \frac{b_2}{a_{23}}\right\} = \min\left\{1, \frac{2}{3}\right\} = \frac{2}{3} = \frac{b_2}{a_{23}},$$

and so the chosen row index is $k = 2$.

Pivoting on the $(2, 3)$ entry produces a model (L_3) from (L_2)

$$(L_3):\ \max y \quad \text{subject to}$$

$$\begin{cases} y + 0 + 0 + 0 + \frac{5}{3}x_4 - \frac{4}{3}x_5 + 0 = -\frac{2}{3}, \\ x_1 + 0 + 0 + \frac{1}{3}x_4 - \frac{2}{3}x_5 + 0 = \frac{2}{3}, \\ 0 - x_2 + x_3 - \frac{2}{3}x_4 + \frac{1}{3}x_5 + 0 = \frac{2}{3}, \\ 0 + 0 + 0 - \frac{4}{3}x_4 + \frac{5}{3}x_5 + x_6 = -\frac{14}{3}, \\ x_1 \geq 0,\ x_2 \geq 0,\ x_3 \geq 0,\ x_4 \geq 0. \end{cases}$$

Looking at the third row of the model (L_3), $b_3 = -\frac{14}{3} < 0$, $a_{34} = -\frac{4}{3} < 0$, the model ($L_3$) is in Case 3. So the chosen column index $l = 4$. Regard the first and second rows as the constraints and the third row as the objective row. Only $a_{14} = \frac{1}{3} > 0$, so the chosen row index $k = 1$.

Pivoting on the $(1, 4)$ entry produces a model (L_4) from the model (L_3)

$$(L_4):\ \max y \quad \text{subject to}$$

$$\begin{cases} y - 5x_1 + 0 + 0 + 0 + 2x_5 + 0 = -4, \\ 3x_1 + 0 + 0 + x_4 - 2x_5 + 0 = 2, \\ 2x_1 - x_2 + x_3 + 0 - x_5 + 0 = 2, \\ 4x_1 + 0 + 0 + 0 - x_5 + x_6 = -2, \\ x_1 \geq 0,\ x_2 \geq 0,\ x_3 \geq 0,\ x_4 \geq 0. \end{cases}$$

Looking at the third row of the model (L_4), $b_3 = -2 < 0$ and $a_{35} = -1 < 0$, the model (L_4) is in Case 3. The chosen column index $l = 5$. Regard the first and second rows as the constraints and the third row as the objective row. All $a_{15} = -2 < 0$ and $a_{25} = -1 < 0$, so the model (L_4) is in the first alternative of Case 3, and so the chosen row index $k = 3$.

Pivoting on the $(3, 5)$ entry in the objective row produces a model (L_5) from the model (L_4)

$$(L_5):\ \max y \quad \text{subject to}$$

$$\begin{cases} y + 3x_1 + 0 + 0 + 0 + 0 + 2x_6 = -8, \\ -5x_1 + 0 + 0 + x_4 + 0 - 2x_6 = 6, \\ -2x_1 - x_2 + x_3 + 0 + 0 - x_6 = 4, \\ -4x_1 + 0 + 0 + 0 + x_5 - x_6 = 2, \\ x_1 \geq 0,\ x_2 \geq 0,\ x_3 \geq 0,\ x_4 \geq 0, \end{cases}$$

where $b_1 = 6$, $b_2 = 4$, $b_3 = 2$. So the model (L_5) is in basic form.

Note that

$$c_1 = 3,$$
$$c_2 = c_3 = c_4 = c_5 = 0,$$
$$c_6 = 2.$$

The model (L_5) is in Case 1 of the simplex algorithm, and basic variables of the model (L_5) are x_3, x_4, x_5 and nonbasic variables are x_1, x_2, x_6. Let nonbasic variables $x_1 = x_2 = x_6 = 0$. Then $x_3 = 4$, $x_4 = 6$, $x_5 = 2$. So the basic solution of the model (L_5)

$$x_1 = 0, \quad x_2 = 0, \quad x_3 = 4, \quad x_4 = 6, \quad x_5 = 2, \quad x_6 = 0$$

is the optimal solution of the model (L^*) and the corresponding value of objective function is $y(00) = -8$. Note that $\min z = -\max y$. Then $\min z = 8$. □

Type 4

The model is a general linear optimization model with ν nonnegative variables and $n-\nu$ free variables subject to μ inequality constraints and $m-\mu$ equality constraints as follows:

$$(L):\ \min z \quad \text{subject to}$$

$$\begin{cases} z = c_1x_1 + \cdots + c_\nu x_\nu + c_{\nu'}x_{\nu'} + \cdots + c_nx_n + d, \\ a_{11}x_1 + \cdots + a_{1\nu}x_\nu + a_{1\nu'}x_{\nu'} + \cdots + a_{1n}x_n \geq b_1, \\ \vdots \\ a_{\mu 1}x_1 + \cdots + a_{\mu\nu}x_\nu + a_{\mu\nu'}x_{\nu'} + \cdots + a_{\mu n}x_n \geq b_\mu, \\ a_{\mu' 1}x_1 + \cdots + a_{\mu'\nu}x_\nu + a_{\mu'\nu'}x_{\nu'} + \cdots + a_{\mu' n}x_n = b_{\mu'}, \\ \vdots \\ a_{m1}x_1 + \cdots + a_{m\nu}x_\nu + a_{\mu'\nu'}x_{\nu'} + \cdots + a_{mn}x_n = b_m, \\ x_1 \geq 0,\ \ldots,\ x_\nu \geq 0, \quad x_{\nu'} \text{ free},\ \ldots,\ x_n \text{ free}, \end{cases}$$

where c_j and b_i have any values and $\mu' = \mu + 1$ and $\nu' = \nu + 1$. Its matrix form is

$$(L):\ \min z \quad \text{subject to}$$

$$\begin{cases} z = \langle \mathbf{c}, \boldsymbol{x} \rangle + d, \\ A_1\boldsymbol{x}^{\mathrm{T}} \geq \boldsymbol{b}_1^{\mathrm{T}}, \\ A_2\boldsymbol{x}^{\mathrm{T}} = \boldsymbol{b}_2^{\mathrm{T}}, \\ \boldsymbol{x} \in \mathbb{R}_+^\nu \times \mathbb{R}^{n-\nu}, \end{cases}$$

where $\boldsymbol{b}_1 = (b_1, \ldots, b_\mu)$, $\boldsymbol{b}_2 = (b_{\mu'}, \ldots, b_m)$, $\mathbf{c} = (c_1, \ldots, c_n)$, $\boldsymbol{x} = (x_1, \ldots, x_n)$, $d \in \mathbb{R}$, and

$$A_1 = \begin{pmatrix} a_{11} & \cdots & a_{1n} \\ \vdots & \ddots & \vdots \\ a_{\mu 1} & \cdots & a_{\mu n} \end{pmatrix}, \quad A_2 = \begin{pmatrix} a_{\mu' 1} & \cdots & a_{\mu' n} \\ \vdots & \ddots & \vdots \\ a_{m1} & \cdots & a_{mn} \end{pmatrix}.$$

For each free variable x_j $(j = \nu', \ldots, n)$, let

$$\begin{aligned} x_j^+ &= x_j, & x_j^- &= 0 & &\text{if } x_j \geq 0, \\ x_j^+ &= 0, & x_j^- &= -x_j & &\text{if } x_j \leq 0. \end{aligned}$$

Then x_j^+ and x_j^- are nonnegative and $x_j = x_j^+ - x_j^-$ $(j = \nu', \ldots, n)$.

Let $y = -z$. Converting minimization of z to maximization of y and multiplying the inequalities by -1, substituting x_j $(j = \nu', \ldots, n)$ by the difference of two nonnegative variable, $x_n = x_n^+ - x_n^-$, where $x_n^+ \geq 0$ and $x_n^- \geq 0$, and then introducing μ slack variables x_{n+i} $(i = 1, \ldots, \mu)$, produces a standard linear optimization model, satisfying the assumption condition given in Section 7.3.2, with $2n - v + \mu$ nonnegative variables subject to m equality constraints as follows:

$$(L^*) \quad \max y \quad \text{subject to}$$

$$\begin{cases} y + c_1x_1 + \cdots + c_vx_v + c_{v'}x_{v'}^+ - c_{v'}x_{v'}^- + \cdots + c_nx_n^+ - c_nx_n^- + 0 + 0 + \cdots + 0 = -d, \\ -a_{11}x_1 - \cdots - a_{1v}x_v - a_{1v'}x_{v'}^+ + a_{1v'}x_{v'}^- - \cdots - a_{1n}x_n^+ + a_{1n}x_n^- + x_{n+1} + 0 + \cdots + 0 = -b_1, \\ \vdots \\ -a_{\mu 1}x_1 - \cdots - a_{\mu v}x_v - a_{\mu v'}x_{v'}^+ + a_{\mu v'}x_{v'}^- - \cdots - a_{\mu n}x_n^+ + a_{\mu n}x_n^- + 0 + \cdots + 0 + x_{n+\mu} = -b_\mu, \\ a_{\mu' 1}x_1 + \cdots + a_{\mu' v}x_v + a_{\mu' v'}x_{v'}^+ - a_{\mu' v'}x_{v'}^- + \cdots + a_{\mu' n}x_n^+ - a_{\mu' n}x_n^- = b_{\mu'}, \\ \vdots \\ a_{m1}x_1 + \cdots + a_{mv}x_v + a_{mv'}x_{v'}^+ - a_{mv'}x_{v'}^- + \cdots + a_{mn}x_n^+ - a_{mn}x_n^- = b_m, \\ x_1 \geq 0, \ \ldots, \ x_v \geq 0, \ x_{v'}^+ \geq 0, \ x_{v'}^- \geq 0, \ \ldots, \ x_n^+ \geq 0, \ x_n^- \geq 0, \ x_{n+1} \geq 0, \ \ldots, \ x_{n+\mu} \geq 0, \end{cases}$$

where c_j and b_i have any values, and $\mu' = \mu + 1$ and $v' = v + 1$. Applying the approach of Type 3, this produces a standard linear optimization model with equality constraints and nonnegative variables in basic form.

Example 7.3.5. Reduce the following linear optimization model:

$$(L): \ \min z \quad \text{subject to}$$

$$\begin{cases} z = -3x_1 + x_2 + x_3, \\ -4x_1 + x_2 + 2x_3 \leq 3, \\ x_1 - 2x_2 + x_3 \leq 11, \\ -2x_1 + 2x_3 = 1, \\ x_1 \geq 0, \ x_2 \geq 0, \ x_3 \text{ free} \end{cases}$$

to a standard linear optimization model with equality constraints and nonnegative variables.

Solution. Let $y = -z$. Converting minimization to maximization and substituting the variable x_3 by $x_3^+ - x_3^-$, where $x_3^+ \geq 0$ and $x_3^- \geq 0$, and then introducing two slack variables x_4, x_5, produces a standard linear optimization model with three equality constraints and six nonnegative variables as follows:

$$(L^*): \ \max y \quad \text{subject to}$$

$$\begin{cases} y - 3x_1 + x_2 + x_3^+ - x_3^- + 0 + 0 = 0, \\ -4x_1 + x_2 + 2x_3^+ - 2x_3^- + x_4 + 0 = 3, \\ x_1 - 2x_2 + x_3^+ - x_3^- + 0 + x_5 = 11, \\ -2x_1 + 0 + 2x_3^+ - 2x_3^- + 0 + 0 = 1, \\ x_1 \geq 0, \ x_2 \geq 0, \ x_3^+ \geq 0, \ x_3^- \geq 0, \ x_4 \geq 0, \ x_5 \geq 0. \end{cases}$$

The model (L^*) is the desired model. □

7.4 Fermat rules

Let $f\colon \mathbb{R} \to \mathbb{R}$ be a function defined on $\mathbb{R}$ and with values on $\mathbb{R}$. If $f(x^*) \leq f(x)$ for all $x \in [x^* - \varepsilon, x^* + \varepsilon]$ $(\varepsilon > 0)$, then we say x^* *minimizes* f on the interval $[x^* - \varepsilon, x^* + \varepsilon]$.

Fermat rule I

Let $f\colon \mathbb{R} \to \mathbb{R}$ be a smooth function. If x^* minimizes f on the interval $[x^* - \varepsilon, x^* + \varepsilon]$ $(\varepsilon > 0)$, then $f'(x^*) = 0$.

The meaning of Fermat rule I is that the tangent to the graph of the smooth function f at the point $(x^*, f(x^*))$ is parallel to the real axis.

Now let $f\colon \mathbb{R}^n \to \mathbb{R}$ be a function defined on $\mathbb{R}^n$ and with values on $\mathbb{R}$. If $f(\boldsymbol{x}^*) \leq f(\boldsymbol{x})$ for all $\boldsymbol{x} \in B(\boldsymbol{x}^*, \varepsilon)$, where $B(\boldsymbol{x}^*, \varepsilon)$ is a closed ball of radius $\varepsilon > 0$ centered at $\boldsymbol{x}^* \in \mathbb{R}^n$, then we say $\boldsymbol{x}^*$ *minimizes* f on $B(\boldsymbol{x}^*, \varepsilon)$. Relying on this concept, Fermat rule I is extended to functions defined on the n-dimensional space $\mathbb{R}^n$.

Fermat rule II

Let $f\colon \mathbb{R}^n \to \mathbb{R}$ be a smooth function. If $\boldsymbol{x}^*$ minimizes f on $B(\boldsymbol{x}^*, \varepsilon)$ $(\varepsilon > 0)$, then

$$\nabla f(\boldsymbol{x}^*) = \left(\frac{\partial f(\boldsymbol{x}^*)}{\partial x_1}, \dots, \frac{\partial f(\boldsymbol{x}^*)}{\partial x_n} \right) = \boldsymbol{0},$$

i.e., the gradient of $f(\boldsymbol{x})$ at $\boldsymbol{x}^*$ is equal to zero, where $\boldsymbol{x} = (x_1, \dots, x_n)$ and $\boldsymbol{x}^* = (x_1^*, \dots, x_n^*)$.

The extension of Fermat rule to convex functions needs the following concepts.

Let X be a subset of $\mathbb{R}^n$. If, for any $\boldsymbol{x}_1, \boldsymbol{x}_2 \in X$, the line segment $[\boldsymbol{x}_1, \boldsymbol{x}_2] \subset X$, i.e.,

$$\boldsymbol{x}_\tau = (1 - \tau)\boldsymbol{x}_1 + \tau \boldsymbol{x}_2 \in X \quad (0 \leq \tau \leq 1),$$

then the set X is said to be a *convex set* of $\mathbb{R}^n$. It is clear by the definition of convex sets that

(a) a set consisting of a single point is a convex set;
(b) the empty set is a convex set;
(c) if $X \subset \mathbb{R}^n$ is a convex set, then the set $S = \{a\boldsymbol{x} + b \mid \boldsymbol{x} \in X, a, b \in \mathbb{R}\}$ is a convex set.

Convex cones are a particularly important subclass of convex sets. For example, $\{\boldsymbol{0}\}$, $\mathbb{R}_+^n = [0, \infty)^n$, $\mathbb{R}^n$, and closed half-spaces are all convex cones.

Let $f\colon \mathbb{R}^n \to \mathbb{R}$ and $X \subset \mathbb{R}^n$ be a convex set. If, for any $\boldsymbol{x}_1, \boldsymbol{x}_2 \in X$,

$$f\left((1 - \tau)\boldsymbol{x}_1 + \tau \boldsymbol{x}_2\right) \leq (1 - \tau) f(\boldsymbol{x}_1) + \tau f(\boldsymbol{x}_2) \quad (0 \leq \tau \leq 1),$$

then f is said to be a *convex function* on X. If, for any $\boldsymbol{x}_1, \boldsymbol{x}_2 \in X$,

$$f\left((1 - \tau)\boldsymbol{x}_1 + \tau \boldsymbol{x}_2\right) < (1 - \tau) f(\boldsymbol{x}_1) + \tau f(\boldsymbol{x}_2) \quad (0 \leq \tau \leq 1),$$

then f is said to be a *strictly convex function* on X. If f is a convex function on X, then $g = -f$ is said to be a *concave function* on X. Clearly, the linear function $f(\boldsymbol{x}) = \langle \boldsymbol{a}, \boldsymbol{x} \rangle$ is both a convex function and a concave function.

The subderivative and the set of subgradients of convex functions are two fundamental concepts of subdifferential calculus. They are defined as follows.

Definition 7.4.1. Let $f : \mathbb{R}^n \to \mathbb{R}$ be convex and its effective domain be $\operatorname{dom} f = \{\boldsymbol{x} \in \mathbb{R}^n \mid f(\boldsymbol{x}) < \infty\}$. The *subderivative* of *f at* $\boldsymbol{x}^* \in \operatorname{dom} f$ *in direction* $\boldsymbol{w}$ is defined as

$$\frac{\partial^+ f(\boldsymbol{x}^*)}{\partial \boldsymbol{w}} = \lim_{\lambda \to 0+} \frac{f(\boldsymbol{x}^* + \lambda \boldsymbol{w}) - f(\boldsymbol{x}^*)}{\lambda} \quad \text{for any } \boldsymbol{w} \in \mathbb{R}^n.$$

The set of *subgradients* of *f at* $\boldsymbol{x}^* \in \operatorname{dom} f$ is defined as

$$\partial f(\boldsymbol{x}^*) = \{\boldsymbol{y} \in \mathbb{R}^n \mid f(\boldsymbol{x}) \geq f(\boldsymbol{x}^*) + \langle \boldsymbol{y}, \boldsymbol{x} - \boldsymbol{x}^* \rangle \text{ for any } \boldsymbol{x} \in \mathbb{R}^n\}. \tag{7.4.1}$$

The set of subgradients depends on the subderivatives. If we set $\boldsymbol{x} - \boldsymbol{x}^* = \lambda \boldsymbol{w}$ $(\lambda > 0)$ in (7.4.1), then the inequality $f(\boldsymbol{x}) \geq f(\boldsymbol{x}^*) + \langle \boldsymbol{y}, \boldsymbol{x} - \boldsymbol{x}^* \rangle$ is equivalent to

$$\frac{f(\boldsymbol{x}^* + \lambda \boldsymbol{w}) - f(\boldsymbol{x}^*)}{\lambda} \geq \langle \boldsymbol{y}, \boldsymbol{w} \rangle.$$

Since f is convex, the ratio on the left-hand side is monotonically nonincreasing, and tends to $\frac{\partial^+ f(\boldsymbol{x}^*)}{\partial \boldsymbol{w}}$ $(\lambda \to 0+)$. Let $\lambda \to 0+$ on both sides of the above inequality. Then

$$\frac{\partial^+ f(\boldsymbol{x}^*)}{\partial \boldsymbol{w}} \geq \langle \boldsymbol{y}, \boldsymbol{w} \rangle.$$

So (7.4.1) has an equivalent form as follows:

$$\partial f(\boldsymbol{x}^*) = \left\{\boldsymbol{y} \in \mathbb{R} \mid \frac{\partial^+ f(\boldsymbol{x}^*)}{\partial \boldsymbol{w}} \geq \langle \boldsymbol{y}, \boldsymbol{w} \rangle \text{ for any } \boldsymbol{w} \in \mathbb{R}^n\right\}. \tag{7.4.2}$$

From this, it is seen that the set of subgradients depends on the subderivatives.

The Fermat rule for convex functions defined on $\mathbb{R}^n$ is as follows.

Fermat rule III

Let $f : \mathbb{R}^n \to \mathbb{R}$ be a convex function. Then $\boldsymbol{x}^* \in \mathbb{R}^n$ minimizes f on $B(\boldsymbol{x}^*, \varepsilon)$ $(\varepsilon > 0)$ if and only if $0 \in \partial f(\boldsymbol{x}^*)$.

Proof. By the definition, $\boldsymbol{x}^*$ minimizes f on $B(\boldsymbol{x}^*, \varepsilon)$ if and only if $f(\boldsymbol{x}) \geq f(\boldsymbol{x}^*)$ for any $\boldsymbol{x} \in B(\boldsymbol{x}^*, \varepsilon)$. Let $\boldsymbol{x} = \boldsymbol{x}^* + \lambda \boldsymbol{w}$, where $\boldsymbol{w} \in \mathbb{R}^n$ and $\lambda > 0$. Then $f(\boldsymbol{x}) \geq f(\boldsymbol{x}^*)$ for any $\boldsymbol{x} \in B(\boldsymbol{x}^*, \varepsilon)$ if and only if $f(\boldsymbol{x}^* + \lambda \boldsymbol{w}) \geq f(\boldsymbol{x}^*) + \lambda \langle \boldsymbol{0}, \boldsymbol{w} \rangle$. By Definition 7.4.1, $f(\boldsymbol{x}^* + \lambda \boldsymbol{w}) \geq f(\boldsymbol{x}^*) + \lambda \langle \boldsymbol{0}, \boldsymbol{w} \rangle$ if and only if $\frac{\partial^+ f(\boldsymbol{x}^*)}{\partial \boldsymbol{w}} \geq \langle \boldsymbol{0}, \boldsymbol{w} \rangle$ $(\boldsymbol{w} \in \mathbb{R}^n)$. Finally, by (7.4.2), it follows that $\frac{\partial^+ f(\boldsymbol{x}^*)}{\partial \boldsymbol{w}} \geq \langle \boldsymbol{0}, \boldsymbol{w} \rangle$ $(\boldsymbol{w} \in \mathbb{R}^n)$ if and only if $\boldsymbol{0} \in \partial f(\boldsymbol{x}^*)$. □

In order to vary Fermat rule III, the following two concepts are required.

Definition 7.4.2. Let Ω be a set and

$$i_\Omega(\boldsymbol{x}) = \begin{cases} 0 & \text{if } \boldsymbol{x} \in \Omega, \\ \infty & \text{otherwise.} \end{cases}$$

Then $i_\Omega(\boldsymbol{x})$ is said to be the *indicator function* of the set Ω.

It is clear by Definition 7.4.2 that the indicator function $i_\Omega(\boldsymbol{x})$ is convex if and only if the set Ω is convex.

Definition 7.4.3. Let $\Omega \subset \mathbb{R}^n$ be a nonempty closed convex set and $\boldsymbol{x}^* \in \Omega$. If $\langle \boldsymbol{v}, \boldsymbol{x} - \boldsymbol{x}^* \rangle \le 0$, where $\boldsymbol{v} \in \mathbb{R}^n$ and for all $\boldsymbol{x} \in \Omega$, then the vector $\boldsymbol{v}$ is said to be *normal* to Ω at $\boldsymbol{x}^*$, denoted by $\boldsymbol{v} \in N_\Omega(\boldsymbol{x}^*)$, where $N_\Omega(\boldsymbol{x}^*)$ is said to be the *normal cone* to Ω at $\boldsymbol{x}^*$.

Assume that $\Omega \subset \mathbb{R}^n$ is a nonempty closed convex set and $\boldsymbol{x}^* \in \Omega$. Then the set of subgradients of the indicator function of Ω is just the normal cone to Ω, i.e.,

$$\partial i_\Omega(\boldsymbol{x}^*) = N_\Omega(\boldsymbol{x}^*). \tag{7.4.3}$$

In fact, from Definition 7.4.3, it follows that

$$N_\Omega(\boldsymbol{x}^*) = \bigcap_{\boldsymbol{x} \in \Omega} \{\boldsymbol{v} \in \mathbb{R}^n \mid \langle \boldsymbol{v}, \boldsymbol{x} - \boldsymbol{x}^* \rangle \le 0\}, \tag{7.4.4}$$

i.e., $N_\Omega(\boldsymbol{x}^*)$ is the intersection of closed half-spaces. Since $\boldsymbol{x}^* \in \Omega$ and $\boldsymbol{x} \in \Omega$, it is clear by Definition 7.4.2 that

$$i_\Omega(\boldsymbol{x}^*) = 0,$$
$$i_\Omega(\boldsymbol{x}) = 0.$$

From this and (7.4.1), the set of subgradients of $i_\Omega(\boldsymbol{x})$ $(\boldsymbol{x} \in \Omega)$ at $\boldsymbol{x}^*$ is

$$\begin{aligned} \partial i_\Omega(\boldsymbol{x}^*) &= \{\boldsymbol{v} \in \mathbb{R}^n \mid i_\Omega(\boldsymbol{x}) \ge i_\Omega(\boldsymbol{x}^*) + \langle \boldsymbol{v}, \boldsymbol{x} - \boldsymbol{x}^* \rangle \text{ for any } \boldsymbol{x} \in \Omega\} \\ &= \{\boldsymbol{v} \in \mathbb{R}^n \mid \langle \boldsymbol{v}, \boldsymbol{x} - \boldsymbol{x}^* \rangle \le 0 \text{ for any } \boldsymbol{x} \in \Omega\} \\ &= \bigcap_{\boldsymbol{x} \in \Omega} \{\boldsymbol{v} \in \mathbb{R}^n \mid \langle \boldsymbol{v}, \boldsymbol{x} - \boldsymbol{x}^* \rangle \le 0\}. \end{aligned}$$

The combination of this and (7.4.4) gives (7.4.3).

The following rule is a variant of Fermat rule III.

Fermat rule IV

Let $f\colon \mathbb{R}^n \to \mathbb{R}$ be a convex function and $\Omega \subset \mathbb{R}^n$ be a nonempty closed convex set. Then $\boldsymbol{x}^*$ minimizes $f(\boldsymbol{x})$ on Ω if and only if $\boldsymbol{0} \in (\partial f(\boldsymbol{x}^*) + N_\Omega(\boldsymbol{x}^*))$, and this is further equivalent to that there is a $\boldsymbol{v} \in \partial f(\boldsymbol{x}^*)$ such that $-\boldsymbol{v} \in N_\Omega(\boldsymbol{x}^*)$.

Proof. By Definition 7.4.2, $i_\Omega(\boldsymbol{x})$ is constant. Since $\boldsymbol{x}^*$ minimizes $f(\boldsymbol{x})$ on Ω, it also minimizes $f(\boldsymbol{x}) + i_\Omega(\boldsymbol{x})$ on Ω. By Fermat rule III, $\boldsymbol{x}^*$ minimizes $f(\boldsymbol{x}) + i_\Omega(\boldsymbol{x})$ on Ω if and only if $\mathbf{0} \in \partial(f + i_\Omega)(\boldsymbol{x}^*)$. By Definition 7.4.1 and (7.4.3), it follows that

$$\partial(f + i_\Omega)(\boldsymbol{x}^*) = \partial f(\boldsymbol{x}^*) + \partial i_\Omega(\boldsymbol{x}^*) = \partial f(\boldsymbol{x}^*) + N_\Omega(\boldsymbol{x}^*).$$

Therefore, $\boldsymbol{x}^*$ minimizes $f(\boldsymbol{x})$ on Ω if and only if $\mathbf{0} \in (\partial f(\boldsymbol{x}^*) + N_\Omega(\boldsymbol{x}^*))$, and this is further equivalent to that there is a $\boldsymbol{\upsilon} \in \partial f(\boldsymbol{x}^*)$ such that $-\boldsymbol{\upsilon} \in N_\Omega(\boldsymbol{x}^*)$. □

An application of Fermat rules is to solve the nonnegative constrained convex optimization model

$$(L_{nc}): \min_{\boldsymbol{x}} f(\boldsymbol{x}) \quad \text{subject to } \boldsymbol{x} \in \mathbb{R}_+^n,$$

where the function $f: \mathbb{R}^n \to \mathbb{R}$ is convex and $\mathbb{R}_+^n = [0, \infty)^n$ is the n-dimensional nonnegative orthant. The following result is given.

Let $\boldsymbol{x}^* = (x_1^*, \ldots, x_n^*) \in \mathbb{R}_+^n$. Then $\boldsymbol{x}^*$ is an optimal solution of the model (L_{nc}) if and only if there is a $\boldsymbol{\upsilon} \in \partial f(\boldsymbol{x}^*)$ such that

$$\begin{aligned} \upsilon_i &\geq 0 \quad (i = 1, \ldots, n), \\ \upsilon_i x_i^* &= 0 \quad (i = 1, \ldots, n), \end{aligned}$$

where $\boldsymbol{\upsilon} = (\upsilon_1, \ldots, \upsilon_n)$.

In fact, according to Fermat rule IV, $\boldsymbol{x}^*$ is an optimal solution of the model (L_{nc}) if and only if there is a $\boldsymbol{\upsilon} \in \partial f(\boldsymbol{x}^*)$ such that $-\boldsymbol{\upsilon} \in N_{\mathbb{R}_+^n}(\boldsymbol{x}^*)$.

However, $-\boldsymbol{\upsilon} \in N_{\mathbb{R}_+^n}(\boldsymbol{x}^*)$ if and only if $-\upsilon_i \in N_{\mathbb{R}_+}(x_i^*)$ $(i = 1, \ldots, n)$, where

$$\begin{aligned} \boldsymbol{\upsilon} &= (\upsilon_1, \ldots, \upsilon_n), \\ \boldsymbol{x}^* &= (x_1^*, \ldots, x_n^*), \\ \mathbb{R}_+^n &= [0, \infty)^n, \quad \mathbb{R}_+ = [0, \infty). \end{aligned}$$

By Definition 7.4.3, a direct computation shows that for $i = 1, \ldots, n$,

$$N_{\mathbb{R}_+}(x_i^*) = \begin{cases} \mathbb{R}_- & \text{if } x_i^* = 0, \\ 0 & \text{if } 0 < x_i^* < \infty, \end{cases}$$

where $\mathbb{R}_- = (-\infty, 0)$. So $-\upsilon_i \in N_{\mathbb{R}_+}(x_i^*)$ $(i = 1, \ldots, n)$ if and only if $\upsilon_i \geq 0$ and $\upsilon_i x_i^* = 0$ $(i = 1, \ldots, n)$. Therefore, $\boldsymbol{x}^*$ is an optimal solution of the model (L_{nc}) if and only if there is a $\boldsymbol{\upsilon} \in \partial f(\boldsymbol{x}^*)$ such that

$$\begin{aligned} \upsilon_i &\geq 0 \quad (i = 1, \ldots, n), \\ \upsilon_i x_i^* &= 0 \quad (i = 1, \ldots, n). \end{aligned}$$

7.5 Karush–Kuhn–Tucker optimality conditions

Karush–Kuhn–Tucker (KKT) conditions are the optimality conditions for solving the optimization models. These conditions are related to KKT multipliers. In order to give the KKT conditions, we first introduce some propositions.

Proposition 7.5.1. *Let Ω be a hyperplane and*

$$\Omega = \{\boldsymbol{x} \in \mathbb{R}^n \mid \langle \boldsymbol{a}, \boldsymbol{x} \rangle = b\} \quad (\boldsymbol{0} \neq \boldsymbol{a} \in \mathbb{R}^n, b \in \mathbb{R}).$$

Then, for $\boldsymbol{x}^* \in \Omega$, the vector $-\boldsymbol{v} \in N_\Omega(\boldsymbol{x}^*)$ if and only if there is a $\lambda^* \in \mathbb{R}$ such that $\boldsymbol{v} = \lambda^* \boldsymbol{a}$.

Proof. Let $U = \Omega - \boldsymbol{x}^*$. For each $\boldsymbol{x} \in U$, there is a $\boldsymbol{x}' \in \Omega$ such that $\boldsymbol{x} = \boldsymbol{x}' - \boldsymbol{x}^*$. Note that

$$\langle \boldsymbol{a}, \boldsymbol{x}' \rangle = b \quad (\boldsymbol{x}' \in \Omega),$$
$$\langle \boldsymbol{a}, \boldsymbol{x}^* \rangle = b \quad (\boldsymbol{x}^* \in \Omega).$$

It follows that for each $\boldsymbol{x} \in U$,

$$\langle \boldsymbol{a}, \boldsymbol{x} \rangle = \langle \boldsymbol{a}, \boldsymbol{x}' - \boldsymbol{x}^* \rangle = \langle \boldsymbol{a}, \boldsymbol{x}' \rangle - \langle \boldsymbol{a}, \boldsymbol{x}^* \rangle = 0.$$

So $U = \{\boldsymbol{x} \mid \langle \boldsymbol{a}, \boldsymbol{x} \rangle = 0\}$.

Let $V = \{\lambda \boldsymbol{a} \mid \lambda \in \mathbb{R}\}$. For any $\boldsymbol{x} \in U$ and any $\lambda \boldsymbol{a} \in V$, $\langle \lambda \boldsymbol{a}, \boldsymbol{x} \rangle = \lambda \langle \boldsymbol{a}, \boldsymbol{x} \rangle = 0$. So $U \perp V$.

Now we prove that $N_\Omega(\boldsymbol{x}^*) = V$.

If $\boldsymbol{v} \in V$, then $\boldsymbol{v} = \lambda \boldsymbol{a}$. For all $\boldsymbol{x} \in \Omega$,

$$\langle \boldsymbol{v}, \boldsymbol{x} - \boldsymbol{x}^* \rangle = \langle \lambda \boldsymbol{a}, \boldsymbol{x} - \boldsymbol{x}^* \rangle = \lambda(\langle \boldsymbol{a}, \boldsymbol{x} \rangle - \langle \boldsymbol{a}, \boldsymbol{x}^* \rangle) = \lambda(b - b) = 0.$$

By Definition 7.4.3, it follows that $\boldsymbol{v} \in N_\Omega(\boldsymbol{x}^*)$. So $V \subset N_\Omega(\boldsymbol{x}^*)$.

Conversely, if $\boldsymbol{v} \in N_\Omega(\boldsymbol{x}^*)$, then $\boldsymbol{v} \in V$.

If $\boldsymbol{v} \notin V$, since $U \perp V$, $\boldsymbol{v} = \lambda \boldsymbol{a} + \hat{\boldsymbol{v}}$ for some $\lambda \in \mathbb{R}$ and $\boldsymbol{0} \neq \hat{\boldsymbol{v}} \in U$. On the one hand, since $\hat{\boldsymbol{v}} \neq 0$,

$$\langle \hat{\boldsymbol{v}}, \hat{\boldsymbol{v}} \rangle > 0. \tag{7.5.1}$$

On the other hand, since $\hat{\boldsymbol{v}} \in U$, there is a $\hat{\boldsymbol{x}} \in \Omega$ such that $\hat{\boldsymbol{v}} = \hat{\boldsymbol{x}} - \boldsymbol{x}^*$. Again, by the assumption $\boldsymbol{v} \in N_\Omega(\boldsymbol{x}^*)$, it follows by Definition 7.4.3 that

$$\langle \boldsymbol{v}, \hat{\boldsymbol{x}} - \boldsymbol{x}^* \rangle \leq 0. \tag{7.5.2}$$

The left-hand side of (7.5.2) becomes

$$\begin{aligned}\langle \boldsymbol{v}, \hat{\boldsymbol{x}} - \boldsymbol{x}^* \rangle &= \langle \lambda \boldsymbol{a} + \hat{\boldsymbol{v}}, \hat{\boldsymbol{x}} - \boldsymbol{x}^* \rangle \\ &= \lambda(\langle \boldsymbol{a}, \hat{\boldsymbol{x}} \rangle - \langle \boldsymbol{a}, \boldsymbol{x}^* \rangle) + \langle \hat{\boldsymbol{v}}, \hat{\boldsymbol{x}} - \boldsymbol{x}^* \rangle \\ &= \lambda(b - b) + \langle \hat{\boldsymbol{v}}, \hat{\boldsymbol{x}} - \boldsymbol{x}^* \rangle \\ &= \langle \hat{\boldsymbol{v}}, \hat{\boldsymbol{v}} \rangle,\end{aligned}$$

and so $\langle \hat{\boldsymbol{v}}, \hat{\boldsymbol{v}} \rangle \leq 0$. This is in contradiction with (7.5.1). Thus, $\boldsymbol{v} \in V$. So $N_\Omega(\boldsymbol{x}^*) \subset V$.

Hence $N_\Omega(\boldsymbol{x}^*) = V$. This implies that $-\boldsymbol{v} \in N_\Omega(\boldsymbol{x}^*)$ if and only if $-\boldsymbol{v} \in V$. Note that $V = \{\lambda \boldsymbol{a} \mid \lambda \in \mathbb{R}\}$. Then, $-\boldsymbol{v} \in V$ if and only if there is a $\lambda^* \in \mathbb{R}$ such that $\boldsymbol{v} = \lambda^* \boldsymbol{a}$. So Proposition 7.5.1 follows. □

Proposition 7.5.2. *Let Ω be a closed half-space and*

$$\Omega = \{\boldsymbol{x} \in \mathbb{R}^n \mid \langle \boldsymbol{a}, \boldsymbol{x}\rangle \geq b\} \quad (\boldsymbol{0} \neq \boldsymbol{a} \in \mathbb{R}^n, b \in \mathbb{R}).$$

Then, for $\boldsymbol{x}^* \in \Omega$, the vector $-\boldsymbol{v} \in N_\Omega(\boldsymbol{x}^*)$ if and only if there is a $\lambda^* \geq 0$ such that

$$\lambda^*(\langle \boldsymbol{a}, \boldsymbol{x}^*\rangle - b) = 0,$$
$$\boldsymbol{v} = \lambda^* \boldsymbol{a}.$$

Proof. There are two cases.

Case 1. $\langle \boldsymbol{a}, \boldsymbol{x}^*\rangle > b$. Then $\boldsymbol{x}^*$ is an interior point of Ω. So $N_\Omega(\boldsymbol{x}^*) = \{\boldsymbol{0}\}$. The conclusion holds clearly.

Case 2. $\langle \boldsymbol{a}, \boldsymbol{x}^*\rangle = b$. Let

$$\Omega_1 = \{\boldsymbol{x} \in \mathbb{R}^n \mid \langle \boldsymbol{a}, \boldsymbol{x}\rangle > b\},$$
$$\Omega_2 = \{\boldsymbol{x} \in \mathbb{R}^n \mid \langle \boldsymbol{a}, \boldsymbol{x}\rangle = b\},$$

where $\boldsymbol{0} \neq \boldsymbol{a} \in \mathbb{R}^n$ and $b \in \mathbb{R}$. Then $\Omega = \Omega_1 \bigcup \Omega_2$ and $\boldsymbol{x}^* \in \Omega_2$.

Let $U = \Omega_2 - \boldsymbol{x}^*$. A similar argument of Proposition 7.5.1 implies that

$$U = \{\boldsymbol{x} \in \mathbb{R}^n \mid \langle \boldsymbol{a}, \boldsymbol{x}\rangle = 0\}.$$

Let $V = \{\lambda \boldsymbol{a} \mid \lambda \leq 0\}$. For any $\boldsymbol{x} \in U$ and any $\lambda \boldsymbol{a} \in V$, $\langle \lambda \boldsymbol{a}, \boldsymbol{x}\rangle = \lambda \langle \boldsymbol{a}, \boldsymbol{x}\rangle = 0$. So $U \perp V$.

Now we prove $N_\Omega(\boldsymbol{x}^*) = V$.

First, we prove that if $\boldsymbol{v} \in N_\Omega(\boldsymbol{x}^*)$, then $\boldsymbol{v} \in V$.

If $\boldsymbol{v} \notin V$, since $U \perp V$, $\boldsymbol{v} = \lambda \boldsymbol{a} + \hat{\boldsymbol{v}}$ for some $\lambda \leq 0$ and $\boldsymbol{0} \neq \hat{\boldsymbol{v}} \in U$. On the one hand, since $\hat{\boldsymbol{v}} \neq 0$,

$$\langle \hat{\boldsymbol{v}}, \hat{\boldsymbol{v}}\rangle > 0. \tag{7.5.3}$$

On the other hand, since $\hat{\boldsymbol{v}} \in U$, there is an $\hat{\boldsymbol{x}} \in \Omega_2$ such that $\hat{\boldsymbol{v}} = \hat{\boldsymbol{x}} - \boldsymbol{x}^*$. Since $\Omega = \Omega_1 \bigcup \Omega_2$, clearly, $\hat{\boldsymbol{x}} \in \Omega$. Again, by the assumption that $\boldsymbol{v} \in N_\Omega(\boldsymbol{x}^*)$, it follows by Definition 7.4.3 that

$$\langle \boldsymbol{v}, \hat{\boldsymbol{x}} - \boldsymbol{x}^*\rangle \leq 0. \tag{7.5.4}$$

Since $\boldsymbol{v} = \lambda \boldsymbol{a} + \hat{\boldsymbol{v}}$, where $\hat{\boldsymbol{v}} = \hat{\boldsymbol{x}} - \boldsymbol{x}^*$ and $\hat{\boldsymbol{x}} \in \Omega_2$, the left-hand side of (7.5.4) becomes

$$\begin{aligned}
\langle \boldsymbol{v}, \hat{\boldsymbol{x}} - \boldsymbol{x}^*\rangle &= \langle \lambda \boldsymbol{a} + \hat{\boldsymbol{v}}, \hat{\boldsymbol{x}} - \boldsymbol{x}^*\rangle \\
&= \lambda(\langle \boldsymbol{a}, \hat{\boldsymbol{x}}\rangle - \langle \boldsymbol{a}, \boldsymbol{x}^*\rangle) + \langle \hat{\boldsymbol{v}}, \hat{\boldsymbol{x}} - \boldsymbol{x}^*\rangle \\
&= \lambda(b - b) + \langle \hat{\boldsymbol{v}}, \hat{\boldsymbol{v}}\rangle \\
&= \langle \hat{\boldsymbol{v}}, \hat{\boldsymbol{v}}\rangle,
\end{aligned}$$

and so $\langle \hat{\boldsymbol{v}}, \hat{\boldsymbol{v}}\rangle \leq 0$. This is in contradiction with (7.5.3). Thus, $\boldsymbol{v} \in V$. So $N_\Omega(\boldsymbol{x}^*) \subset V$.

Next, we prove $V \subset N_\Omega(\boldsymbol{x}^*)$.

Let $v \in V$. Then $v = \lambda \boldsymbol{a}$, where $\lambda \leq 0$. Note that $\boldsymbol{x}^* \in \Omega_2$. For all $\boldsymbol{x} \in \Omega$,

$$\langle v, \boldsymbol{x} - \boldsymbol{x}^*\rangle = \langle \lambda \boldsymbol{a}, \boldsymbol{x} - \boldsymbol{x}^*\rangle = \lambda(\langle \boldsymbol{a}, \boldsymbol{x}\rangle - \langle \boldsymbol{a}, \boldsymbol{x}^*\rangle) = \lambda(\langle \boldsymbol{a}, \boldsymbol{x}\rangle - b) \leq 0.$$

By Definition 7.4.3, $v \in N_\Omega(\boldsymbol{x}^*)$. So $V \subset N_\Omega(\boldsymbol{x}^*)$.

Hence $V = N_\Omega(\boldsymbol{x}^*)$. This implies that $-\boldsymbol{v} \in N_\Omega(\boldsymbol{x}^*)$ if and only if $-\boldsymbol{v} \in V$. Note that $V = \{\lambda \boldsymbol{a} \mid \lambda \le 0\}$. Then, $-\boldsymbol{v} \in V$ if and only if there is a $\lambda^* \ge 0$ such that $\boldsymbol{v} = \lambda^* \boldsymbol{a}$ and $\lambda^*(\langle \boldsymbol{a}, \boldsymbol{x}^*\rangle - b) = 0$ since the assumption in Case 2 is $\langle \boldsymbol{a}, \boldsymbol{x}^*\rangle = b$. So Proposition 7.5.2 follows. □

Propositions 7.5.1 and 7.5.2 can be generalized as follows.

Proposition 7.5.3. *Let Ω be a nonempty polyhedral set and*

$$\Omega = \left\{\boldsymbol{x} \in \mathbb{R}^n \mid \langle \boldsymbol{a}_i, \boldsymbol{x}\rangle \ge b_i \ (i = 1, \dots, s),\ \langle \boldsymbol{a}_i, \boldsymbol{x}\rangle = b_i \ (i = s+1, \dots, m)\right\},$$

where $\boldsymbol{0} \ne \boldsymbol{a}_i \in \mathbb{R}^n$ and $b_i \in \mathbb{R}$. Then for $\boldsymbol{x}^* \in \Omega$, the vector $-\boldsymbol{v} \in N_\Omega(\boldsymbol{x}^*)$ if and only if there are $\lambda_i^* \ge 0$ $(i = 1, \dots, s)$ and $\lambda_i^* \in \mathbb{R}$ $(i = s+1, \dots, m)$ such that

$$\lambda_i^*(\langle \boldsymbol{a}_i, \boldsymbol{x}^*\rangle - b_i) = 0 \quad (i = 1, \dots, s),$$
$$\boldsymbol{v} = \lambda_1^* \boldsymbol{a}_1 + \cdots + \lambda_s^* \boldsymbol{a}_s + \lambda_{s+1}^* \boldsymbol{a}_{s+1} + \cdots + \lambda_m^* \boldsymbol{a}_m.$$

Proof. Let

$$\Omega_1^i = \{\boldsymbol{x} \in \mathbb{R}^n \mid \langle \boldsymbol{a}_i, \boldsymbol{x}\rangle \ge b_i\} \quad (i = 1, \dots, s),$$
$$\Omega_2^i = \{\boldsymbol{x} \in \mathbb{R}^n \mid \langle \boldsymbol{a}_i, \boldsymbol{x}\rangle = b_i\} \quad (i = s+1, \dots, m).$$

and

$$\Omega_1 = \bigcap_1^s \Omega_1^i, \quad \Omega_2 = \bigcap_{s+1}^m \Omega_2^i.$$

Then both Ω_1 and Ω_2 are polyhedral sets. Let $\Omega = \Omega_1 \bigcap \Omega_2$. Then Ω is also a polyhedral set. It is well known that the normal cones of polyhedral sets have the following property:

$$N_\Omega(\boldsymbol{x}^*) = N_{\Omega_1}(\boldsymbol{x}^*) + N_{\Omega_2}(\boldsymbol{x}^*) = N_{\Omega_1^1}(\boldsymbol{x}^*) + \cdots + N_{\Omega_1^s}(\boldsymbol{x}^*) + N_{\Omega_2^{s+1}}(\boldsymbol{x}^*) + \cdots + N_{\Omega_2^m}(\boldsymbol{x}^*).$$

So $-\boldsymbol{v} \in N_\Omega(\boldsymbol{x}^*)$ if and only if there exist $\boldsymbol{v}_1^i$ $(i = 1, \dots, s)$ and $\boldsymbol{v}_2^i$ $(i = s+1, \dots, m)$ such that

$$-\boldsymbol{v}_1^i \in N_{\Omega_1^i}(\boldsymbol{x}^*) \quad (i = 1, \dots, s),$$
$$-\boldsymbol{v}_2^i \in N_{\Omega_2^i}(\boldsymbol{x}^*) \quad (i = s+1, \dots, m),$$

where $\boldsymbol{v} = \boldsymbol{v}_1^1 + \cdots + \boldsymbol{v}_1^s + \boldsymbol{v}_2^{s+1} + \cdots + \boldsymbol{v}_2^m$.

By Proposition 7.5.2, $-\boldsymbol{v}_1^i \in N_{\Omega_1^i}(\boldsymbol{x}^*)$ $(i = 1, \dots, s)$ if and only if there is a $\lambda_i^* \ge 0$ $(i = 1, \dots, s)$ such that

$$\lambda_i^*(\langle \boldsymbol{a}_i, \boldsymbol{x}^*\rangle - b) = 0 \qquad (i = 1, \dots, s),$$
$$\boldsymbol{v}_1^i = \lambda_i^* \boldsymbol{a}_i \quad (i = 1, \dots, s).$$

By Proposition 7.5.1, $-\boldsymbol{v}_2^i \in N_{\Omega_2^i}(\boldsymbol{x}^*)$ $(i = s+1, \dots, m)$ if and only if there is a $\lambda_i^* \in \mathbb{R}(i = s+1, \dots, m)$ such that

$$\boldsymbol{v}_2^i = \lambda_i^* \boldsymbol{a}_i \quad (i = s+1, \dots, m).$$

Therefore, $-\boldsymbol{v} \in N_\Omega(\boldsymbol{x}^*)$ if and only if there are $\lambda_i^* \ge 0$ $(i = 1, \dots, s)$ and $\lambda_i^* \in \mathbb{R}$ $(i = s+1, \dots, m)$ such that

$$\lambda_i^*(\langle \boldsymbol{a}_i, \boldsymbol{x}^* \rangle - b) = 0 \quad (i = 1, \dots, s),$$
$$\boldsymbol{v} = \lambda_1^* \boldsymbol{a}_1 + \cdots + \lambda_s^* \boldsymbol{a}_s + \lambda_{s+1}^* \boldsymbol{a}_{s+1} + \cdots + \lambda_m^* \boldsymbol{a}_m. \qquad \square$$

Now we solve the convex optimization models by Propositions 7.5.1, 7.5.2, and 7.5.3.

The *simplest convex optimization model with an equality constraint* is the form

$$(L_{ce}): \min_{\boldsymbol{x}} f(\boldsymbol{x}) \quad \text{subject to} \quad \langle \boldsymbol{a}, \boldsymbol{x} \rangle = b,$$

where $f: \mathbb{R}^n \to \mathbb{R}$ is convex and $\boldsymbol{0} \ne \boldsymbol{a} \in \mathbb{R}^n$, $\boldsymbol{x} \in \mathbb{R}^n$, $b \in \mathbb{R}$. Its feasible set is

$$\Omega_{ce} = \{\boldsymbol{x} \in \mathbb{R}^n \mid \langle \boldsymbol{a}, \boldsymbol{x} \rangle = b\}.$$

Theorem 7.5.4. *$\boldsymbol{x}^* \in \Omega_{ce}$ is the optimal solution of the model (L_{ce}) if and only if there is a $\lambda^* \in \mathbb{R}$ such that $\lambda^* \boldsymbol{a} \in \partial f(\boldsymbol{x}^*)$, where $\langle \boldsymbol{a}, \boldsymbol{x}^* \rangle = b$.*

Proof. It is seen that the feasible set Ω_{ce} is a hyperplane and $\boldsymbol{x}^*$ minimizes f on Ω_{ce}, where $\langle \boldsymbol{a}, \boldsymbol{x}^* \rangle = b$. Fermat rule IV shows that $\boldsymbol{x}^*$ minimizes f on Ω_{ce} if and only if there is a $\boldsymbol{v} \in \partial f(\boldsymbol{x}^*)$ such that $-\boldsymbol{v} \in N_{\Omega_{ce}}(\boldsymbol{x}^*)$. Proposition 7.5.1 shows that when the feasible set Ω_{ce} is a hyperplane, $-\boldsymbol{v} \in N_{\Omega_{ce}}(\boldsymbol{x}^*)$ if and only if there is a $\lambda^* \in \mathbb{R}$ such that $\boldsymbol{v} = \lambda^* \boldsymbol{a}$. Note that $\boldsymbol{v} \in \partial f(\boldsymbol{x}^*)$. Therefore, $\boldsymbol{x}^* \in \Omega_{ce}$ is the optimal solution of the model (L_{ce}) if and only if there is a $\lambda^* \in \mathbb{R}$ such that $\lambda^* \boldsymbol{a} \in \partial f(\boldsymbol{x}^*)$, where $\langle \boldsymbol{a}, \boldsymbol{x}^* \rangle = b$. $\square$

The *simplest convex optimization model with an inequality constraint* is the form

$$(L_{ci}): \min_{\boldsymbol{x}} f(\boldsymbol{x}) \quad \text{subject to} \quad \langle \boldsymbol{a}, \boldsymbol{x} \rangle \ge b,$$

where $f: \mathbb{R}^n \to \mathbb{R}$ is convex and $\boldsymbol{0} \ne \boldsymbol{a} \in \mathbb{R}^n$, $\boldsymbol{x} \in \mathbb{R}^n$, $b \in \mathbb{R}$. Its feasible set is

$$\Omega_{ci} = \{\boldsymbol{x} \in \mathbb{R}^n \mid \langle \boldsymbol{a}, \boldsymbol{x} \rangle \ge b\}.$$

Theorem 7.5.5. *$\boldsymbol{x}^* \in \Omega_{ci}$ is the optimal solution of the model (L_{ci}) if and only if there is a $\lambda^* \ge 0$ such that $\lambda^* \boldsymbol{a} \in \partial f(\boldsymbol{x}^*)$ and $\lambda^*(\langle \boldsymbol{a}, \boldsymbol{x}^* \rangle - b) = 0$, where $\langle \boldsymbol{a}, \boldsymbol{x}^* \rangle \ge b$.*

Proof. It is seen that the feasible set Ω_{ci} is a half-space and $\boldsymbol{x}^*$ minimizes f on Ω_{ci}, where $\langle \boldsymbol{a}, \boldsymbol{x}^* \rangle \ge b$. Fermat rule IV says that $\boldsymbol{x}^*$ minimizes f on Ω_{ci} if and only if there is a $\boldsymbol{v} \in \partial f(\boldsymbol{x}^*)$ such that $-\boldsymbol{v} \in N_{\Omega_{ci}}(\boldsymbol{x}^*)$. Proposition 7.5.2 says that when the feasible set Ω_{ci} is a half-space, $-\boldsymbol{v} \in N_{\Omega_{ci}}(\boldsymbol{x}^*)$ if and only if there is a $\lambda^* \ge 0$ such that

$$\lambda^*(\langle \boldsymbol{a}, \boldsymbol{x}^* \rangle - b) = 0,$$
$$\boldsymbol{v} = \lambda^* \boldsymbol{a}.$$

From this and $\boldsymbol{v} \in \partial f(\boldsymbol{x}^*)$, it follows that $\boldsymbol{x}^*$ is the optimal solution of the model (L_{ci}) if and only if there is a $\lambda^* \ge 0$ such that $\lambda^*(\langle \boldsymbol{a}, \boldsymbol{x}^* \rangle - b) = 0$ and $\lambda^* \boldsymbol{a} \in \partial f(\boldsymbol{x}^*)$. $\square$

The *linearly constrained convex optimization model* is the form

$$
\begin{aligned}
&(L_{lc}):\ \min_{\boldsymbol{x}} f(\boldsymbol{x}) \quad \text{subject to}\\
&\text{inequality constraints:}\quad \langle \boldsymbol{a}_i, \boldsymbol{x}\rangle \geq b_i \quad (i = 1, \ldots, s),\\
&\text{and equality constraints:}\quad \langle \boldsymbol{a}_i, \boldsymbol{x}\rangle = b_i \quad (i = s+1, \ldots, m),\\
&\text{and the restriction:}\quad \boldsymbol{x} \in X \subset \mathbb{R}^n,
\end{aligned}
$$

where $f: \mathbb{R}^n \to \mathbb{R}$ is convex and X is a polyhedral set, and $\boldsymbol{a}_i, \boldsymbol{x} \in \mathbb{R}^n$, $b_i \in \mathbb{R}$ $(i = 1, \ldots, m)$. Let

$$X_{lc} = \{\boldsymbol{x} \in \mathbb{R}^n \mid \langle \boldsymbol{a}_i, \boldsymbol{x}\rangle \geq b_i\ (i = 1, \ldots, s);\ \langle \boldsymbol{a}_i, \boldsymbol{x}\rangle = b_i\ (i = s+1, \ldots, m)\}.$$

Its feasible set is

$$\Omega_{lc} = X_{lc} \bigcap X.$$

Both X and X_{lc} are polyhedral sets, so is Ω_{lc}. The assumption that the function $f: \mathbb{R}^n \to \mathbb{R}$ is convex and the set X is a polyhedral set is called the *blanket assumption*. The convex property of the function ensures that the function is continuous.

Theorem 7.5.6. *$\boldsymbol{x}^* \in \Omega_{lc}$ is the optimal solution of the model (L_{lc}) if and only if the following three conditions hold:*
(a) *$\langle \boldsymbol{a}_i, \boldsymbol{x}^*\rangle \geq b_i$ $(i = 1, \ldots, s)$ and $\langle \boldsymbol{a}_i, \boldsymbol{x}^*\rangle = b_i$ $(i = s+1, \ldots, m)$;*
(b) *there are m multipliers $\lambda_1^*, \ldots, \lambda_m^*$ such that*

$$
\begin{aligned}
\lambda_i^* &\geq 0 \quad (i = 1, \ldots, s),\\
\lambda_i^* &\in \mathbb{R} \quad (i = s+1, \ldots, m),\\
\lambda_i^*(\langle \boldsymbol{a}_i, \boldsymbol{x}^*\rangle - b_i) &= 0 \quad (i = 1, \ldots, s);
\end{aligned}
$$

(c) *$\boldsymbol{x}^*$ minimizes $f(\boldsymbol{x}) + \langle \lambda_1^* \boldsymbol{a}_1 + \cdots + \lambda_m^* \boldsymbol{a}_m, \boldsymbol{x}\rangle$ on X.*
Here (c) is equivalent to that there is a $\boldsymbol{v} \in \partial f(\boldsymbol{x}^*)$ such that $-(\boldsymbol{v} + \lambda_1^* \boldsymbol{a}_1 + \cdots + \lambda_m^* \boldsymbol{a}_m) \in N_X(\boldsymbol{x}^*)$.

The multipliers $\lambda_1^*, \ldots, \lambda_m^*$ in Theorem 7.5.6 are called *KKT multipliers* of the linearly constrained convex optimization model. The conditions (a)–(c) in Theorem 7.5.6 are called *KKT-optimality conditions* of the linearly constrained convex optimization model.

Proof. By $\boldsymbol{x}^* \in \Omega_{lc}$, it is clear that

$$
\begin{aligned}
\langle \boldsymbol{a}_i, \boldsymbol{x}^*\rangle &\geq b_i \quad (i = 1, \ldots, s),\\
\langle \boldsymbol{a}_i, \boldsymbol{x}^*\rangle &= b_i \quad (i = s+1, \ldots, m).
\end{aligned}
$$

By Fermat rule IV, $\boldsymbol{x}^*$ minimizes f on Ω_{lc} if and only if $\boldsymbol{0} \in (\partial f(\boldsymbol{x}^*) + N_{\Omega_{lc}}(\boldsymbol{x}^*))$. Note that

$$\Omega_{lc} = X_{lc} \bigcap X.$$

It is well known that normal cones of polyhedral sets have the following property:

$$N_{\Omega_{lc}}(\boldsymbol{x}^*) = N_{X_{lc}}(\boldsymbol{x}^*) + N_X(\boldsymbol{x}^*).$$

So $\mathbf{0} \in (\partial f(\boldsymbol{x}^*) + N_{\Omega_{lc}}(\boldsymbol{x}^*))$ if and only if

$$\mathbf{0} \in (\partial f(\boldsymbol{x}^*) + N_{X_{lc}}(\boldsymbol{x}^*) + N_X(\boldsymbol{x}^*)) . \tag{7.5.5}$$

By Fermat rule IV, this is further equivalent to that there is a $\boldsymbol{v} \in (\partial f(\boldsymbol{x}^*) + N_X(\boldsymbol{x}^*))$ such that $-\boldsymbol{v} \in N_{X_{lc}}(\boldsymbol{x}^*)$. By Proposition 7.5.3, $-\boldsymbol{v} \in N_{X_{lc}}(\boldsymbol{x}^*)$ if and only if there are $\lambda_i^* \geq 0$ $(i = 1, \dots, s)$ and $\lambda_i^* \in \mathbb{R}$ $(i = s+1, \dots, m)$ such that

$$\begin{aligned} \lambda_i^*(\langle \boldsymbol{a}_i, \boldsymbol{x}^* \rangle - b_i) = 0 \quad (i = 1, \dots, s), \\ \boldsymbol{v} = \lambda_1^* \boldsymbol{a}_1 + \dots + \lambda_m^* \boldsymbol{a}_m. \end{aligned}$$

Note that $\boldsymbol{v} \in (\partial f(\boldsymbol{x}^*) + N_X(\boldsymbol{x}^*))$. Then $(\lambda_1^* \boldsymbol{a}_1 + \dots + \lambda_m^* \boldsymbol{a}_m) \in (\partial f(\boldsymbol{x}^*) + N_X(\boldsymbol{x}^*))$ or

$$\mathbf{0} \in (\partial f(\boldsymbol{x}^*) + N_X(\boldsymbol{x}^*) - \lambda_1^* \boldsymbol{a}_1 - \dots - \lambda_m^* \boldsymbol{a}_m). \tag{7.5.6}$$

Note that $\partial \langle \lambda_i^* \boldsymbol{a}_i, \boldsymbol{x}^* \rangle = \lambda_i^* \boldsymbol{a}_i$ $(i = 1, \dots, m)$. Then $\mathbf{0} \in (\partial f(\boldsymbol{x}^*) + N_X(\boldsymbol{x}^*) - \partial \langle \lambda_1^* \boldsymbol{a}_1 + \dots + \lambda_m^* \boldsymbol{a}_m, \boldsymbol{x}^* \rangle)$ or

$$\mathbf{0} \in (\partial [f(\boldsymbol{x}^*) - \langle \lambda_1^* \boldsymbol{a}_1 + \dots + \lambda_m^* \boldsymbol{a}_m, \boldsymbol{x}^* \rangle] + N_X(\boldsymbol{x}^*)) . \tag{7.5.7}$$

By Fermat rule IV, (7.5.7) holds if and only if $\boldsymbol{x}^*$ minimizes $f(\boldsymbol{x}) - \langle \lambda_1^* \boldsymbol{a}_1 + \dots + \lambda_m^* \boldsymbol{a}_m, \boldsymbol{x} \rangle$ on X, and (7.5.6) holds if and only if there is $\boldsymbol{v} \in \partial f(\boldsymbol{x}^*)$ such that $(-\boldsymbol{v} + \lambda_1^* \boldsymbol{a}_1 + \dots + \lambda_m^* \boldsymbol{a}_m) \in N_X(\boldsymbol{x}^*)$. □

In the linearly constrained convex optimization model, if $X = \mathbb{R}_+^r \times \mathbb{R}^{n-r}$, the corresponding optimization model is called a *linearly nonnegative constrained convex optimization model*. Again, if $f(\boldsymbol{x}) = \langle \boldsymbol{c}, \boldsymbol{x} \rangle$, where $\boldsymbol{c}$ is an n-dimensional constant vector, the corresponding optimization model is a linear optimization model. For these two models, two corollaries of Theorem 7.5.6 are given as follows.

Corollary 7.5.7. *Let $X = \mathbb{R}_+^r \times \mathbb{R}^{n-r}$. Then $\boldsymbol{x}^* \in \Omega_{lc}$ is an optimal solution of the linearly nonnegative constrained convex optimization model if and only if the following KKT-optimality conditions* (a)–(d) *hold:*

(a) $x_j^* \geq 0$ $(j = 1, \dots, r)$ *and* $x_j^* \in \mathbb{R}$ $(j = r+1, \dots, n)$*, where* $\boldsymbol{x}^* = (x_1^*, \dots, x_n^*)$*;*

(b) $\langle \boldsymbol{a}_i, \boldsymbol{x}^* \rangle \geq b_i$ $(i = 1, \dots, s)$ *and* $\langle \boldsymbol{a}_i, \boldsymbol{x}^* \rangle = b_i$ $(i = s+1, \dots, m)$*, where* $\boldsymbol{a}_i = (a_{i1}, \dots, a_{in})$*;*

(c) *there are m KKT multipliers $\lambda_1^*, \dots, \lambda_m^*$ such that*

$$\begin{aligned} \lambda_i^* &\geq 0 \quad (i = 1, \dots, s), \\ \lambda_i^* &\in \mathbb{R} \quad (i = s+1, \dots, m), \\ \lambda_i^*(\langle \boldsymbol{a}_i, \boldsymbol{x}^* \rangle - b_i) &= 0 \quad (i = 1, \dots, s); \end{aligned}$$

(d) *there is a* $\boldsymbol{v} \in \partial f(\boldsymbol{x}^*)$ *such that*

$$\begin{aligned} v_j - \lambda_1^* a_{1j} - \cdots - \lambda_m^* a_{mj} &\geq 0 \quad (j = 1, \ldots, r), \\ (v_j - \lambda_1^* a_{1j} - \cdots - \lambda_m^* a_{mj})\, x_j^* &= 0 \quad (j = 1, \ldots, r), \\ v_j - \lambda_1^* a_{1j} - \cdots - \lambda_m^* a_{mj} &= 0 \quad (j = r+1, \ldots, n), \end{aligned}$$

where $\boldsymbol{v} = (v_1, \ldots, v_n)$.

Proof. Theorem 7.5.6 gives that $\boldsymbol{x}^* \in \Omega_{lc}$ is the optimal solution of the model (L_{lc}) if and only if the following KKT-optimality conditions (a)–(c) hold:

(a) $\langle \boldsymbol{a}_i, \boldsymbol{x}^* \rangle \geq b_i$ $(i = 1, \ldots, s)$ and $\langle \boldsymbol{a}_i, \boldsymbol{x}^* \rangle = b_i$ $(i = s+1, \ldots, m)$;

(b) there are m KKT multipliers $\lambda_1^*, \ldots, \lambda_m^*$ such that

$$\begin{aligned} \lambda_i^* &\geq 0 \quad (i = 1, \ldots, s), \\ \lambda_i^* &\in \mathbb{R} \quad (i = s+1, \ldots, m), \\ \lambda_i^* (\langle \boldsymbol{a}_i, \boldsymbol{x}^* \rangle - b_i) &= 0 \quad (i = 1, \ldots, s); \end{aligned}$$

(c) there is a $\boldsymbol{v} \in \partial f(\boldsymbol{x}^*)$ such that $(-\boldsymbol{v} + \lambda_1^* \boldsymbol{a}_1 + \cdots + \lambda_m^* \boldsymbol{a}_m) \in N_{\mathbb{R}_+^r \times \mathbb{R}^{n-r}}(\boldsymbol{x}^*)$.

Let $\boldsymbol{v} = (v_1, \ldots, v_n)$ and $\boldsymbol{a}_i = (a_{i1}, \ldots, a_{in})$ $(i = 1, \ldots, m)$ in the KKT-optimality condition (c). Then

$$(-\boldsymbol{v} + \lambda_1^* \boldsymbol{a}_1 + \cdots + \lambda_m^* \boldsymbol{a}_m) \in N_{\mathbb{R}_+^r \times \mathbb{R}^{n-r}}(\boldsymbol{x}^*)$$

if and only if

$$\begin{aligned} (-v_j + \lambda_1^* a_{1j} + \cdots + \lambda_m^* a_{mj}) &\in N_{\mathbb{R}_+}(x_j^*) \quad (j = 1, \ldots, r), \\ (-v_j + \lambda_1^* a_{1j} + \cdots + \lambda_m^* a_{mj}) &\in N_{\mathbb{R}}(x_j^*) \quad (j = r+1, \ldots, n). \end{aligned} \tag{7.5.8}$$

By Definition 7.4.3, for $j = 1, \ldots, r$,

$$N_{\mathbb{R}_+}(x_j^*) = \begin{cases} \mathbb{R}_- & \text{if } x_j^* = 0, \\ 0 & \text{if } 0 < x_j^* < \infty, \end{cases}$$

where $\mathbb{R}_+ = [0, \infty)$ and $\mathbb{R}_- = (-\infty, 0)$, and for $j = r+1, \ldots, n$,

$$N_{\mathbb{R}}(x_j^*) = \{0\}.$$

From this and (7.5.8), it is seen that the following (a) and (b) hold:

$$(-v_j + \lambda_1^* a_{1j} + \cdots + \lambda_m^* a_{mj}) \in N_{\mathbb{R}_+}(x_j^*) \quad (j = 1, \ldots, r) \tag{a}$$

if and only if

$$\begin{aligned} v_j - \lambda_1^* a_{1j} - \cdots - \lambda_m^* a_{mj} &\geq 0 \quad (j = 1, \ldots, r), \\ (v_j - \lambda_1^* a_{1j} - \cdots - \lambda_m^* a_{mj})\, x_j^* &= 0 \quad (j = 1, \ldots, r); \end{aligned}$$

$$(-v_j + \lambda_1^* a_{1j} + \cdots + \lambda_m^* a_{mj}) \in N_{\mathbb{R}}(x_j^*) \quad (j = r+1, \ldots, n) \tag{b}$$

if and only if

$$v_j - \lambda_1^* a_{1j} - \cdots - \lambda_m^* a_{mj} = 0 \quad (j = r+1, \ldots, n).$$

Therefore $(-\boldsymbol{v} + \lambda_1^* \boldsymbol{a}_1 + \cdots + \lambda_m^* \boldsymbol{a}_m) \in N_{\mathbb{R}_+^r \times \mathbb{R}^{n-r}}(\boldsymbol{x}^*)$ if and only if

$$\begin{aligned} v_j - \lambda_1^* a_{1j} - \cdots - \lambda_m^* a_{mj} &\geq 0 \quad (j = 1, \ldots, r), \\ (v_j - \lambda_1^* a_{1j} - \cdots - \lambda_m^* a_{mj}) x_j^* &= 0 \quad (j = 1, \ldots, r), \\ v_j - \lambda_1^* a_{1j} - \cdots - \lambda_m^* a_{mj} &= 0 \quad (j = r+1, \ldots, n). \end{aligned}$$

Corollary 7.5.7 follows. □

Corollary 7.5.8. *Let $X = \mathbb{R}_+^r \times \mathbb{R}^{n-r}$ and $f(\boldsymbol{x}) = \langle \boldsymbol{c}, \boldsymbol{x} \rangle$. Then $\boldsymbol{x}^* \in \Omega_{lc}$ is the optimal solution of the linear optimization model if and only if the following KKT-optimality conditions hold:*

(a) *$x_j^* \geq 0$ $(j = 1, \ldots, r)$ and $x_j^* \in \mathbb{R}$ $(j = r+1, \ldots, n)$, where $\boldsymbol{x}^* = (x_1^*, \ldots, x_n^*)$;*

(b) *$\langle \boldsymbol{a}_i, \boldsymbol{x}^* \rangle \geq b_i$ $(i = 1, \ldots, s)$ and $\langle \boldsymbol{a}_i, \boldsymbol{x}^* \rangle = b_i$ $(i = s+1, \ldots, m)$, where $\boldsymbol{a}_i = (a_{i1}, \ldots, a_{in})$;*

(c) *there are m KKT multipliers $\lambda_1^*, \ldots, \lambda_m^*$ such that*

$$\begin{aligned} \lambda_i^* &\geq 0 \quad (i = 1, \ldots, s), \\ \lambda_i^* &\in \mathbb{R} \quad (i = s+1, \ldots, m), \\ \lambda_i^*(\langle \boldsymbol{a}_i, \boldsymbol{x}^* \rangle - b_i) &= 0 \quad (i = 1, \ldots, s); \end{aligned}$$

(d) *for $\boldsymbol{c} = (c_1, \ldots, c_n)$,*

$$\begin{aligned} c_j - \lambda_1^* a_{1j} - \cdots - \lambda_m^* a_{mj} &\geq 0 \quad (j = 1, \ldots, r), \\ (c_j - \lambda_1^* a_{1j} - \cdots - \lambda_m^* a_{mj}) x_j^* &= 0 \quad (j = 1, \ldots, r), \\ c_j - \lambda_1^* a_{1j} - \cdots - \lambda_m^* a_{mj} &= 0 \quad (j = r+1, \ldots, n). \end{aligned}$$

Proof. Corollary 7.5.8 follows immediately from Corollary 7.5.7 and $\partial f(\boldsymbol{x}^*) = \{\boldsymbol{c}\}$. □

7.6 Primal and dual pairs of linear optimization

Assume that a bivariate function $S(\boldsymbol{x}, \boldsymbol{y})$ defined on $\mathbb{R}^n \times \mathbb{R}^m$ is a convex-concave bivariate function, i.e.,

- for all $\boldsymbol{y} \in \mathbb{R}^m$, the function $S(\boldsymbol{x}, \boldsymbol{y})$ is convex with respect to $\boldsymbol{x} \in \mathbb{R}^n$,
- for all $\boldsymbol{x} \in \mathbb{R}^n$, the function $S(\boldsymbol{x}, \boldsymbol{y})$ is concave with respect to $\boldsymbol{y} \in \mathbb{R}^m$,

then $S(\boldsymbol{x}, \boldsymbol{y})$ is called a *saddle function* on $\mathbb{R}^n \times \mathbb{R}^m$. Assume further that there is a point $(\boldsymbol{x}^*, \boldsymbol{y}^*)$ such that $\boldsymbol{y}^*$ maximizes $S(\boldsymbol{x}^*, \boldsymbol{y})$ on $\mathbb{R}^m$ and $\boldsymbol{x}^*$ minimizes $S(\boldsymbol{x}, \boldsymbol{y}^*)$ on $\mathbb{R}^n$. Then the point $(\boldsymbol{x}^*, \boldsymbol{y}^*)$ is called a *saddle point* of the saddle function $S(\boldsymbol{x}, \boldsymbol{y})$.

Since $\boldsymbol{y}^*$ maximizes $S(\boldsymbol{x}^*, \boldsymbol{y})$ on $\mathbb{R}^m$ and $\boldsymbol{x}^*$ minimizes $S(\boldsymbol{x}, \boldsymbol{y}^*)$ on $\mathbb{R}^n$,

$$S(\boldsymbol{x}^*, \boldsymbol{y}) \le S(\boldsymbol{x}^*, \boldsymbol{y}^*),$$
$$S(\boldsymbol{x}, \boldsymbol{y}^*) \ge S(\boldsymbol{x}^*, \boldsymbol{y}^*).$$

From these two inequalities, it follows that if $(\boldsymbol{x}^*, \boldsymbol{y}^*)$ is a saddle point of a saddle function $S(\boldsymbol{x}, \boldsymbol{y})$, then

$$S(\boldsymbol{x}^*, \boldsymbol{y}) \le S(\boldsymbol{x}^*, \boldsymbol{y}^*) \le S(\boldsymbol{x}, \boldsymbol{y}^*).$$

Definition 7.6.1. The following two optimization models:

$$(L^p): \min_{\boldsymbol{x}\in\mathbb{R}^n} \sup_{\boldsymbol{y}\in\mathbb{R}^m} S(\boldsymbol{x}, \boldsymbol{y}),$$
$$(L^d): \max_{\boldsymbol{y}\in\mathbb{R}^m} \inf_{\boldsymbol{x}\in\mathbb{R}^n} S(\boldsymbol{x}, \boldsymbol{y})$$

are called a *primal/dual pair*, where the model (L^p) is called the *primal model* and the model (L^d) is called the *dual model.*

Let

$$f(\boldsymbol{x}) = \sup_{\boldsymbol{y}\in\mathbb{R}^m} S(\boldsymbol{x}, \boldsymbol{y}),$$
$$g(\boldsymbol{y}) = \inf_{\boldsymbol{x}\in\mathbb{R}^n} S(\boldsymbol{x}, \boldsymbol{y}).$$

By the convex-concave property of the saddle function $S(\boldsymbol{x}, \boldsymbol{y})$, it is easy to deduce that $f(\boldsymbol{x})$ is a convex function and $g(\boldsymbol{y})$ is a concave function.

Dual theorem. *The following two conditions are equivalent:*
(a) $(\boldsymbol{x}^*, \boldsymbol{y}^*)$ *is a saddle point of the saddle function* $S(\boldsymbol{x}, \boldsymbol{y})$,
(b) $\boldsymbol{x}^*$ *solves the primal model* (L^p) *and* $\boldsymbol{y}^*$ *solves the dual model* (L^d), *and* $f(\boldsymbol{x}^*) = g(\boldsymbol{y}^*)$, *and the optimal values of the primal model* (L^p) *and the dual model* (L^d) *are both equal to* $S(\boldsymbol{x}^*, \boldsymbol{y}^*)$.

Now consider the linearly constrained convex optimization model (L) given in Section 7.5

$$\min_{\boldsymbol{x}} f(\boldsymbol{x}) \quad \text{subject to}$$
$$\begin{aligned} \text{inequality constraints:} \quad & \langle \boldsymbol{a}_i, \boldsymbol{x}\rangle \ge b_i \quad (i = 1, \ldots, s), \\ \text{and equality constraints:} \quad & \langle \boldsymbol{a}_i, \boldsymbol{x}\rangle = b_i \quad (i = s+1, \ldots, m), \\ \text{and the restriction:} \quad & \boldsymbol{x} \in X \subset \mathbb{R}^n, \end{aligned}$$

where $f: \mathbb{R}^n \to \mathbb{R}$ is convex and X is a polyhedral set. Define its Lagrangian function as

$$L(\boldsymbol{x}, \boldsymbol{\lambda}) = \begin{cases} f(\boldsymbol{x}) + \sum_1^m \lambda_i (b_i - \langle \boldsymbol{a}_i, \boldsymbol{x}\rangle) & \text{if } \boldsymbol{x} \in X, \boldsymbol{\lambda} \in S, \\ \infty & \text{if } \boldsymbol{x} \notin X, \\ -\infty & \text{if } \boldsymbol{x} \in X, \boldsymbol{\lambda} \notin S, \end{cases}$$

where $S = \mathbb{R}_+^s \times \mathbb{R}^{m-s}$ and $\lambda = (\lambda_1, \ldots, \lambda_m)$.

7.6.1 Saddle function

It is clear that the Lagrangian function $L(\boldsymbol{x}, \lambda)$ is a bivariate function on $\mathbb{R}^n \times \mathbb{R}^m$. We will prove that $L(\boldsymbol{x}, \boldsymbol{\lambda})$ is a convex-concave bivariate function on $\mathbb{R}^n \times \mathbb{R}^m$.

Without loss of generality, consider $\boldsymbol{x} \in X$ and $\lambda \in S$. Then

$$L(\boldsymbol{x}, \boldsymbol{\lambda}) = f(\boldsymbol{x}) + \sum_1^m \lambda_i (b_i - \langle \boldsymbol{a}_i, \boldsymbol{x} \rangle).$$

On the one hand, for all $\lambda \in S$ and any $\boldsymbol{x}_1, \boldsymbol{x}_2 \in X$,

$$\begin{aligned} &L((1-\tau)\boldsymbol{x}_1 + \tau \boldsymbol{x}_2, \boldsymbol{\lambda}) \\ &\quad = f((1-\tau)\boldsymbol{x}_1 + \tau \boldsymbol{x}_2) + \sum_1^m \lambda_i \left(b_i - \langle \boldsymbol{a}_i, (1-\tau)\boldsymbol{x}_1 + \tau \boldsymbol{x}_2 \rangle\right) \quad (0 \le \tau \le 1). \end{aligned}$$

Since f is a convex function and $b_i = (1-\tau)b_i + \tau b_i$, it follows that

$$\begin{aligned} f((1-\tau)\boldsymbol{x}_1 + \tau \boldsymbol{x}_2) &\le (1-\tau) f(\boldsymbol{x}_1) + \tau f(\boldsymbol{x}_2), \\ b_i - \langle \boldsymbol{a}_i, (1-\tau)\boldsymbol{x}_1 + \tau \boldsymbol{x}_2 \rangle &= (1-\tau)(b_i - \langle \boldsymbol{a}_i, \boldsymbol{x}_1 \rangle) + \tau (b_i - \langle \boldsymbol{a}_i, \boldsymbol{x}_2 \rangle). \end{aligned}$$

Substituting them into the above equality gives

$$\begin{aligned} L((1-\tau)\boldsymbol{x}_1 + \tau \boldsymbol{x}_2, \boldsymbol{\lambda}) &\le (1-\tau) \left(f(\boldsymbol{x}_1) + \sum_1^m \lambda_i (b_i - \langle \boldsymbol{a}_i, \boldsymbol{x}_1 \rangle) \right) \\ &\quad + \tau \left(f(\boldsymbol{x}_2) + \sum_1^m \lambda_i (b_i - \langle \boldsymbol{a}_i, \boldsymbol{x}_2 \rangle) \right) \\ &= (1-\tau) L(\boldsymbol{x}_1, \lambda) + \tau L(\boldsymbol{x}_2, \lambda) \qquad (0 \le \tau \le 1), \end{aligned}$$

i.e., for all $\lambda \in S$, the Lagrangian function is convex with respect to $\boldsymbol{x} \in X$.

On the other hand, for all $\boldsymbol{x} \in X$ and any $\boldsymbol{\lambda}_i^{(1)}, \boldsymbol{\lambda}_i^{(2)} \in S$,

$$L(\boldsymbol{x}, (1-\tau)\boldsymbol{\lambda}_i^{(1)} + \tau \boldsymbol{\lambda}_i^{(2)}) = f(\boldsymbol{x}) + \sum_1^m \left((1-\tau)\lambda_i^{(1)} + \tau \lambda_i^{(2)} \right) (b_i - \langle \boldsymbol{a}_i, \boldsymbol{x} \rangle) \quad (0 \le \tau \le 1).$$

Note that $f(\boldsymbol{x}) = (1-\tau) f(\boldsymbol{x}) + \tau f(\boldsymbol{x})$. Substituting it into the above equality, we get

$$\begin{aligned} L(\boldsymbol{x}, (1-\tau)\boldsymbol{\lambda}_i^{(1)} + \tau \boldsymbol{\lambda}_i^{(2)}) &= (1-\tau) \left(f(\boldsymbol{x}) + \sum_1^m \lambda_i^{(1)} (b_i - \langle \boldsymbol{a}_i, \boldsymbol{x} \rangle) \right) \\ &\quad + \tau \left(f(\boldsymbol{x}) + \sum_1^m \lambda_i^{(2)} (b_i - \langle \boldsymbol{a}_i, \boldsymbol{x} \rangle) \right) \\ &= (1-\tau) L(\boldsymbol{x}, \lambda^{(1)}) + \tau L(\boldsymbol{x}, \lambda^{(2)}) \qquad (0 \le \tau \le 1), \end{aligned}$$

i.e., for all $\boldsymbol{x} \in X$, the Lagrangian function is concave with respect to $\lambda \in S$.

Thus Lagrangian function $L(\boldsymbol{x}, \boldsymbol{\lambda})$ is a saddle function.

7.6.2 Saddle point

Consider the KKT-optimization conditions in Theorem 7.5.6. Let

$$\begin{aligned}\boldsymbol{a}_i &= (a_{i1}, \ldots, a_{in}) \quad (i = 1, \ldots, m),\\ \boldsymbol{x} &= (x_1, \ldots, x_n).\end{aligned}$$

Then

$$\begin{aligned}\langle \lambda_1^* \boldsymbol{a}_1 + \lambda_2^* \boldsymbol{a}_2 + \cdots + \lambda_m^* \boldsymbol{a}_m, \boldsymbol{x}\rangle &= (\lambda_1^* a_{11} + \lambda_2^* a_{21} + \cdots + \lambda_m^* a_{m1}) x_1\\ &\quad + \cdots + (\lambda_1^* a_{1n} + \lambda_2^* a_{2n} + \cdots + \lambda_m^* a_{mn}) x_n\\ &= \lambda_1^* (a_{11} x_1 + \cdots + a_{1n} x_n) + \lambda_2^* (a_{21} x_1 + \cdots + a_{2n} x_n)\\ &\quad + \cdots + \lambda_m^* (a_{m1} x_1 + \cdots + a_{mn} x_n)\\ &= \sum_1^m \lambda_i^* (a_{i1} x_1 + \cdots + a_{in} x_n) = \sum_1^m \lambda_i^* \langle \boldsymbol{a}_i, \boldsymbol{x}\rangle,\end{aligned}$$

and so

$$f(\boldsymbol{x}) - \langle \lambda_1^* \boldsymbol{a}_1 + \cdots + \lambda_m^* \boldsymbol{a}_m, \boldsymbol{x}\rangle = f(\boldsymbol{x}) - \sum_1^m \lambda_i^* \langle \boldsymbol{a}_i, \boldsymbol{x}\rangle.$$

Note that $\sum_{i=1}^m \lambda_i b_i$ is constant. Therefore, the KKT-optimization condition (c) of Theorem 7.5.6 means that $\boldsymbol{x}^*$ minimizes $L(\boldsymbol{x}, \lambda^*)$.

The KKT-optimization conditions (a) and (b) in Theorem 7.5.6 imply that

$$\sum_{i=1}^m \lambda_i^* (b_i - \langle \boldsymbol{a}_i, \boldsymbol{x}^*\rangle) = 0.$$

On the other hand, since $\lambda = (\lambda_1, \ldots, \lambda_m) \in \mathbb{R}_+^s \times \mathbb{R}^{m-s}$, it is clear that $\lambda_i \ge 0$ $(i = 1, \ldots, s)$. Note that

$$\begin{aligned}\langle \boldsymbol{a}_i, \boldsymbol{x}\rangle &\ge b_i \quad (i = 1, \ldots, s),\\ \langle \boldsymbol{a}_i, \boldsymbol{x}\rangle &= b_i \quad (i = s+1, \ldots, m).\end{aligned}$$

Then

$$\sum_1^m \lambda_i (b_i - \langle \boldsymbol{a}_i, \boldsymbol{x}\rangle) = \sum_1^s \lambda_i (b_i - \langle \boldsymbol{a}_i, \boldsymbol{x}\rangle) + \sum_{s+1}^m \lambda_i (b_i - \langle \boldsymbol{a}_i, \boldsymbol{x}\rangle) \le 0.$$

So λ^* maximizes $L(\boldsymbol{x}^*, \lambda)$.

Thus, $(\boldsymbol{x}^*, \lambda^*)$ is a saddle point of the Lagrange function $L(\boldsymbol{x}, \lambda)$ for the model (L).

7.6.3 Dual pair

Theorem 7.6.2. *Let*

$$\begin{aligned} \boldsymbol{c} &= (c_1, \ldots, c_n), & \boldsymbol{x} &= (x_1, \ldots, x_n), & \boldsymbol{a}_i &= (a_{i1}, \ldots, a_{in}) & (i &= 1, \ldots, m), \\ \boldsymbol{b} &= (b_1, \ldots, b_m), & \lambda &= (\lambda_1, \ldots, \lambda_m), & \boldsymbol{a}^j &= (a_{1j}, \ldots, a_{mj}) & (j &= 1, \ldots, n). \end{aligned}$$

Then the linear optimization model

$$\begin{aligned} (L^p):\ \min_{\boldsymbol{x}} \langle \boldsymbol{c}, \boldsymbol{x}\rangle \quad & \textit{subject to} \\ \textit{the inequality constraints:}\quad & \langle \boldsymbol{a}_i, \boldsymbol{x}\rangle \geq b_i \qquad (i = 1, \ldots, s) \\ \textit{and the equality constraints:}\quad & \langle \boldsymbol{a}_i, \boldsymbol{x}\rangle = b_i \qquad (i = s+1, \ldots, m), \\ \textit{and the restriction:}\quad & \boldsymbol{x} \in \mathbb{R}_+^r \times \mathbb{R}^{n-r} \end{aligned}$$

and the linear optimization model

$$\begin{aligned} (L^d):\ \max_{\boldsymbol{\lambda}} \langle \boldsymbol{b}, \boldsymbol{\lambda}\rangle \quad & \textit{subject to} \\ \textit{the inequality constraints:}\quad & \langle \boldsymbol{a}^j, \boldsymbol{\lambda}\rangle \leq c_j \qquad (j = 1, \ldots, r) \\ \textit{and the equality constraints:}\quad & \langle \boldsymbol{a}^j, \boldsymbol{\lambda}\rangle = c_j \qquad (j = r+1, \ldots, n), \\ \textit{and the restriction:}\quad & \boldsymbol{\lambda} \in \mathbb{R}_+^s \times \mathbb{R}^{m-s} \end{aligned}$$

are a primal/dual pair.

Proof. In fact, let

$$\begin{aligned} X &= \mathbb{R}_+^r \times \mathbb{R}^{n-r}, \\ S &= \mathbb{R}_+^s \times \mathbb{R}^{m-s}. \end{aligned}$$

Then the Lagrangian function associated with the primal model (L^p) is given by

$$L(\boldsymbol{x}, \boldsymbol{\lambda}) = \begin{cases} \langle \boldsymbol{c}, \boldsymbol{x}\rangle + \sum_1^m \lambda_i (b_i - \langle \boldsymbol{a}_i, \boldsymbol{x}\rangle) & \text{if } \boldsymbol{x} \in X, \boldsymbol{\lambda} \in S, \\ \infty & \text{if } \boldsymbol{x} \notin X, \\ -\infty & \text{if } \boldsymbol{x} \in X, \boldsymbol{\lambda} \notin S, \end{cases}$$

where $\lambda = (\lambda_1, \ldots, \lambda_m)$. Note that for $\boldsymbol{x} \in X$ and $\lambda \in S$,

$$L(\boldsymbol{x}, \lambda) = \langle \boldsymbol{c}, \boldsymbol{x}\rangle + \sum_1^m \lambda_i (b_i - \langle \boldsymbol{a}_i, \boldsymbol{x}\rangle). \tag{7.6.1}$$

The right-hand side of (7.6.1) is computed as follows:

$$\begin{aligned} \langle \boldsymbol{c}, \boldsymbol{x}\rangle + \sum_1^m \lambda_i (b_i - \langle \boldsymbol{a}_i, \boldsymbol{x}\rangle) &= \langle \boldsymbol{c}, \boldsymbol{x}\rangle + \sum_1^m b_i \lambda_i - \sum_1^m \lambda_i \langle \boldsymbol{a}_i, \boldsymbol{x}\rangle \\ &= \sum_1^n c_j x_j + \langle \boldsymbol{b}, \lambda\rangle - \sum_1^m \lambda_i \langle \boldsymbol{a}_i, \boldsymbol{x}\rangle. \end{aligned}$$

However,

$$\begin{aligned}\sum_1^m \lambda_i\langle \boldsymbol{a}_i, \boldsymbol{x}\rangle &= \sum_1^m \lambda_i(a_{i1}x_1 + \cdots + a_{in}x_n)\\ &= \lambda_1(a_{11}x_1 + \cdots + a_{1n}x_n) + \cdots + \lambda_m(a_{m1}x_1 + \cdots + a_{mn}x_n)\\ &= (\lambda_1 a_{11} + \cdots + \lambda_m a_{m1})x_1 + \cdots + (\lambda_1 a_{1n} + \cdots + \lambda_m a_{mn})x_n\\ &= \langle \boldsymbol{a}^1, \lambda\rangle x_1 + \cdots + \langle \boldsymbol{a}^n, \lambda\rangle x_n\\ &= \sum_1^n \langle \boldsymbol{a}^j, \lambda\rangle x_j\end{aligned}$$

and

$$\begin{aligned}\sum_1^n c_j x_j + \langle \boldsymbol{b}, \lambda\rangle - \sum_1^m \lambda_i\langle \boldsymbol{a}_i, \boldsymbol{x}\rangle &= \langle \boldsymbol{b}, \lambda\rangle + \sum_1^n c_j x_j - \sum_1^n \langle \boldsymbol{a}^j, \lambda\rangle x_j\\ &= \langle \boldsymbol{b}, \lambda\rangle + \sum_1^n (c_j - \langle \boldsymbol{a}^j, \lambda\rangle) x_j.\end{aligned}$$

From this and (7.6.1), it follows that for $\boldsymbol{x} \in X$ and $\lambda \in S$,

$$L(\boldsymbol{x}, \lambda) = \langle \boldsymbol{b}, \lambda\rangle + \sum_1^n (c_j - \langle \boldsymbol{a}^j, \lambda\rangle)\, x_j. \tag{7.6.2}$$

The combination of (7.6.1) and (7.6.2) gives

$$L(\boldsymbol{x}, \lambda) = \langle \boldsymbol{c}, \boldsymbol{x}\rangle + \sum_1^m \lambda_i(b_i - \langle \boldsymbol{a}_i, \boldsymbol{x}\rangle) = \langle \boldsymbol{b}, \lambda\rangle + \sum_1^n (c_j - \langle \boldsymbol{a}^j, \lambda\rangle)\, x_j.$$

From this, by Definition 7.6.1 of primal/dual pairs, it follows that

$$\begin{aligned}(L^p):\ \min_{\boldsymbol{x}\in X}\sup_{\lambda\in S} L(\boldsymbol{x}, \lambda) &= \min_{\boldsymbol{x}\in X}\sup_{\lambda\in S}\left(\langle \boldsymbol{c}, \boldsymbol{x}\rangle + \sum_1^m \lambda_i(b_i - \langle \boldsymbol{a}_i, \boldsymbol{x}\rangle)\right)\\ &= \min_{\boldsymbol{x}\in X}\left(\langle \boldsymbol{c}, \boldsymbol{x}\rangle + \sup_{\lambda\in S}\sum_1^m \lambda_i(b_i - \langle \boldsymbol{a}_i, \boldsymbol{x}\rangle)\right),\end{aligned} \tag{7.6.3}$$

$$\begin{aligned}(L^d):\ \max_{\lambda\in S}\inf_{\boldsymbol{x}\in X} L(\boldsymbol{x}, \lambda) &= \max_{\lambda\in S}\inf_{\boldsymbol{x}\in X}\left(\langle \boldsymbol{b}, \lambda\rangle + \sum_1^n (c_j - \langle \boldsymbol{a}^j, \lambda\rangle)\, x_j\right)\\ &= \max_{\lambda\in S}\left(\langle \boldsymbol{b}, \lambda\rangle + \inf_{\boldsymbol{x}\in X}\sum_1^n (c_j - \langle \boldsymbol{a}^j, \lambda\rangle)\, x_j\right).\end{aligned} \tag{7.6.4}$$

Note that

$$\begin{aligned}\boldsymbol{x} &= (x_1, \ldots, x_n) \in X = \mathbb{R}^r \times \mathbb{R}^{n-r},\\ \lambda &= (\lambda_1, \ldots, \lambda_m) \in S = \mathbb{R}^s \times \mathbb{R}^{m-s}.\end{aligned}$$

It is seen that

$$\begin{aligned}&x_1 \geq 0, \quad \ldots, \quad x_r \geq 0, \quad x_{r+1}, \ldots, x_n \in \mathbb{R},\\ &\lambda_1 \geq 0, \quad \ldots, \quad \lambda_s \geq 0, \quad \lambda_{s+1}, \ldots, \lambda_m \in \mathbb{R}.\end{aligned} \tag{7.6.5}$$

By the assumption that $\langle \boldsymbol{a}_i, \boldsymbol{x}\rangle \geq b_i$ $(i = 1, \ldots, s)$ and $\langle \boldsymbol{a}_i, \boldsymbol{x}\rangle = b_i$ $(i = s+1, \ldots, m)$, and (7.6.5),

$$\sum_1^m \lambda_i (b_i - \langle \boldsymbol{a}_i, \boldsymbol{x}\rangle) = \sum_1^s \lambda_i (b_i - \langle \boldsymbol{a}_i, \boldsymbol{x}\rangle) \leq 0.$$

By the assumption that $\langle \boldsymbol{a}^j, \lambda\rangle \leq c_j$ $(j = 1, \ldots, r)$ and $\langle \boldsymbol{a}^j, \lambda\rangle = c_j$ $(j = r+1, \ldots, n)$, and (7.6.5),

$$\sum_1^n (c_j - \langle \boldsymbol{a}^j, \lambda\rangle)\, x_j = \sum_1^r (c_j - \langle \boldsymbol{a}^j, \lambda\rangle)\, x_j \geq 0.$$

Thus,

$$\sup_{\lambda\in S} \sum_1^m \lambda_i(b_i - \langle \boldsymbol{a}_i, \boldsymbol{x}\rangle) = 0,$$

$$\inf_{\boldsymbol{x}\in X} \sum_1^n (c_j - \langle \boldsymbol{a}^j, \lambda\rangle)\, x_j = 0.$$

From this and (7.6.3), and (7.6.4), it follows that

$$(L^p):\ \min_{\boldsymbol{x}\in X} \sup_{\lambda\in S} L(\boldsymbol{x}, \lambda) = \min_{\boldsymbol{x}\in X} \langle \boldsymbol{c}, \boldsymbol{x}\rangle,$$

$$(L^d):\ \max_{\lambda\in S} \inf_{\boldsymbol{x}\in X} L(\boldsymbol{x}, \lambda) = \max_{\lambda\in S} \langle \boldsymbol{b}, \lambda\rangle.$$

Therefore, the linear optimization models (L^p) and (L^d) are a primal/dual pair. □

Theorem 7.6.2 shows that two linear optimization models (L^p) and (L^d) are a primal/dual pair. This result will be applied in Chapter 8.

Example 7.6.3. The linear optimization model

$$(L^p):\ \min\{ f(x_1, x_2, x_3) = 5x_1 - 6x_2 + 8x_3 \}\quad \text{subject to}$$

$$\begin{cases} 2x_1 + 3x_2 + 2x_3 \geq 2, \\ 4x_1 + 4x_2 + 3x_3 \geq 1, \\ 5x_1 - 8x_2 + x_3 = -3, \\ 7x_1 + 9x_2 - 7x_3 = 1, \\ x_1 \geq 0, \quad x_2 \geq 0, \quad x_3 \in \mathbb{R}. \end{cases}$$

and the linear optimization model

$$(L^d):\ \max\{ g(\lambda_1, \lambda_2, \lambda_3, \lambda_4) = 2\lambda_1 + \lambda_2 - 3\lambda_3 + \lambda_4 \}\quad \text{subject to}$$

$$\begin{cases} 2\lambda_1 + 4\lambda_2 + 5\lambda_3 + 7\lambda_4 \leq 5, \\ 3\lambda_1 + 4\lambda_2 - 8\lambda_3 + 9\lambda_4 \leq -6, \\ 2\lambda_1 + 3\lambda_2 + \lambda_3 - 7\lambda_4 = 8, \\ \lambda_1 \geq 0, \quad \lambda_2 \geq 0, \quad \lambda_3 \in \mathbb{R}, \quad \lambda_4 \in \mathbb{R} \end{cases}$$

are a primal/dual pair, and the model (L^p) is the primal model and the model (L^d) is the dual model.

7.7 Case studies

Optimization problems arise in almost all areas of environmental science. Optimization can used to reduce overall product cost, minimize negative environment impacts and maximize the probability of making a correct decision. Here we give some case studies to explain how the algorithms in this chapter are applied in environmental science.

7.7.1 Building design

Buildings have considerable impacts on the environment and it has become necessary to pay more attention to environmental performance in building design. Recent progress in the design of greener buildings advances the research and applications of various optimization methods in the building sector. In these optimization problems, main variables include building orientation, window type, window-to-wall ratio, roof type, etc. The lower and upper boundary values of each variable are viewed as constraints. The objective is to achieve cost-effective green building design under these constraints.

7.7.2 Supply chain planning

The efficiency of a company can often be constrained by the efficiency of its supply chain management procedures. The supply chain planning problem consists of determining the optimal production, storage, backorder, subcontracting and distribution variables associated with a supply chain network, where the objective is to achieve minimization of the overall total cost, carbon emissions, total delivery time, tardiness, or maximization of the benefit, etc. and the main constraints includes mass balances, capacities (production, distribution, budget, storage), time restrictions, etc.

7.7.3 Coal mining

Significant quantities of groundwater are discharged in underground coal mining. It brings further damage to local water environment. Xu et al. (2016) developed an equilibrium strategy-based optimization method to solve the coal-water conflict in China. Each colliery pursues the largest possible profit and minimizes the cost which consists of production costs, sewage treatment costs and punitive fees. The main constraints include production capacity, mining quota limitations, and environmental protection constraints. At the leader level, the government has the obligation to promote financial revenue so as to ensure an environmental self-repair capacity. In this case, financial revenue is set to the objective function and environmental protection is the constraint.

7.7.4 Iron and steel industry

The rapid development of the iron and steel industry directly needs large amount of energy. Because energy shortages and environmental degradation problems have become increasingly prominent, energy conservation and emission reduction have become more important. The objective function of the optimization is the energy intensity in the iron and steel industry, while the constraints include production unit ferrite balance, process ferrite balance, market order as well as production capacity.

7.7.5 Electric energy generating systems

Economic operation of electric energy generating systems is one of the prevailing problems in energy systems. The objective function includes cost reduction, voltage profile improvement, voltage stability enhancement, emission reduction, as well as their combinations. The constraints can be classified into equality and inequality constraints. Equality constraints consist of real power constraints and reactive power constraints. Inequality constraints consists of generator constraints, transformer constraints, shunt VAR compensator constraints, and security constraints.

Further reading

[1] Ahmadi P, Dincer I, Rosen MA. Multi-objective optimization of an ocean thermal energy conversion system for hydrogen production. *International Journal of Hydrogen Energy*. 2015(40): 7601–7608.
[2] Cauchy A. Methode génerale pour la résolution des systéms d'equations simultanées, *C R Acad Sci Paris*. 1847(25):536–538.
[3] Chaib AE, Bouchekara HREH, Mehasni R, Abido MA. Optimal power flow with emission and non-smooth cost functions using backtracking search optimization algorithm. *International Journal of Electrical Power & Energy Systems*. 2016(81):64–77.
[4] Cooper FC, Zanna L. Optimisation of an idealised ocean model, stochastic parameterisation of sub-grid eddies. *Ocean Modelling*. 2015(88):38–53.
[5] Feng L, Mears L, Beaufort C, Schulte J. Energy, economy, and environment analysis and optimization on manufacturing plant energy supply system. *Energy Conversion and Management*. 2016(117):454–465.
[6] Fergani Z, Touil D, Morosuk T. Multi-criteria exergy based optimization of an organic rankine cycle for waste heat recovery in the cement industry. *Energy Conversion and Management*. 2016(112):81–90.
[7] Harver CM. *Operations Research: An Introduction to Linear Optimization and Decision Analysis*, Elsevier North Holland, Inc., New York, 1979.

[8] Kucukmehmetoglu M, Geymen A. Optimization models for urban land readjustment practices in Turkey. *Habitat International*. 2016(53):517–533.
[9] Kuhn HW, Tucker AW. *Nonlinear Programming*. Proceedings of Second Berkeley Symposium on Mathematical Statistics and Probability, pp. 481–492, University of California Press, Berkeley, CA, 1950.
[10] Lemke CE. The dual method of solving a linear programming problem. *Naval Res Logistics Q*. 1954(1):48–54.
[11] Lendering KT, Jonkman SN, van Gelder PHAJM, Peters DJ. Risk-based optimization of land reclamation. *Reliability Engineering & System Safety*. 2015(144):193–203.
[12] Ma D, Wang S, Zhang Z. Hybrid algorithm of minimum relative entropy-particle swarm optimization with adjustment parameters for gas source term identification in atmosphere. *Atmospheric Environment*. 2014(94):637–646.
[13] Nguyen AT, Reiter S, Rigo P. A review on simulation-based optimization methods applied to building performance analysis. *Applied Energy*. 2014(113):1043–1058.
[14] Wang W, Zmeureanu R, Rivard H. Applying multi-objective genetic algorithms in green building design optimization. *Building and Environment*. 2005(40):1512–1525.
[15] Xu J, Lv C, Zhang M, Yao L, Zeng Z. Equilibrium strategy-based optimization method for the coal-water conflict: A perspective from China. *Journal of Environmental Management*. 2015(160):312–323.
[16] Zamarripa MA, Aguirre AM, Mendez CA, Espuna A. Mathematical programming and game theory optimization-based tool for supply chain planning in cooperative/competitive environments. *Chemical Engineering Research and Design*. 2013(91):1588–1600.

8 Data envelopment analysis

Data Envelopment Analysis (DEA) is a very powerful service management and benchmarking technique to evaluate organizations (i.e. so-called Decision Making Units (DMUs)) such as companies, schools, hospitals, shops, bank branches and similar instances where there is a relatively homogeneous set of units. Each DMU has a varying level of inputs and gives a varying level of outputs. DEA uses a linear optimization technique to measure the relative performance of DMUs where the presence of multiple inputs and outputs makes comparisons difficult. The main DEA models can be classified in three groups according to orientation, returns to scale, and distance function. The oriented models include input-oriented, output-oriented, and nonoriented envelopment models. The returns to scale models include constant, variable, nonincreasing, nondecreasing, and generalized returns to scale models. The distance function models include traditional radial DEA, nonradial slack-based measure, and radial and hybrid slack based measure models. In this chapter we will introduce these models.

8.1 Charnes–Cooper–Rhodes DEA models

The Charnes–Cooper–Rhodes (CCR) DEA model is the origin of DEA models. The CCR DEA model is referred to as the Constant Returns to Scale (CRS) model or the synthetic technical efficiency model.

Assume that there are n DMUs to be evaluated, denoted by DMU_j $(j = 1, \dots, n)$. For each DMU_j, the input $\boldsymbol{x}_j$ has m components and the output $\boldsymbol{y}_j$ has q components

$$\boldsymbol{x}_j = (x_{1j}, x_{2j}, \dots, x_{mj}), \quad \boldsymbol{y}_j = (y_{1j}, y_{2j}, \dots, y_{qj})$$

and each DMU_j consumes amount x_{ij} of input i $(i = 1, \dots, m)$ and produces amount y_{rj} of output r $(r = 1, \dots, q)$. Assume that $x_{ij} \geq 0$ and $y_{rj} \geq 0$ and that each DMU_j has at least one positive input value and one positive output value. For the particular DMU_0 being evaluated, its input and output are, respectively,

$$\boldsymbol{x}_0 = (x_{10}, x_{20}, \dots, x_{m0}), \quad \boldsymbol{y}_0 = (y_{10}, y_{20}, \dots, y_{q0}),$$

and it consumes amounts x_{i0} of input i $(i = 1, \dots, m)$ and produces amount y_{r0} of output r $(r = 1, \dots, q)$. Assume that $x_{i0} \geq 0$ and $y_{r0} \geq 0$.

8.1.1 Input-oriented CCR DEA models

Let

$$\boldsymbol{u} = (u_1, \dots, u_q),$$
$$\boldsymbol{v} = (v_1, \dots, v_m).$$

DOI 10.1515/9783110424904-009

The ratio of outputs $\boldsymbol{y}_0$ with weight $\boldsymbol{u}$ to inputs $\boldsymbol{x}_0$ with weight $\boldsymbol{v}$ forms the objective function

$$\frac{\langle \boldsymbol{u}, \boldsymbol{y}_0\rangle}{\langle \boldsymbol{v}, \boldsymbol{x}_0\rangle} = \frac{u_1 y_{10} + u_2 y_{20} + \cdots + u_q y_{q0}}{v_1 x_{10} + v_2 x_{20} + \cdots + v_m x_{m0}},$$

where both $\boldsymbol{u} \geq 0$ and $\boldsymbol{v} \geq 0$ are variables, and $\boldsymbol{x}_0, \boldsymbol{y}_0$ are the observed input and output values, respectively. The objective function is a measure of efficiency. An additional constraint is to guarantee that the efficiency of a DMU must be less than or equal to unity, i.e.,

$$\frac{\langle \boldsymbol{u}, \boldsymbol{y}_j\rangle}{\langle \boldsymbol{v}, \boldsymbol{x}_j\rangle} = \frac{u_1 y_{1j} + u_2 y_{2j} + \cdots + u_q y_{qj}}{v_1 x_{1j} + v_2 x_{2j} + \cdots + v_m x_{mj}} \leq 1 \quad (j = 1, \ldots, n).$$

The first input-oriented CCR DEA model presented by Charnes, Cooper, and Rhodes is the form

$$(L_{\mathrm{CCR}}^{\mathrm{ip}})\colon \max\left\{\frac{\langle \boldsymbol{u}, \boldsymbol{y}_0\rangle}{\langle \boldsymbol{v}, \boldsymbol{x}_0\rangle}\right\} \quad \text{subject to}$$

$$\begin{cases} \frac{\langle \boldsymbol{u}, \boldsymbol{y}_j\rangle}{\langle \boldsymbol{v}, \boldsymbol{x}_j\rangle} \leq 1 \quad (j = 1, \ldots, n), \\ \boldsymbol{u} \geq 0, \quad \boldsymbol{v} \geq 0. \end{cases}$$

If $(\boldsymbol{u}^*, \boldsymbol{v}^*)$ is the optimal solution of the model $(L_{\mathrm{CCR}}^{\mathrm{ip}})$, then $(t\boldsymbol{u}^*, t\boldsymbol{v}^*)$ $(t > 0)$ is also its optimal solution. Hence, the above ratio form yields an infinite number of solutions. Using the Charnes–Cooper transformation

$$t = \frac{1}{\langle \boldsymbol{v}, \boldsymbol{x}_0\rangle} \quad \text{or} \quad \langle t\boldsymbol{v}, \boldsymbol{x}_0\rangle = 1, \tag{8.1.1}$$

the objective function becomes

$$\frac{\langle \boldsymbol{u}, \boldsymbol{y}_0\rangle}{\langle \boldsymbol{v}, \boldsymbol{x}_0\rangle} = t\langle \boldsymbol{u}, \boldsymbol{y}_0\rangle = \langle t\boldsymbol{u}, \boldsymbol{y}_0\rangle \tag{8.1.2}$$

and the additional constraint becomes

$$\frac{\langle \boldsymbol{u}, \boldsymbol{y}_j\rangle}{\langle \boldsymbol{v}, \boldsymbol{x}_j\rangle} = \frac{t\langle \boldsymbol{u}, \boldsymbol{y}_j\rangle}{t\langle \boldsymbol{v}, \boldsymbol{x}_j\rangle} = \frac{\langle t\boldsymbol{u}, \boldsymbol{y}_j\rangle}{\langle t\boldsymbol{v}, \boldsymbol{x}_j\rangle} \leq 1 \quad (j = 1, \ldots, n).$$

By the assumption, $\langle t\boldsymbol{v}, \boldsymbol{x}_j\rangle = t\langle \boldsymbol{v}, \boldsymbol{x}_j\rangle > 0$. So the additional constraint is equivalent to $\langle t\boldsymbol{u}, \boldsymbol{y}_j\rangle \leq \langle t\boldsymbol{v}, \boldsymbol{x}_j\rangle$ or

$$\langle t\boldsymbol{u}, \boldsymbol{y}_j\rangle - \langle t\boldsymbol{v}, \boldsymbol{x}_j\rangle \leq 0. \tag{8.1.3}$$

Let

$$\boldsymbol{\mu} = (\mu_1, \ldots, \mu_q),$$
$$\boldsymbol{\nu} = (\nu_1, \ldots, \nu_m).$$

Using change of variables $\boldsymbol{\mu} = t\boldsymbol{u}$ and $\boldsymbol{\nu} = t\boldsymbol{v}$ in (8.1.1), (8.1.2), and (8.1.3), the input-oriented CCR DEA model has the following equivalent form

$$(L_{\mathrm{CCR}}^{\mathrm{ip}})\colon \max\langle \boldsymbol{\mu}, \boldsymbol{y}_0\rangle \quad \text{subject to}$$

$$\begin{cases} \langle \boldsymbol{\mu}, \boldsymbol{y}_j\rangle - \langle \boldsymbol{\nu}, \boldsymbol{x}_j\rangle \leq 0 \quad (j = 1, \ldots, n), \\ \langle \boldsymbol{\nu}, \boldsymbol{x}_0\rangle = 1, \\ \boldsymbol{\mu} \geq 0, \ \boldsymbol{\nu} \geq 0. \end{cases}$$

The input-oriented CCR DEA model is also called the *multiplier model* since $\boldsymbol{\mu}$ and $\mathbf{v}$ are multipliers.

Example 8.1.1. Assume that there are three DMUs with two inputs x_1, x_2 and one output y to be evaluated. The data for the input-oriented CCR DEA model are listed in the following table:

DMU	x_1	x_2	y	x_1/y	x_2/y
DMU_1	15	25	10	1.50	2.50
DMU_2	32	24	16	2.00	1.50
DMU_3	50	60	20	2.50	3.00

Using the model (L_{CCR}^{ip}), three input-oriented CCR DEA models are used to assess DMUs, respectively

$$
\begin{aligned}
\mathbf{DMU}_1: \ & \max 10\, W_y \quad \text{subject to} \\
& 10\, W_y - 15\, W_{x_1} - 25\, W_{x_2} \le 0 \\
& 16\, W_y - 32\, W_{x_1} - 24\, W_{x_2} \le 0 \\
& 20\, W_y - 50\, W_{x_1} - 60\, W_{x_2} \le 0 \\
& 15\, W_{x_1} + 25\, W_{x_2} = 1.
\end{aligned}
$$

$$
\begin{aligned}
\mathbf{DMU}_2: \ & \max 16\, W_y \quad \text{subject to} \\
& 10\, W_y - 15\, W_{x_1} - 25\, W_{x_2} \le 0 \\
& 16\, W_y - 32\, W_{x_1} - 24\, W_{x_2} \le 0 \\
& 20\, W_y - 50\, W_{x_1} - 60\, W_{x_2} \le 0 \\
& 32\, W_{x_1} + 24\, W_{x_2} = 1.
\end{aligned}
$$

$$
\begin{aligned}
\mathbf{DMU}_3: \ & \max 20\, W_y \quad \text{subject to} \\
& 10\, W_y - 15\, W_{x_1} - 25\, W_{x_2} \le 0 \\
& 16\, W_y - 32\, W_{x_1} - 24\, W_{x_2} \le 0 \\
& 20\, W_y - 50\, W_{x_1} - 60\, W_{x_2} \le 0 \\
& 50\, W_{x_1} + 60\, W_{x_2} = 1.
\end{aligned}
$$

The three evaluated DMUs may be represented by the following three points, respectively,

$$D_1(1.5, 2.5), \quad D_2(2, 1.5), \quad D_3(2.5, 3).$$

Set a rectangular coordinate system with the origin $O(0,0)$ and horizontal axis x_1/y, and vertical axis x_2/y on a sheet of graph paper and plot these three points D_1, D_2, and D_3. The curve through points D_1, D_2, and its extension is called the *frontier curve* which looks like a *convex envelopment* towards the origin and envelops these three points. Denote by D_i' the intersection point of the straight line OD_i and the frontier curve. The intersection point D_i' is called the *projection* from the point D_i to the frontier curve. Define the efficiency value of D_i as $\frac{OD_i'}{OD_i}$. Clearly,

$$\frac{OD_1'}{OD_1} = 1, \quad \frac{OD_2'}{OD_2} = 1, \quad \frac{OD_3'}{OD_3} < 1,$$

i.e., the efficiency values of D_1 and D_2 are both 1, and the efficiency value of D_3 is less than 1.

The detailed version of the input-oriented CCR DEA model is

$$(L_{\text{CCR}}^{\text{ip}}): \max\{y_{10}\mu_1 + y_{20}\mu_2 + \cdots + y_{q0}\mu_q\} \quad \text{subject to}$$

$$\begin{cases} y_{11}\mu_1 + y_{21}\mu_2 + \cdots + y_{q1}\mu_q - x_{11}\nu_1 - x_{21}\nu_2 - \cdots - x_{m1}\nu_m \le 0, \\ \vdots \\ y_{1n}\mu_1 + y_{2n}\mu_2 + \cdots + y_{qn}\mu_q - x_{1n}\nu_1 - x_{2n}\nu_2 - \cdots - x_{mn}\nu_n \le 0, \\ x_{10}\nu_1 + x_{20}\nu_2 + \cdots + x_{m0}\nu_m = 1, \\ \mu_1 \ge 0, \ \ldots, \ \mu_q \ge 0, \quad \nu_1 \ge 0, \ \ldots, \ \nu_m \ge 0. \end{cases}$$

By Theorem 7.6.2, the *dual model* of the input-oriented CCR DEA model is

$$(L_{\text{CCR}}^{\text{id}}): \min \theta \quad \text{subject to}$$

$$\begin{cases} y_{11}\lambda_1 + y_{12}\lambda_2 + \cdots + y_{1n}\lambda_n \ge y_{10}, \\ \vdots \\ y_{q1}\lambda_1 + y_{q2}\lambda_2 + \cdots + y_{qn}\lambda_n \ge y_{q0}, \\ -x_{11}\lambda_1 - x_{12}\lambda_2 - \cdots - x_{1n}\lambda_n + \theta x_{10} \ge 0, \\ \vdots \\ -x_{m1}\lambda_1 - x_{m2}\lambda_2 - \cdots - x_{mn}\lambda_n + \theta x_{m0} \ge 0, \\ \lambda_1 \ge 0, \ \lambda_2 \ge 0, \ \ldots, \ \lambda_n \ge 0. \end{cases}$$

Note that

$$x_{11}\lambda_1 + x_{12}\lambda_2 + \cdots + x_{1n}\lambda_n \le \theta x_{10}.$$

From $x_{11}\lambda_1 + x_{12}\lambda_2 + \cdots + x_{1n}\lambda_n > 0$ and $x_{10} \ge 0$, it follows that $\theta > 0$.

Let $\boldsymbol{\lambda} = (\lambda_1, \ldots, \lambda_n)$ and

$$\mathbf{x}^i = (x_{i1}, \ldots, x_{in}) \quad (i = 1, \ldots, m),$$
$$\mathbf{y}^r = (y_{r1}, \ldots, y_{rn}) \quad (r = 1, \ldots, q).$$

Then the contraction of the dual model is as follows:

$$(L_{\mathrm{CCR}}^{\mathrm{id}}): \ \min \theta \quad \text{subject to}$$
$$\begin{cases} \langle \boldsymbol{\lambda}, \mathbf{y}^r \rangle \geq y_{r0} & (r = 1, \ldots, q), \\ \langle \boldsymbol{\lambda}, \mathbf{x}^i \rangle \leq \theta x_{i0} & (i = 1, \ldots, m), \\ \lambda \geq 0, \ \theta > 0. \end{cases}$$

If the input-oriented CCR DEA model has the optimal solution z^* and its dual model has the optimal solution θ^*, the dual theorem says that $z^* = \theta^*$. Thus, the problem of finding the optimal solution of the input-oriented CCR DEA model is reduced to the problem of finding the optimal solution of the dual model. One can solve the dual model to obtain the optimal solution because when $\theta = 1$, $\lambda_0 = \lambda_k = 1$, and all other $\lambda_j = 0$, the constraints of the dual model hold. So the solution satisfies $\theta^* = \min \theta \leq 1$. So the optimal solution of the dual model satisfies that $0 < \theta^* \leq 1$. A DMU is *efficient* if the optimal solution $\theta^* = 1$ and the optimal multipliers $\mu_r^* > 0$ and $v_i^* > 0$. All efficient points representing DMUs lie on the frontier curve which looks like a *concave envelopment* towards the origin and envelops all DMUs. So the dual model is called the *envelopment model.*

Example 8.1.2. Assume that there are five DMUs. Each DMU consumes a single input to produce a single output. The data are given in the following table:

DMUs	DMU_1	DMU_2	DMU_3	DMU_4	DMU_5
Input	$x_{11} = 2$	$x_{12} = 3$	$x_{13} = 6$	$x_{14} = 9$	$x_{15} = 5$
Output	$y_{11} = 1$	$y_{12} = 4$	$y_{13} = 6$	$y_{14} = 7$	$y_{15} = 3$

and the observed input and output values are, respectively, $x_{10} = x_{15} = 5$ and $y_{10} = y_{15} = 3$. To evaluate the efficiency of DMU_5, by the dual theorem, we solve the following dual model:

$$(L_{\mathrm{CCR}}^{\mathrm{id}}): \ \min \theta \quad \text{subject to}$$
$$\begin{cases} 2\lambda_1 + 3\lambda_2 + 6\lambda_3 + 9\lambda_4 + 5\lambda_5 \leq 5\theta \\ \lambda_1 + 4\lambda_2 + 6\lambda_3 + 7\lambda_4 + 3\lambda_5 \geq 3 \\ \lambda_1 \geq 0, \ \lambda_2 \geq 0, \ \lambda_3 \geq 0, \ \lambda_4 \geq 0, \ \lambda_5 \geq 0, \ \theta > 0. \end{cases}$$

Let $\theta = -\xi$. Converting minimization of θ to maximization of ξ, and then introducing two slack variables $s_1 \geq 0$ and $s_2 \geq 0$, this produces a new model (L_0) from the dual model $(L_{\mathrm{CCR}}^{\mathrm{id}})$

$$(L_0):\ \max \xi \quad \text{subject to}$$

$$\begin{cases} 5\xi - 2\lambda_1 - 3\lambda_2 - 6\lambda_3 - 9\lambda_4 - 5\lambda_5 + s_1 = 0, \\ \lambda_1 + 4\lambda_2 + 6\lambda_3 + 7\lambda_4 + 3\lambda_5 + s_2 = 3 \\ \lambda_1 \geq 0,\ \lambda_2 \geq 0,\ \lambda_3 \geq 0,\ \lambda_4 \geq 0,\ \lambda_5 \geq 0. \end{cases}$$

Consider the first row as the objective and the second row as the only constraint, we will solve the above dual model (L_0) using the simplex method given in Section 7.3. There is only one constraint, so the computation process becomes very simple.

Pivoting on the $(1, 1)$ entry, this produces a model (L_1) from the model (L_0)

$$(L_1):\ \max \xi \quad \text{subject to}$$

$$\begin{cases} 5\xi + 5\lambda_2 + 6\lambda_3 + 5\lambda_4 + \lambda_5 + s_1 + 2s_2 = 6, \\ \lambda_1 + 4\lambda_2 + 6\lambda_3 + 7\lambda_4 + 3\lambda_5 + s_2 = 3 \\ \lambda_1 \geq 0,\ \lambda_2 \geq 0,\ \lambda_3 \geq 0,\ \lambda_4 \geq 0,\ \lambda_5 \geq 0. \end{cases} \qquad \phi(1) = 1,$$

The basic variable is λ_1. Other variables are nonbasic variables. Let $\lambda_i = 0\ (i \neq 1)$ and $s_i = 0\ (i = 1, 2)$. Then $\lambda_1 = 3$ and $\xi = -\frac{6}{5}$.

Pivoting on the $(1, 2)$ entry, this produces a model (L_2) from the model (L_0)

$$(L_2):\ \max \xi \quad \text{subject to}$$

$$\begin{cases} 5\xi + \frac{5}{4}\lambda_1 + \frac{3}{2}\lambda_3 + \frac{15}{4}\lambda_4 + \frac{11}{4}\lambda_5 - s_1 - \frac{3}{4}s_2 = -\frac{9}{4}, \\ \frac{1}{4}\lambda_1 + \lambda_2 + \frac{3}{2}\lambda_3 + \frac{7}{4}\lambda_4 + \frac{3}{4}\lambda_5 + \frac{1}{4}s_2 = \frac{3}{4}, \\ \lambda_1 \geq 0,\ \lambda_2 \geq 0,\ \lambda_3 \geq 0,\ \lambda_4 \geq 0,\ \lambda_5 \geq 0. \end{cases} \qquad \phi(1) = 2,$$

The basic variable is λ_2. Other variables are nonbasic variables. Let $\lambda_i = 0\ (i \neq 2)$ and $s_i = 0\ (i = 1, 2)$. Then $\lambda_2 = \frac{3}{4}$ and $\xi = -\frac{9}{20}$.

Pivoting on the $(1, 3)$ entry, this produces a model (L_3) from the model (L_0)

$$(L_3):\ \max \xi \quad \text{subject to}$$

$$\begin{cases} 5\xi + \lambda_1 - \lambda_2 + 2\lambda_4 + 2\lambda_5 - s_1 - s_2 = -3, \\ \frac{1}{6}\lambda_1 + \frac{2}{3}\lambda_2 + \lambda_3 + \frac{7}{6}\lambda_4 + \frac{1}{2}\lambda_5 + \frac{1}{6}s_2 = \frac{1}{2}, \\ \lambda_1 \geq 0,\ \lambda_2 \geq 0,\ \lambda_3 \geq 0,\ \lambda_4 \geq 0,\ \lambda_5 \geq 0. \end{cases} \qquad \phi(1) = 3,$$

The basic variable is λ_3. Other variables are nonbasic variables. Let $\lambda_i = 0\ (i \neq 3)$ and $s_i = 0\ (i = 1, 2)$. Then $\lambda_3 = \frac{1}{2}$ and $\xi = -\frac{3}{5}$.

Pivoting on the $(1, 4)$ entry, this produces a model (L_4) from the model (L_0)

$$(L_4):\ \max \xi \quad \text{subject to}$$

$$\begin{cases} \frac{5}{9}\xi + \frac{5}{63}\lambda_1 - \frac{15}{63}\lambda_2 - \frac{12}{63}\lambda_3 + \frac{8}{63}\lambda_5 - \frac{1}{9}s_1 - \frac{1}{7}s_2 = -\frac{3}{7}, \\ \frac{1}{7}\lambda_1 + \frac{4}{7}\lambda_2 + \frac{6}{7}\lambda_3 + \lambda_4 + \frac{3}{7}\lambda_5 + \frac{1}{7}s_2 = \frac{3}{7}, \\ \lambda_1 \geq 0,\ \lambda_2 \geq 0,\ \lambda_3 \geq 0,\ \lambda_4 \geq 0,\ \lambda_5 \geq 0. \end{cases} \qquad \phi(1) = 4,$$

The basic variable is λ_4. Other variables are nonbasic variables. Let $\lambda_i = 0$ $(i \neq 4)$ and $s_i = 0$ $(i = 1, 2)$. Then $\lambda_4 = \frac{3}{7}$ and $\xi = -\frac{27}{35}$.

Pivoting on the $(1, 5)$ entry, this produces a model (L_5) from the model (L_0)

$$(L_5):\ \max \xi \quad \text{subject to}$$

$$\begin{cases} \xi + \frac{1}{15}\lambda_1 - \frac{11}{15}\lambda_2 - \frac{4}{5}\lambda_3 - \frac{8}{15}\lambda_4 - \frac{1}{5}s_1 - \frac{1}{3}s_2 = -1, \\ \frac{1}{3}\lambda_1 + \frac{4}{3}\lambda_2 + 2\lambda_3 + \frac{7}{3}\lambda_4 + \lambda_5 + \frac{1}{3}s_2 = 1, & \phi(1) = 5, \\ \lambda_1 \geq 0,\ \lambda_2 \geq 0,\ \lambda_3 \geq 0,\ \lambda_4 \geq 0,\ \lambda_5 \geq 0. \end{cases}$$

The basic variable is λ_5. Other variables are nonbasic variables. Let $\lambda_i = 0$ $(i \neq 5)$ and $s_i = 0$ $(i = 1, 2)$. Then $\lambda_5 = 1$ and $\xi = -1$.

Therefore,

$$\max \xi = \max \left\{ -\frac{6}{5}, -\frac{9}{20}, -\frac{3}{5}, -\frac{27}{35}, -1 \right\} = -\frac{9}{20}$$

and the corresponding $\lambda_2 = \frac{3}{4}$ and $\lambda_i = 0$ $(i \neq 2)$.

Note that $\min \theta = -\max \xi$. So when $\lambda_2 = \frac{3}{4}$ and $\lambda_i = 0$ $(i \neq 2)$, $\min \theta = -\max \xi = \frac{9}{20}$.

8.1.2 Output-oriented CCR DEA models

Let $\boldsymbol{x}_0, \boldsymbol{y}_0, \boldsymbol{x}_j, \boldsymbol{y}_j$ $(j = 1, \ldots, n)$, $\boldsymbol{u}$, $\boldsymbol{v}$, and $\boldsymbol{\mu}$, $\boldsymbol{\nu}$ be stated as in Section 8.1.1, i.e.,

$$\begin{aligned} &\boldsymbol{x}_0 = (x_{10}, \ldots, x_{m0}), && \boldsymbol{y}_0 = (y_{10}, \ldots, y_{q0}), \\ &\boldsymbol{x}_j = (x_{1j}, \ldots, x_{mj}), && \boldsymbol{y}_j = (y_{1j}, \ldots, y_{qj}) \qquad (j = 1, \ldots, n), \\ &\boldsymbol{u} = (u_1, \ldots, u_q), && \boldsymbol{v} = (v_1, \ldots, v_m), \\ &\boldsymbol{\mu} = (\mu_1, \ldots, \mu_q), && \boldsymbol{\nu} = (\nu_1, \ldots, \nu_m). \end{aligned}$$

The ratio of inputs $\boldsymbol{x}_0$ with weight $\boldsymbol{v}$ to outputs $\boldsymbol{y}_0$ with weight $\boldsymbol{u}$ forms the objective function

$$\frac{\langle \boldsymbol{v}, \boldsymbol{x}_0 \rangle}{\langle \boldsymbol{u}, \boldsymbol{y}_0 \rangle} = \frac{v_1 x_{10} + v_2 x_{20} + \cdots + v_m x_{m0}}{u_1 y_{10} + u_2 y_{20} + \cdots + u_q y_{q0}},$$

where $\boldsymbol{u} \geq 0$ and $\boldsymbol{v} \geq 0$ are variables, and $\boldsymbol{x}_0, \boldsymbol{y}_0$ are the observed input and output values, respectively. Using the Charnes–Cooper transformation

$$t = \frac{1}{\langle \boldsymbol{u}, \boldsymbol{y}_0 \rangle} \quad \text{or} \quad \langle t\boldsymbol{u}, \boldsymbol{y}_0 \rangle = 1, \tag{8.1.4}$$

the objective function becomes

$$\frac{\langle \boldsymbol{v}, \boldsymbol{x}_0 \rangle}{\langle \boldsymbol{u}, \boldsymbol{y}_0 \rangle} = t\langle \boldsymbol{v}, \boldsymbol{x}_0 \rangle = \langle t\boldsymbol{v}, \boldsymbol{x}_0 \rangle. \tag{8.1.5}$$

An additional constraint is to guarantee that the efficiency of DMU must be less than or equal to unity, i.e.,

$$\frac{\langle \boldsymbol{u}, \boldsymbol{y}_j \rangle}{\langle \boldsymbol{v}, \boldsymbol{x}_j \rangle} = \frac{t\langle \boldsymbol{u}, \boldsymbol{y}_j \rangle}{t\langle \boldsymbol{v}, \boldsymbol{x}_j \rangle} = \frac{\langle t\boldsymbol{u}, \boldsymbol{y}_j \rangle}{\langle t\boldsymbol{v}, \boldsymbol{x}_j \rangle} \leq 1 \quad (j = 1, \ldots, n). \tag{8.1.6}$$

Using change of variable $\boldsymbol{\mu} = t\boldsymbol{u}$ and $\boldsymbol{\nu} = t\boldsymbol{v}$ in (8.1.4), (8.1.5), and (8.1.6), the output-oriented CCR DEA model becomes the following form

$$(L_{\text{CCR}}^{\text{op}}):\ \min\langle \boldsymbol{\nu}, \boldsymbol{x}_0\rangle \quad \text{subject to}$$
$$\begin{cases} \langle \boldsymbol{\nu}, \boldsymbol{x}_j\rangle - \langle \boldsymbol{\mu}, \boldsymbol{y}_j\rangle \geq 0 \quad (j = 1, \ldots, n), \\ \langle \boldsymbol{\mu}, \boldsymbol{y}_0\rangle = 1, \\ \boldsymbol{\nu} \geq 0, \quad \boldsymbol{\mu} \geq 0. \end{cases}$$

Example 8.1.3. Assume that there are three DMUs with two outputs y_1, y_2 and one input x to be evaluated. The data for the output-oriented CCR DEA model are listed in the following table:

DMU	y_1	y_2	x	y_1/x	y_2/x
DMU_1	10	40	10	1.00	4.00
DMU_2	20	40	20	1.00	2.00
DMU_3	25	35	10	2.50	3.50

Using the model $(L_{\text{CCR}}^{\text{op}})$, three output-oriented CCR DEA models are used to assess DMUs, respectively

$$\mathbf{DMU}_1:\ \max 10\,W_x \quad \text{subject to}$$
$$\begin{aligned} 10\,W_x - 10\,W_{y_1} - 40\,W_{y_2} &\geq 0 \\ 20\,W_x - 20\,W_{y_1} - 40\,W_{y_2} &\geq 0 \\ 10\,W_x - 25\,W_{y_1} - 35\,W_{y_2} &\geq 0 \\ 10\,W_{y_1} + 40\,W_{y_2} &= 1. \end{aligned}$$

$$\mathbf{DMU}_2:\ \max 20\,W_x \quad \text{subject to}$$
$$\begin{aligned} 10\,W_x - 10\,W_{y_1} - 40\,W_{y_2} &\geq 0 \\ 20\,W_x - 20\,W_{y_1} - 40\,W_{y_2} &\geq 0 \\ 10\,W_x - 25\,W_{y_1} - 35\,W_{y_2} &\geq 0 \\ 20\,W_{y_1} + 40\,W_{y_2} &= 1. \end{aligned}$$

$$\mathbf{DMU}_3:\ \max 10\,W_x \quad \text{subject to}$$
$$\begin{aligned} 10\,W_x - 10\,W_{y_1} - 40\,W_{y_2} &\geq 0 \\ 20\,W_x - 20\,W_{y_1} - 40\,W_{y_2} &\geq 0 \\ 10\,W_x - 25\,W_{y_1} - 35\,W_{y_2} &\geq 0 \\ 25\,W_{y_1} + 35\,W_{y_2} &= 1. \end{aligned}$$

The three evaluated DMUs may be represented by the following three points, respectively,

$$D_1(1, 4), \quad D_2(1, 2), \quad D_3(2.5, 3.5).$$

Set a rectangular coordinate system with the origin $O(0,0)$, horizontal axis y_1/x, and vertical axis y_2/x on a sheet of graph paper and plot these three points D_1, D_2, D_3. The curve through points D_1, D_3, and its extension is called the *frontier curve* which looks like a *concave envelopment* towards the origin and envelops these three points. Denote by D_i' the intersection point of the straight line OD_i and the frontier curve. The intersection point D_i' is called the *projection* from the point D_i to the frontier curve.

Define the efficiency value of D_i as $\frac{OD_i}{OD_i'}$. Clearly,

$$\frac{OD_1}{OD_1'} = 1, \quad \frac{OD_2}{OD_2'} < 1, \quad \frac{OD_3}{OD_3'} = 1,$$

i.e., the efficiency values of D_1 and D_3 are both 1, and the efficiency values of D_2 is less than 1.

The detailed version of the output-oriented CCR DEA model is

$$(L_{\text{CCR}}^{\text{op}}): \min\{x_{10}v_1 + x_{20}v_2 + \cdots + x_{m0}v_m\} \quad \text{subject to}$$

$$\begin{cases} x_{11}v_1 + x_{21}v_2 + \cdots + x_{m1}v_m - y_{11}\mu_1 - y_{21}\mu_2 - \cdots - y_{q1}\mu_q \ge 0, \\ x_{12}v_1 + x_{22}v_2 + \cdots + x_{m2}v_m - y_{12}\mu_1 - y_{22}\mu_2 - \cdots - y_{q2}\mu_q \ge 0, \\ \vdots \\ x_{1n}v_1 + x_{2n}v_2 + \cdots + x_{mn}v_n - y_{1n}\mu_1 - y_{2n}\mu_2 - \cdots - y_{qn}\mu_q \ge 0, \\ y_{10}\mu_1 + y_{20}\mu_2 + \cdots + y_{q0}\mu_q = 1, \\ v_1 \ge 0, \ \ldots, \ v_m \ge 0, \quad \mu_1 \ge 0, \ \ldots, \ \mu_q \ge 0. \end{cases}$$

By Theorem 7.6.2, the *dual model* of the output-oriented CCR DEA model is

$$(L_{\text{CCR}}^{\text{od}}): \max \varphi \quad \text{subject to}$$

$$\begin{cases} x_{11}\lambda_1 + x_{12}\lambda_2 + \cdots + x_{1n}\lambda_n \le x_{10}, \\ \vdots \\ x_{m1}\lambda_1 + x_{m2}\lambda_2 + \cdots + x_{mn}\lambda_n \le x_{m0}, \\ -y_{11}\lambda_1 - y_{12}\lambda_2 - \cdots - y_{1n}\lambda_n + y_{10}\varphi \le 0, \\ \vdots \\ -y_{q1}\lambda_1 - y_{q2}\lambda_2 - \cdots - y_{qn}\lambda_n + y_{q0}\varphi \le 0, \\ \lambda_1 \ge 0, \ \lambda_2 \ge 0, \ \ldots, \ \lambda_n \ge 0. \end{cases}$$

Let $\boldsymbol{x}^i$, $\boldsymbol{y}^r$, and $\boldsymbol{\lambda}$ be stated as in Section 8.1.1, i.e.,

$$\begin{aligned} \boldsymbol{x}^i &= (x_{i1}, \ldots, x_{in}) \quad (i = 1, \ldots, m), \\ \boldsymbol{y}^r &= (y_{r1}, \ldots, y_{rn}) \quad (r = 1, \ldots, q), \\ \boldsymbol{\lambda} &= (\lambda_1, \ldots, \lambda_n). \end{aligned}$$

The contraction of the dual model of the output-oriented CCR DEA model is as follows:

$$(L^{\mathrm{od}}_{\mathrm{CCR}}):\ \max\varphi \quad \text{subject to}$$
$$\begin{cases} \langle \boldsymbol{\lambda}, \mathbf{x}^i\rangle \le x_{i0} & (i = 1,\dots,m),\\ \langle \boldsymbol{\lambda}, \mathbf{y}^r\rangle \ge \varphi y_{r0} & (r = 1,\dots,q),\\ \boldsymbol{\lambda} \ge 0. \end{cases}$$

8.2 Banker–Charnes–Cooper DEA models

The Banker–Charnes–Cooper (BCC) DEA model was presented by Banker, Charnes, and Cooper in 1984. This model is referred to as the Variable Returns to Scale (VRS) model or the pure technical efficiency model. BCC DEA models are constructed by adding an additional constraint to the dual models $(L^{\mathrm{id}}_{\mathrm{CCR}})$ and $(L^{\mathrm{od}}_{\mathrm{CCR}})$.

8.2.1 Input-oriented BCC DEA models

Consider the dual model $(L^{\mathrm{id}}_{\mathrm{CCR}})$ introduced in Section 8.1.1. Adding an additional constraint

$$\lambda_1 + \cdots + \lambda_n = 1 \quad (\lambda \ge 0)$$

to the dual model $(L^{\mathrm{id}}_{\mathrm{CCR}})$ yields the *input-oriented BCC DEA model* $(L^{\mathrm{ip}}_{\mathrm{BCC}})$.

The detailed version of the input-oriented BCC DEA model $(L^{\mathrm{ip}}_{\mathrm{BCC}})$ is as follows

$$(L^{\mathrm{ip}}_{\mathrm{BCC}}):\ \min\theta \quad \text{subject to}$$
$$\begin{cases} y_{11}\lambda_1 + y_{12}\lambda_2 + \cdots + y_{1n}\lambda_n \ge y_{10},\\ \quad\vdots\\ y_{q1}\lambda_1 + y_{q2}\lambda_2 + \cdots + y_{qn}\lambda_n \ge y_{q0},\\ -x_{11}\lambda_1 - x_{12}\lambda_2 - \cdots - x_{1n}\lambda_n + \theta x_{10} \ge 0,\\ \quad\vdots\\ -x_{m1}\lambda_1 - x_{m2}\lambda_2 - \cdots - x_{mn}\lambda_n + \theta x_{m0} \ge 0,\\ \lambda_1 + \lambda_2 + \cdots + \lambda_n = 1,\\ \lambda_1 \ge 0,\ \lambda_2 \ge 0,\ \dots,\ \lambda_n \ge 0, \end{cases}$$

where $\theta > 0$.

By Theorem 7.6.2, the dual model $(L^{\rm id}_{\rm BCC})$ of the input-oriented BCC DEA model $(L^{\rm ip}_{\rm BCC})$ is

$$(L^{\rm id}_{\rm BCC}):\ \max\{y_{10}\mu_1+y_{20}\mu_2+\cdots+y_{q0}\mu_q+\mu_0\}\quad \text{subject to}$$
$$\begin{cases} y_{11}\mu_1+y_{21}\mu_2+\cdots+y_{q1}\mu_q-x_{11}\nu_1-x_{21}\nu_2-\cdots-x_{m1}\nu_m+\mu_0\le 0,\\ \quad\vdots\\ y_{1n}\mu_1+y_{2n}\mu_2+\cdots+y_{qn}\mu_q-x_{1n}\nu_1-x_{2n}\nu_2-\cdots-x_{mn}\nu_m+\mu_0\le 0,\\ x_{10}\nu_1+x_{20}\nu_2+\cdots+x_{m0}\nu_m=1,\\ \mu_1\ge 0,\ \ldots,\ \mu_n\ge 0,\quad \nu_1\ge 0,\ \ldots,\ \nu_m\ge 0,\quad \mu_0\in\mathbb{R}. \end{cases}$$

Let $\boldsymbol{x}_0$, $\boldsymbol{y}_0$, $\boldsymbol{x}_j$, $\boldsymbol{y}_j$, $\boldsymbol{x}^i$, $\boldsymbol{y}^r$, and $\boldsymbol{\mu}$, $\boldsymbol{\nu}$, $\boldsymbol{\lambda}$ be stated in Section 8.1, and let $\boldsymbol{e}=(1,1,\ldots,1)$. The contraction of the input-oriented BCC DEA model is the form

$$(L^{\rm ip}_{\rm BCC}):\ \min\theta\quad \text{subject to}$$
$$\begin{cases} \langle\boldsymbol{\lambda},\boldsymbol{y}^r\rangle\ge y_{r0} & (r=1,\ldots,q),\\ \langle\boldsymbol{\lambda},\boldsymbol{x}^i\rangle\le\theta x_{i0} & (i=1,\ldots,m),\\ \langle\boldsymbol{e},\boldsymbol{\lambda}\rangle=1,\ \boldsymbol{\lambda}\ge 0,\ \theta>0, \end{cases}$$

and the contraction of the dual model of the input-oriented BCC DEA model is the form

$$(L^{\rm id}_{\rm BCC}):\ \max\{\langle\boldsymbol{\mu},\boldsymbol{y}_0\rangle+\mu_0\}\quad \text{subject to}$$
$$\begin{cases} \langle\boldsymbol{\mu},\boldsymbol{y}_j\rangle-\langle\boldsymbol{\nu},\boldsymbol{x}_j\rangle+\mu_0\le 0 & (j=1,\ldots,n),\\ \langle\boldsymbol{\nu},\boldsymbol{x}_0\rangle=1,\\ \boldsymbol{\mu}\ge 0,\ \boldsymbol{\nu}\ge 0,\ \mu_0\in\mathbb{R}, \end{cases}$$

where μ_0 is a free variable.

It has been seen that the added constraint in the input-oriented BCC DEA model introduces an additional variable μ_0 in its dual model. This extra variable μ_0 makes it possible to effect returns-to-scale evaluations. Thus the input-oriented BCC DEA model is refereed to as the Variable Returns to Scale (VRS) model.

8.2.2 Output-oriented BCC DEA models

Similarly, adding the additional constraint $\lambda_1+\lambda_2+\cdots+\lambda_n=1$ ($\boldsymbol{\lambda}\ge 0$) to the dual model $(L^{\rm od}_{\rm CCR})$ yields the *output-oriented BCC DEA model*. Its detail version is as follows:

$$(L_{\text{BCC}}^{\text{op}}):\ \max\varphi \quad \text{subject to}$$
$$\begin{cases} x_{11}\lambda_1 + x_{12}\lambda_2 + \cdots + x_{1n}\lambda_n \le x_{10}, \\ \vdots \\ x_{m1}\lambda_1 + x_{m2}\lambda_2 + \cdots + x_{mn}\lambda_n \le x_{m0}, \\ -y_{11}\lambda_1 - y_{12}\lambda_2 - \cdots - y_{1n}\lambda_n + y_{10}\varphi \le 0, \\ \vdots \\ -y_{q1}\lambda_1 - y_{q2}\lambda_2 - \cdots - y_{qn}\lambda_n + y_{q0}\varphi \le 0, \\ \lambda_1 + \lambda_2 + \cdots + \lambda_n = 1, \\ \lambda_1 \ge 0,\ \lambda_2 \ge 0,\ \ldots,\ \lambda_n \ge 0. \end{cases}$$

By Theorem 7.6.2, its dual model $(L_{\text{BCC}}^{\text{od}})$ is as follows:

$$(L_{\text{BCC}}^{\text{od}}):\ \max\{x_{10}v_1 + x_{20}v_2 + \cdots + x_{m0}v_m + v_0\} \quad \text{subject to}$$
$$\begin{cases} x_{11}v_1 + x_{21}v_2 + \cdots + x_{m1}v_m - y_{11}\mu_1 - y_{21}\mu_2 - \cdots - y_{q1}\mu_q + v_0 \ge 0, \\ \vdots \\ x_{1n}v_1 + x_{2n}v_2 + \cdots + x_{mn}v_m - y_{1n}\mu_1 - y_{2n}\mu_2 - \cdots - y_{qn}\mu_q + v_0 \ge 0, \\ y_{10}\mu_1 + y_{20}\mu_2 + \cdots + y_{q0}\mu_q = 1, \\ v_1 \ge 0,\ \ldots,\ v_m \ge 0,\quad \mu_1 \ge 0,\ \ldots,\ \mu_q \ge 0,\quad v_0 \in \mathbb{R}, \end{cases}$$

where v_0 is a free variable.

Let $\boldsymbol{x}_0, \boldsymbol{y}_0, \boldsymbol{x}_j, \boldsymbol{y}_j, \boldsymbol{x}^i, \boldsymbol{y}^r$, and $\boldsymbol{\mu}, \boldsymbol{v}, \boldsymbol{\lambda}$ be stated in Section 8.1 and $\boldsymbol{e} = (1, 1, \ldots, 1)$. The contraction of the output-oriented BCC DEA model is as follows:

$$(L_{\text{BCC}}^{\text{op}}):\ \min\varphi \quad \text{subject to}$$
$$\begin{cases} \langle \boldsymbol{\lambda}, \boldsymbol{x}^i \rangle \le x_{i0} & (i = 1, \ldots, m), \\ \langle \boldsymbol{\lambda}, \boldsymbol{y}^r \rangle \ge \varphi y_{r0} & (r = 1, \ldots, q), \\ \langle \boldsymbol{e}, \boldsymbol{\lambda} \rangle = 1, \\ \boldsymbol{\lambda} \ge 0. \end{cases}$$

The contraction of its dual model is as follows:

$$(L_{\text{BCC}}^{\text{od}}):\ \max\{\langle \boldsymbol{v}, \boldsymbol{x}_0 \rangle + v_0\} \quad \text{subject to}$$
$$\begin{cases} \langle \boldsymbol{\mu}, \boldsymbol{y}_j \rangle - \langle \boldsymbol{v}, \boldsymbol{x}_j \rangle + v_0 \le 0 \quad (j = 1, \ldots, n), \\ \langle \boldsymbol{\mu}, \boldsymbol{y}_0 \rangle = 1, \\ \boldsymbol{v} \ge 0,\ \boldsymbol{\mu} \ge 0,\ v_0 \in \mathbb{R}, \end{cases}$$

where v_0 is a free variable. Similarly, the added constraint $\langle \boldsymbol{e}, \boldsymbol{\lambda} \rangle = 1$ introduces an extra variable v_0.

8.3 One-stage and two-stage methods

The one-stage and two stage methods are used for solving optimization models. Cite the dual model ($L^{\text{id}}_{\text{CCR}}$) of the input-oriented CCR DEA model to introduce these two methods.

8.3.1 One-stage method

Slack variables are used to convert the inequalities in the model ($L^{\text{id}}_{\text{CCR}}$) to equivalent equations. The one-stage method is to add only slack variables to constraints. The dual model ($L^{\text{id}}_{\text{CCR}}$) is the form

$$(L^{\text{id}}_{\text{CCR}}): \ \min \theta \quad \text{subject to}$$
$$\begin{cases} \langle \boldsymbol{\lambda}, \boldsymbol{y}^r \rangle \geq y_{r0} & (r = 1, \ldots, q), \\ \langle \boldsymbol{\lambda}, \boldsymbol{x}^i \rangle \leq \theta x_{i0} & (i = 1, \ldots, m), \\ \boldsymbol{\lambda} \geq 0, \ \theta > 0. \end{cases}$$

Introducing slack variables to constraints, the dual model has the following equivalent form

$$(L^{\text{id}}_{\text{CCR}}): \ \min \theta \quad \text{subject to}$$
$$\begin{cases} \langle \boldsymbol{\lambda}, \boldsymbol{y}^r \rangle - s_r^- = y_{r0}, & s_r^- \geq 0 \quad (r = 1, \ldots, q), \\ \langle \boldsymbol{\lambda}, \boldsymbol{x}^i \rangle + s_i^+ = \theta x_{i0}, & s_i^+ \geq 0 \quad (i = 1, \ldots, m), \\ \boldsymbol{\lambda} \geq 0, \ \theta > 0. \end{cases}$$

In the previous Section 8.1.1, the one-stage method was used for solving Example 8.1.2.

8.3.2 Two-stage method

Nonzero slacks will be such that some boundary points may be weakly efficient which is defined as follows.

Definition 8.3.1. Let the particular DMU_0 being evaluated consume amounts x_{i0} of input i ($i = 1, \ldots, m$) and produce amounts y_{r0} of output r ($r = 1, \ldots, q$), and let θ^* be the optimal solution of the dual model ($L^{\text{id}}_{\text{CCR}}$). Then

(a) DMU_0 is efficient if and only if $\theta^* = 1$ and the slack variables $s_i^{-*} = s_r^{+*} = 0$ for all i and r.

(b) DMU_0 is weakly efficient if and only if $\theta^* = 1$ and the slack variables $s_i^{-*} \neq 0$ and/or $s_r^{+*} \neq 0$ for some i and r.

To avoid weak efficiency, the two-stage method is to add the slack variables to both constraints and the objective, i.e.,

$$(L_{\text{CCR}}^{\text{id}}):\ \min\left\{\theta-\varepsilon\left(\sum_1^m s_i^-+\sum_1^q s_r^+\right)\right\}\quad\text{subject to}$$

$$\begin{cases}\langle\boldsymbol{\lambda},\mathbf{y}^r\rangle-s_r^-=y_{r0}, & s_r^-\ge 0\quad (r=1,\dots,q),\\ \langle\boldsymbol{\lambda},\mathbf{x}^i\rangle+s_i^+=\theta x_{i0}, & s_i^+\ge 0\quad (i=1,\dots,m),\\ \boldsymbol{\lambda}\ge 0,\ \theta>0,\ \varepsilon>0, & \end{cases}$$

where ε is a so-called non-Archimedean element defined to be smaller than any positive real number.

This model is solved in two stages.

The first stage is to find the optimal solution θ^* of the dual model $(L_{\text{CCR}}^{\text{id}})$

$$(L_{\text{CCR}}^{\text{id}}):\ \min\theta\quad\text{subject to}$$

$$\begin{cases}\langle\boldsymbol{\lambda},\mathbf{y}^r\rangle\ge y_{r0} & (r=1,\dots,q),\\ \langle\boldsymbol{\lambda},\mathbf{x}^i\rangle\le\theta x_{i0} & (i=1,\dots,m),\\ \boldsymbol{\lambda}\ge 0,\ \theta>0 & \end{cases}$$

or its equivalent model

$$(L_{\text{CCR}}^{\text{id}}):\ \min\theta\quad\text{subject to}$$

$$\begin{cases}\langle\boldsymbol{\lambda},\mathbf{y}^r\rangle-s_r^-=y_{r0}, & s_r^-\ge 0\quad (r=1,\dots,q),\\ \langle\boldsymbol{\lambda},\mathbf{x}^i\rangle+s_i^+=\theta x_{i0}, & s_i^+\ge 0\quad (i=1,\dots,m),\\ \boldsymbol{\lambda}\ge 0,\ \theta>0. & \end{cases}$$

The second stage is to solve the following model:

$$\max\left\{\sum_1^m s_i^-+\sum_1^q s_r^+\right\}\quad\text{subject to}$$

$$\begin{cases}\langle\boldsymbol{\lambda},\mathbf{y}^r\rangle-s_r^-=y_{r0} & (r=1,\dots,q),\\ \langle\boldsymbol{\lambda},\mathbf{x}^i\rangle+s_i^+=\theta^* x_{i0} & (i=1,\dots,m),\\ \boldsymbol{\lambda}\ge 0, & \end{cases}$$

where θ^* is the optimal solution obtained in the first stage.

8.4 Advanced DEA models

Let $\boldsymbol{x}_0, \boldsymbol{y}_0, \boldsymbol{x}_j, \boldsymbol{y}_j, \boldsymbol{x}^i, \boldsymbol{y}^r$ and $\boldsymbol{\mu}, \boldsymbol{\nu}, \boldsymbol{\lambda}$ be stated in Section 8.1, and let $\boldsymbol{e} = (1, 1, \dots, 1)$.

8.4.1 Free disposal hull DEA models

The Free Disposal Hull (FDH) DEA model presented by Tulkens (1993) is a *mixed integer linear programming model*. The FDH DEA model is obtained by changing the constraint $\boldsymbol{\lambda} \geq 0$ into $\boldsymbol{\lambda} = (\lambda_1, \dots, \lambda_n)$ and $\lambda_j = 0$ or 1 $(j = 1, \dots, n)$ in DEA models. The input-oriented FDH DEA model is the form

$$\min \theta \quad \text{subject to}$$

$$\begin{cases} \langle \boldsymbol{\lambda}, \boldsymbol{y}^r \rangle \geq y_{r0} & (r = 1, \dots, q), \\ \langle \boldsymbol{\lambda}, \boldsymbol{x}^i \rangle \leq \theta x_{i0} & (i = 1, \dots, m), \\ \langle \boldsymbol{e}, \boldsymbol{\lambda} \rangle = 1 & (\lambda_j = 0 \text{ or } 1 \ (j = 1, \dots, n)), \ \theta > 0. \end{cases}$$

8.4.2 Slack-based measure DEA models

Slack-Based Measure (SBM) DEA models were presented by Tone (2001). The SBM DEA models include the input-oriented, the output-oriented, and the nonoriented SBM DEA model.

- The input-oriented SBM DEA model is the form

$$\min \left\{ 1 - \frac{1}{m} \sum_1^m \frac{s_i^-}{x_{i0}} \right\} \quad \text{subject to}$$

$$\begin{cases} \langle \boldsymbol{\lambda}, \boldsymbol{y}^r \rangle \geq y_{r0} & & (r = 1, \dots, q), \\ \langle \boldsymbol{\lambda}, \boldsymbol{x}^i \rangle + s_i^- = x_{i0}, & s_i^- \geq 0 & (i = 1, \dots, m), \\ \boldsymbol{\lambda} \geq 0, & & \end{cases}$$

  where s_i^- $(i = 1, \dots, m)$ are m slack variables.
- The output-oriented SBM DEA model is the form

$$\min \left\{ \frac{1}{1 + \frac{1}{q} \sum_1^q \frac{s_r^+}{y_{r0}}} \right\} \quad \text{subject to}$$

$$\begin{cases} \langle \boldsymbol{\lambda}, \boldsymbol{y}^r \rangle - s_r^+ = y_{r0}, & s_r^+ \geq 0 & (r = 1, \dots, q), \\ \langle \boldsymbol{\lambda}, \boldsymbol{x}^i \rangle \leq x_{i0} & & (i = 1, \dots, m), \\ \boldsymbol{\lambda} \geq 0, & & \end{cases}$$

  where s_r^+ $(r = 1, \dots, q)$ are q slack variables.

- The nonoriented SBM DEA model is the form

$$\min \left\{ \frac{1 - \frac{1}{m}\sum_1^m \frac{s_i^-}{x_{i0}}}{1 + \frac{1}{q}\sum_1^q \frac{s_r^+}{y_{r0}}} \right\} \quad \text{subject to}$$

$$\begin{cases} \langle \boldsymbol{\lambda}, \mathbf{y}^r \rangle - s_r^+ = y_{r0}, & s_r^+ \geq 0 \quad (r = 1, \dots, q), \\ \langle \boldsymbol{\lambda}, \mathbf{x}^i \rangle + s_i^- = x_{i0}, & s_i^- \geq 0 \quad (i = 1, \dots, m), \\ \boldsymbol{\lambda} \geq 0, \end{cases}$$

where s_i^- $(i = 1, \dots, m)$ and s_r^+ $(r = 1, \dots, q)$ are $m + q$ slack variables.

The above input-oriented, output-oriented, and nonoriented SBM models are referred to as *CRS models*. If the additional constraint $\langle \mathbf{e}, \boldsymbol{\lambda} \rangle = 1$ is added to the above models, the obtained models are referred to as *VRS models*. If the additional constraints

$$\begin{aligned} \langle \mathbf{e}, \boldsymbol{\lambda} \rangle &\geq 1, \\ \langle \mathbf{e}, \boldsymbol{\lambda} \rangle &\leq 1, \\ L \leq \langle \mathbf{e}, \boldsymbol{\lambda} \rangle &\leq U, \end{aligned}$$

are added, respectively, to the above models, the obtained models are referred to as *Non-Decreasing Returns to Scale* (NDRS) models, *Non-Increasing Returns to Scale* (NIRS) models, and *Generalized Returns to Scale* (GRS) models, respectively.

8.4.3 MSBM, WSBM, and WMSBM DEA models

- The Modified Slack-Based Measure (MSBM) model was presented by Sharp et al. (2007). Its form is

$$\min \left\{ \frac{1 - \frac{1}{m}\sum_1^m \frac{s_i^-}{R_{i0}}}{1 + \frac{1}{q}\sum_1^q \frac{s_r^+}{R_{r0}}} \right\} \quad \text{subject to}$$

$$\begin{cases} \langle \boldsymbol{\lambda}, \mathbf{y}^r \rangle - s_r^+ = y_{r0}, & s_r^+ \geq 0 \quad (r = 1, \dots, q), \\ \langle \boldsymbol{\lambda}, \mathbf{x}^i \rangle + s_i^- = x_{i0}, & s_i^- \geq 0 \quad (i = 1, \dots, m), \\ \langle \mathbf{e}, \boldsymbol{\lambda} \rangle = 1, \\ \boldsymbol{\lambda} \geq 0, \end{cases}$$

where s_r^+ $(r = 1, \dots, q)$ and s_i^- $(i = 1, \dots, m)$ are slack variables, and

$$\begin{aligned} R_{i0} &= x_{i0} - \min_{1 \leq j \leq n} \{ x_{ij} \}, \\ R_{r0} &= \max_{1 \leq j \leq n} \{ y_{rj} \} - y_{r0}. \end{aligned}$$

– The nonoriented Weighted Slack-Based Measure (WSBM) model is the form

$$\min \left\{ \frac{1 - \frac{1}{\sum_1^m w_i^I} \sum_1^m \frac{w_i^I s_i^-}{x_{i0}}}{1 + \frac{1}{\sum_1^q w_r^o} \sum_1^q \frac{w_r^o s_r^+}{y_{r0}}} \right\} \quad \text{subject to}$$

$$\begin{cases} \langle \boldsymbol{\lambda}, \mathbf{y}^r \rangle - s_r^+ = y_{r0}, & s_r^+ \geq 0 \quad (r = 1, \ldots, q), \\ \langle \boldsymbol{\lambda}, \mathbf{x}^i \rangle + s_i^- = x_{i0}, & s_i^- \geq 0 \quad (i = 1, \ldots, m), \\ \boldsymbol{\lambda} \geq 0, \end{cases}$$

where w_i^I $(i = 1, \ldots, m)$ and w_r^o $(r = 1, \ldots, q)$ are the input weights and the output weights, respectively.

– The Weighted Modified Slack-Based Measure (WMSBM) model is the form

$$\min \left\{ \frac{1 - \frac{1}{\sum_1^m w_i^I} \sum_1^m \frac{w_i^I s_i^-}{R_{i0}}}{1 + \frac{1}{\sum_1^q w_r^o} \sum_1^q \frac{w_r^o s_r^+}{R_{r0}}} \right\} \quad \text{subject to}$$

$$\begin{cases} \langle \boldsymbol{\lambda}, \mathbf{y}^r \rangle - s_r^+ = y_{r0}, & s_r^+ \geq 0 \quad (r = 1, \ldots, q), \\ \langle \boldsymbol{\lambda}, \mathbf{x}^i \rangle + s_i^- = x_{i0}, & s_i^- \geq 0 \quad (i = 1, \ldots, m), \\ \langle \mathbf{e}, \boldsymbol{\lambda} \rangle = 1, \\ \boldsymbol{\lambda} \geq 0, \end{cases}$$

where s_r^+, s_i^-, w_i^I, w_r^o, R_{i0}, and R_{r0} are stated as above.

8.4.4 Hybrid distance function DEA models

The main hybrid distance function DEA models include two groups. The first group was presented by Tone and Tsutsui (2010). Due to the parameter ε, Tone and Tsutsui call the first group *Epsilon-Based Measure (EBM) models*. The second group was presented by Cooper, Seiford, and Tone (2007). Due to the mix of the radial model and SBM model, Cooper, Seiford, and Tone call the second group *Hybrid Models*.

(a) EBM models

- The input-oriented EBM model is the form

$$\min\left\{\theta - \frac{\varepsilon^-}{\sum_1^m w_i^-}\sum_1^m \frac{w_i^- s_i^-}{x_{i0}}\right\} \quad \text{subject to}$$

$$\begin{cases} \langle \boldsymbol{\lambda}, \boldsymbol{x}^i\rangle + s_i^- = \theta x_{i0}, & s_i^- \geq 0 \quad (i = 1, \dots, m),\\ \langle \boldsymbol{\lambda}, \boldsymbol{y}^r\rangle \geq y_{r0} & \quad (r = 1, \dots, q),\\ \boldsymbol{\lambda} \geq 0, \end{cases}$$

where s_i^- and w_i^- are stated as above, and $\varepsilon^- \in [0, 1]$.

- The output-oriented EBM model is the form

$$\min\left\{\frac{1}{\varphi + \frac{\varepsilon^+}{\sum_1^q w_r^+}\sum_1^q \frac{w_r^+ s_r^+}{y_{r0}}}\right\} \quad \text{subject to}$$

$$\begin{cases} \langle \boldsymbol{\lambda}, \boldsymbol{y}^r\rangle - s_r^+ = \varphi y_{r0}, & s_r^+ \geq 0 \quad (r = 1, \dots, q),\\ \langle \boldsymbol{\lambda}, \boldsymbol{x}^i\rangle \leq x_{i0} & \quad (i = 1, \dots, m),\\ \boldsymbol{\lambda} \geq 0, \end{cases}$$

where s_r^+ and w_r^+ are stated as above, and $\varepsilon^+ \in [0, 1]$.

- The non-oriented EBM model is the form

$$\min\left\{\frac{\theta - \frac{\varepsilon^-}{\sum_1^m w_i^-}\sum_1^m \frac{w_i^- s_i^-}{x_{i0}}}{\varphi + \frac{\varepsilon^+}{\sum_1^q w_r^+}\sum_1^q \frac{w_r^+ s_r^+}{y_{r0}}}\right\} \quad \text{subject to}$$

$$\begin{cases} \langle \boldsymbol{\lambda}, \boldsymbol{x}^i\rangle + s_i^- = \theta x_{i0}, & s_i^- \geq 0 \quad (i = 1, \dots, m),\\ \langle \boldsymbol{\lambda}, \boldsymbol{y}^r\rangle - s_r^+ = \varphi y_{r0}, & s_r^+ \geq 0 \quad (r = 1, \dots, q),\\ \boldsymbol{\lambda} \geq 0, \end{cases}$$

where s_i^-, s_r^+, w_i^-, w_r^+ are stated as above, and $\varepsilon^-, \varepsilon^+ \in [0, 1]$. If $\varepsilon^- = \varepsilon^+ = 0$ and $\varphi = 1$, the EBM model is just the dual model of the input-oriented CCR DEA model. If $\varepsilon^- = \varepsilon^+ = 1$ and $\theta = \varphi = 1$, the EBM model is just the nonoriented weighted SBM model.

(b) Hybrid models

- The input-oriented hybrid model is the form

$$\min\left\{1-\frac{m_1}{m}(1-\theta)-\frac{1}{m}\sum_{1}^{m_2}\frac{s_i^{N-}}{x_{i0}^{N}}\right\}\quad\text{subject to}$$

$$\begin{cases}\langle\boldsymbol{\lambda},\boldsymbol{x}_R^i\rangle+s_i^{R-}=\theta x_{i0}^R, & s_i^{R-}\ge 0\quad(i=1,\ldots,m),\\ \langle\boldsymbol{\lambda},\boldsymbol{x}_N^i\rangle+s_i^{N-}=x_{i0}^N, & s_i^{N-}\ge 0\quad(i=1,\ldots,m),\\ \langle\boldsymbol{\lambda},\boldsymbol{y}_N^r\rangle\ge y_{r0}^N & (r=1,\ldots,q),\\ \boldsymbol{\lambda}\ge 0,\end{cases}$$

where R is radial and N is nonradial, and m is the input, where m_1 is the radial input and m_2 is the nonradial input.

- The output-oriented hybrid model is the form

$$\min\left\{\frac{1}{1-\frac{q_1}{q}(\varphi-1)+\frac{1}{q}\sum_{1}^{q_2}\frac{s_r^{N+}}{y_{i0}^{N}}}\right\}\quad\text{subject to}$$

$$\begin{cases}\langle\boldsymbol{\lambda},\boldsymbol{y}_R^r\rangle-s_r^{R+}=\varphi y_{r0}^R, & s_r^{R+}\ge 0\quad(r=1,\ldots,q),\\ \langle\boldsymbol{\lambda},\boldsymbol{y}_N^r\rangle-s_r^{N+}=y_{r0}^N, & s_r^{N+}\ge 0\quad(r=1,\ldots,q),\\ \langle\boldsymbol{\lambda},\boldsymbol{x}_N^i\rangle\le x_{i0}^N & (i=1,\ldots,m),\\ \boldsymbol{\lambda}\ge 0,\end{cases}$$

where R is radial and N is nonradial, and q is the output, where q_1 is the radial output and q_2 is the nonradial output.

- Hybrid model mixing the radial and SMB models is the form

$$\min\left\{\frac{1-\frac{m_1}{m}(1-\theta)-\frac{1}{m}\sum_{1}^{m_2}\frac{s_i^{N-}}{x_{i0}^{N}}}{1+\frac{q_1}{q}(\varphi-1)+\frac{1}{q}\sum_{1}^{q_2}\frac{s_r^{N+}}{y_{r0}^{N}}}\right\}\quad\text{subject to}$$

$$\begin{cases}\langle\boldsymbol{\lambda},\boldsymbol{x}_R^i\rangle+s_i^{R-}=\theta x_{i0}^R, & s_i^{R-}\ge 0\quad(i=1,\ldots,m),\\ \langle\boldsymbol{\lambda},\boldsymbol{y}_R^r\rangle-s_r^{R+}=\varphi y_{r0}^R, & s_r^{R+}\ge 0\quad(r=1,\ldots,q),\\ \langle\boldsymbol{\lambda},\boldsymbol{x}_N^i\rangle+s_i^{N-}=x_{i0}^N, & s_i^{N-}\ge 0\quad(i=1,\ldots,m),\\ \langle\boldsymbol{\lambda},\boldsymbol{y}_N^r\rangle-s_r^{N+}=y_{r0}^N, & s_r^{N+}\ge 0\quad(r=1,\ldots,q),\\ \boldsymbol{\lambda}\ge 0,\end{cases}$$

where R is radial and N is nonradial, m is the input and q is the output, where m_1 is the radial input and m_2 is the nonradial input, and q_1 is the radial output and q_2 is the nonradial output.

8.4.5 Super efficiency models

The main super efficiency models include three groups. The first group is the radial super efficiency models presented by Andersen and Petersen (1993). The second group is the directional distance function super efficiency models presented by Ray (2008). The third group is the SMB super efficiency models presented by Tone (2002).

(a) Radial super efficiency models

- The input-oriented CRS radial super efficiency model is the form

$$\min\theta \quad \text{subject to}$$
$$\begin{cases} \langle \boldsymbol{\lambda}, \mathbf{x}^i\rangle - \lambda_k x_{ik} \le \theta x_{i0} & (i = 1, \dots, m), \\ \langle \boldsymbol{\lambda}, \mathbf{y}^r\rangle - \lambda_k y_{rk} \ge y_{r0} & (r = 1, \dots, q), \\ \boldsymbol{\lambda} \ge 0, \ \theta > 0. \end{cases}$$

- The output-oriented CRS radial super efficiency model is the form

$$\max\varphi \quad \text{subject to}$$
$$\begin{cases} \langle \boldsymbol{\lambda}, \mathbf{x}^i\rangle - \lambda_k x_{ik} \le x_{i0} & (i = 1, \dots, m), \\ \langle \boldsymbol{\lambda}, \mathbf{y}^r\rangle - \lambda_k y_{rk} \ge \varphi y_{r0} & (r = 1, \dots, q), \\ \boldsymbol{\lambda} \ge 0. \end{cases}$$

- The input-oriented VRS radial super efficiency model is the form

$$\min\theta \quad \text{subject to}$$
$$\begin{cases} \langle \boldsymbol{\lambda}, \mathbf{x}^i\rangle - \lambda_k x_{ik} \le \theta x_{i0} & (i = 1, \dots, m), \\ \langle \boldsymbol{\lambda}, \mathbf{y}^r\rangle - \lambda_k y_{rk} \ge y_{r0} & (r = 1, \dots, q), \\ \langle \mathbf{e}, \boldsymbol{\lambda}\rangle - \lambda_k = 1, \\ \boldsymbol{\lambda} \ge 0, \ \theta > 0. \end{cases}$$

- The output-oriented VRS radial super efficiency model is the form

$$\max\varphi \quad \text{subject to}$$
$$\begin{cases} \langle \boldsymbol{\lambda}, \mathbf{x}^i\rangle - \lambda_k x_{ik} \le x_{i0} & (i = 1, \dots, m), \\ \langle \boldsymbol{\lambda}, \mathbf{y}^r\rangle - \lambda_k y_{rk} \ge \varphi y_{r0} & (r = 1, \dots, q), \\ \langle \mathbf{e}, \boldsymbol{\lambda}\rangle - \lambda_k = 1, \\ \boldsymbol{\lambda} \ge 0. \end{cases}$$

(b) Directional distance function super efficiency models

– The form is as follows:

$$\max \beta \quad \text{subject to}$$

$$\begin{cases} \langle \boldsymbol{\lambda}, \mathbf{x}^i \rangle - \lambda_k x_{ik} + \beta g_x^i \le x_{i0} & (i = 1, \ldots, m), \\ \langle \boldsymbol{\lambda}, \mathbf{y}^r \rangle - \lambda_k y_{rk} - \beta g_y^r \ge y_{r0} & (r = 1, \ldots, q), \\ \langle \boldsymbol{\lambda}, \mathbf{b}^t \rangle - \lambda_k b_{tk} - \beta g_y^t \le b_{t0}, \\ \boldsymbol{\lambda} \ge 0, \end{cases}$$

where $\boldsymbol{b}^t = (b_{t1}, \ldots, b_{tn})$.

(c) SBM super efficiency models

– The input-oriented SBM super efficiency model is the form

$$\min \left\{ 1 + \frac{1}{m} \sum_1^m \frac{s_i^-}{x_{i0}} \right\} \quad \text{subject to}$$

$$\begin{cases} \langle \boldsymbol{\lambda}, \mathbf{x}^i \rangle - \lambda_k x_{ik} + s_i^- = x_{i0}, \quad s_i^- \ge 0 & (i = 1, \ldots, m), \\ \langle \boldsymbol{\lambda}, \mathbf{y}^r \rangle - \lambda_k y_{rk} \ge y_{r0} & (r = 1, \ldots, q), \\ \boldsymbol{\lambda} \ge 0. \end{cases}$$

– The output-oriented SBM super efficiency model is the form

$$\min \left\{ \frac{1}{1 - \frac{1}{q} \sum_1^q \frac{s_r^+}{y_{r0}}} \right\} \quad \text{subject to}$$

$$\begin{cases} \langle \boldsymbol{\lambda}, \mathbf{x}^i \rangle - \lambda_k x_{ik} \le x_{i0} & (i = 1, \ldots, m), \\ \langle \boldsymbol{\lambda}, \mathbf{y}^r \rangle - \lambda_k y_{rk} - s_r^+ = y_{r0}, \quad s_r^+ \ge 0 & (r = 1, \ldots, q), \\ \boldsymbol{\lambda} \ge 0. \end{cases}$$

– The nonoriented SBM super efficiency model is the form

$$\max \left\{ \frac{1 + \frac{1}{m} \sum_1^m \frac{s_i^-}{x_{i0}}}{1 - \frac{1}{q} \sum_1^q \frac{s_r^+}{y_{r0}}} \right\} \quad \text{subject to}$$

$$\begin{cases} \langle \boldsymbol{\lambda}, \mathbf{x}^i \rangle - \lambda_k x_{ik} + s_i^- = x_{i0}, \quad s_i^- \ge 0 & (i = 1, \ldots, m), \\ \langle \boldsymbol{\lambda}, \mathbf{y}^r \rangle - \lambda_k y_{rk} - s_r^+ = y_{r0}, \quad s_r^+ \ge 0 & (r = 1, \ldots, q), \\ \boldsymbol{\lambda} \ge 0. \end{cases}$$

8.4.6 Directional distance function models

The Directional Distance Function (DDF) model was presented by Chung et al. in 1997.

- Directional distance function CRS model is

$$\max \beta \quad \text{subject to}$$

$$\begin{cases} \langle \boldsymbol{\lambda}, \boldsymbol{x}^i \rangle + \beta g_x^i \le x_{i0} & (i = 1, \dots, m), \\ \langle \boldsymbol{\lambda}, \boldsymbol{y}^r \rangle + \beta g_y^r \ge y_{r0} & (r = 1, \dots, q), \\ \boldsymbol{\lambda} \ge 0, \end{cases}$$

where $\boldsymbol{g}_x = (g_x^1, \dots, g_x^m)$ and $\boldsymbol{g}_y = (g_y^1, \dots, g_y^q)$. By Theorem 7.6.2, its dual model is

$$\min(\langle \boldsymbol{v}, \boldsymbol{x}_0 \rangle - \langle \boldsymbol{\mu}, \boldsymbol{y}_0 \rangle) \quad \text{subject to}$$

$$\begin{cases} \langle \boldsymbol{v}, \boldsymbol{x}_j \rangle - \langle \boldsymbol{\mu}, \boldsymbol{y}_j \rangle \le 0 & (j = 1, \dots, n), \\ \langle \boldsymbol{v}, \boldsymbol{g}_x \rangle - \langle \boldsymbol{\mu}, \boldsymbol{g}_y \rangle = 1, \\ \boldsymbol{v} \ge 0, \ \boldsymbol{\mu} \ge 0. \end{cases}$$

- Directional distance function VRS model is

$$\max \beta \quad \text{subject to}$$

$$\begin{cases} \langle \boldsymbol{\lambda}, \boldsymbol{x}^i \rangle + \beta g_x^i \le x_{i0} & (i = 1, \dots, m), \\ \langle \boldsymbol{\lambda}, \boldsymbol{y}^r \rangle + \beta g_y^r \ge y_{r0} & (r = 1, \dots, q), \\ \langle \boldsymbol{e}, \boldsymbol{\lambda} \rangle = 1, \\ \lambda \ge 0, \end{cases}$$

where $\boldsymbol{g}_x = (g_x^1, \dots, g_x^m)$ and $\boldsymbol{g}_y = (g_y^1, \dots, g_y^q)$. By Theorem 7.6.2, its dual model is

$$\min\{ \langle \boldsymbol{v}, \boldsymbol{x}_0 \rangle - \langle \boldsymbol{\mu}, \boldsymbol{y}_0 \rangle + \mu_0 \} \quad \text{subject to}$$

$$\begin{cases} \langle \boldsymbol{v}, \boldsymbol{x}_j \rangle - \langle \boldsymbol{\mu}, \boldsymbol{y}_j \rangle + \mu_0 \le 0 & (j = 1, \dots, n), \\ \langle \boldsymbol{v}, \boldsymbol{g}_x \rangle - \langle \boldsymbol{\mu}, \boldsymbol{g}_y \rangle = 1, \\ \boldsymbol{v} \ge 0, \ \boldsymbol{\mu} \ge 0, \ \mu_0 \in \mathbb{R}. \end{cases}$$

8.5 Software and case studies

Data Envelopment Analysis (DEA) is a very powerful service management and benchmarking technique to evaluate organizations. MaxDEA is an easy-to-use DEA software and can be downloaded from http://www.maxdea.cn/. MaxDEA includes most DEA models and can import data from Excel, Access, and text files. Below we give some case studies that use DEA algorithms.

8.5.1 Carbon emissions reduction

DEA can be used to make environmental efficiency analysis of several regional industries. Capital stock, population and energy consumption are three main inputs, GDP is a single desirable output and carbon emissions are an undesirable output. Miao et al. (2016) use DEA to quantify the efficient allocation of carbon emissions between different provinces using China's provincial data in 2006–2010. Results showed that the actual carbon emissions in some provinces were higher than their maximal carbon emission allowances calculated from the DEA model, indicating that these provinces are facing great pressures on carbon emission reduction.

8.5.2 Coal mining industry

For the coal mining industry, the main inputs are capital and labour, as well as fossil-fuels; the main output is coal production and some undesirable outputs such as waste water, gas and solids. With the help of DEA, Liu et al. compared two major coal producing areas in China, Shanxi and Inner Mongolia provinces, and showed that the market-oriented policy package performs better than the nationalization regulations, because not only can the former policy instruments reduce the economic shock on industry productivity, but also improve to a larger extent the unified environmental and operational efficiency.

8.5.3 Coal-fired power plants

DEA can be used to assess energy efficiency of coal-fired plants. Song et al. (2015) use input-oriented CCR/BCC DEA models to measure overall technical efficiency and pure technical efficiency of 34 coal-fired power plants in China. The generalized energy efficiency is calculated based on four input parameters: coal consumption, oil consumption, water consumption and auxiliary power consumption by power units. The special energy efficiency is only based on two input parameters: coal consumption and auxiliary power consumption. Results showed that electricity saving potential exists for 14 out of 34 power plants.

Further reading

[1] Andersen P, Petersen NC. A procedure for ranking efficient units in data envelopment analysis. *Management Science*. 1993(39):1261–1265.

[2] Banker RD, Charnes A, Cooper WW. Some models for estimating technical and scale inefficiencies in data envelopment analysis. *Management Science*. 1984(30):1078–1092.

[3] Cauchy A. Methode génerale pour la résolution des systéms d'equations simultanées. *C R Acad Sci Paris*. 1847(25):536–538.

[4] Charnes A , Cooper WW, Rhodes E. Measuring the efficiency of decision making units. *European Journal of Operational Research*. 1978(2):429–444.

[5] Chen L, Jia G. Environmental efficiency analysis of China's regional industry: a data envelopment analysis (DEA) based approach. *Journal of Cleaner Production*. 2016: in press.

[6] Cooper WW, Seiford LM, Tone K. *Data Envelopment Analysis*. A comprehensive text with models, applications, references and DEA-solver software, Kluwer Academic Publishers, Boston, 2000.

[7] Cooper WW, Seiford LM, Tone K. *Data Envelopment Analysis*. A comprehensive text with models, applications, references and DEA-Solver software, 2nd ed. New York: Springer Science & Business Media, 2007.

[8] Chung YH, Färe R, Grosskopf S. Productivity and undesirable outputs: A directional distance function approach. *Journal of Environmental Management*. 1997(51):229–240.

[9] Harver CM. *Operations Research: An Introduction to Linear Optimization and Decision Analysis*, Elsevier North Holland, Inc., New York, 1979.

[10] Kuhn HW, Tucker AW. *Nonlinear Programming*. Proceedings of Second Berkeley Symposium on Mathematical Statistics and Probability, pp. 481–492, University of California Press, Berkeley, CA, 1950.

[11] Lemke CE. The dual method of solving a linear programming problem. *Naval Res Logistics Q*. 1954(1):48–54.

[12] Liu J, Liu H, Yao XL, Liu Y. Evaluating the sustainability impact of consolidation policy in China's coal mining industry: A data envelopment analysis. *Journal of Cleaner Production*. 2016(112): 2969–2976.

[13] Miao Z, Geng Y, Sheng J. Efficient allocation of CO_2 emissions in China: A zero sum gains data envelopment model. *Journal of Cleaner Production*. 2016(112):4144–4150.

[14] Ray SC. The directional distance function and measurement of super-efficiency: An application to airlines data. *Journal of the Operational Research Society*. 2008(59):788–797.

[15] Sharp JA, Meng W, Liu W. A modified slacks-based measure model for data envelopment analysis with natural negative outputs and inputs. *Journal of the Operational Research Society*. 2007(58):1672–1677.

[16] Song C, Li M, Zhang F, He YL, Tao WQ. A data envelopment analysis for energy efficiency of coal-fired power units in China. *Energy Conversion and Management*. 2015(102):121–130.

[17] Tone K. A slacks-based measure of efficiency in data envelopment analysis. *European Journal of Operational Research*. 2001(130):498–509.

[18] Tone K. A slacks-based measure of super-efficiency in data envelopment analysis. *European Journal of Operational Research*. 2002(143):32–41.

[19] Tone K. Variations on the theme of slacks-based measure of efficiency in DEA. *European Journal of Operational Research*. 2010(200):901–907.

[20] Tulkens H. On FDH efficiency analysis: Some methodological issues and applications to retail banking, courts, and urban transit. *Journal of Productivity Analysis*. 1993(4):183–210.

9 Risk assessments

Risk assessment provides a systematic procedure for estimating the probability of harm to, or from, the environment, the severity of harm, and uncertainty. Lowrance defined a risk is a measure of the probability and severity of adverse effects. This definition is harmonious with the mathematical formula used to calculate the expected value of risk. Later, Haimes showed that a risk assessment is a process consisting of a set of logical, systematic, and well-defined activities that provide the decision maker with a sound identification, measurement, quantification, and evaluation of the risk associated with certain natural phenomena or man-made activities. The main process of risk assessment includes formulating the problem; assessment of the risk; appraising various management options; and determining the optimal risk management strategy. In this chapter we will introduce fundamental methods in risk assessment processes.

9.1 Decision rules under uncertainty

Denote by a_i $(i = 1, \ldots, I)$ the i-th decision adopted by the decision maker and denote by s_j $(j = 1, \ldots, J)$ the j-th scenario, and denote by P_{ij} $(i = 1, \ldots, I; j = 1, \ldots, J)$ the payoff associated with the pair (a_i, s_j).

The common decision rules include the pessimistic rule, the optimistic rule, and the Hurwitz rule. The pessimistic rule includes both the maximin criteria and the minimax criterion. The optimistic rule is a maximax criterion. The Hurwitz rule is a linear combination between the maximin and the maximax criteria.

The pessimistic rule

The rule includes two criteria as follows.

- Let P_{ij} represent a payoff. The maximin criteria is represented by

$$\max_{1\le i\le I}\left(\min_{1\le j\le J} P_{ij}\right).$$

- Let P_{ij} represent a loss or a risk. The minimax criterion is represented by

$$\min_{1\le i\le I}\left(\max_{1\le j\le J} P_{ij}\right).$$

The pessimistic rule is also called the *maximin criterion* or the *minimax criterion*. This rule ensures that the decision makers will at least realize the minimum gain or avoid maximum loss.

DOI 10.1515/9783110424904-010

The optimistic rule

The rule is represented mathematically by

$$\max_{1\le i\le I}\left(\max_{1\le j\le J} P_{ij}\right).$$

The decision maker is most optimistic and seeks to maximize the maximum gain using the optimistic rule. The optimistic rule is sometimes called the *maximax criterion.*

The Hurwitz rule

The rule is represented mathematically by

$$\max_{1\le i\le I} P_i(\alpha) = \max_{1\le i\le I}\left(\alpha \min_{1\le j\le J} P_{ij} + (1-\alpha)\max_{1\le j\le J} P_{ij}\right) \quad (0\le\alpha\le 1).$$

From the Hurwitz rule, it is seen that $\max_{1\le i\le I} P_i(0)$ is the maximax criterion and $\max_{1\le i\le I} P_i(1)$ is the maximin criterion. Therefore, the Hurwitz rule is a compromise between two extreme criteria through an α-index and the decision maker's degree of optimism is specified by the parameter α.

Example 9.1.1. Assume that a_i $(i = 1, 2)$ are two kinds of carbon-emission-reduction technology adopted potentially by the decision maker and s_j $(j = 1, 2, 3)$ are three scenarios for future carbon price, the corresponding payoff P_{ij} $(i = 1, 2; j = 1, 2, 3)$ for production with value \$ 1000 associated with pair (a_i, s_j) $(i = 1, 2; j = 1, 2, 3)$ is

$$\begin{pmatrix} P_{11} & P_{12} & P_{13} \\ P_{21} & P_{22} & P_{23} \end{pmatrix} = \begin{pmatrix} 200 & 110 & -15 \\ 100 & 50 & 5 \end{pmatrix}.$$

Try to find the optimal decision by using the pessimistic rule, the optimistic rule, and the Hurwitz rule, respectively.

Solution. First, consider the pessimistic rule. From

$$\min_{1\le j\le 3} P_{1j} = \min(P_{11}, P_{12}, P_{13}) = \min(200, 110, -15) = -15 \quad \text{for decision } a_1,$$

$$\min_{1\le j\le 3} P_{2j} = \min(P_{21}, P_{22}, P_{23}) = \min(100, 50, 5) = 5 \quad \text{for decision } a_2,$$

it follows that

$$\max_{1\le i\le 2}\left(\min_{1\le j\le 3} P_{ij}\right) = \max\left(\min_{1\le j\le 3} P_{1j}, \min_{1\le j\le 3} P_{2j}\right) = \max(-15, 5) = 5,$$

i.e., the maximin criterion implies a gain of at least \$ 5 and the corresponding decision is a_2.

Secondly, consider the optimistic rule. Since

$$\max_{1\le j\le 3} P_{1j} = \max(P_{11}, P_{12}, P_{13}) = \max(200, 110, -15) = 200 \quad \text{for decision } a_1,$$

$$\max_{1\le j\le 3} P_{2j} = \max(P_{21}, P_{22}, P_{23}) = \max(100, 50, 5) = 100 \quad \text{for decision } a_2,$$

it follows that

$$\max_{1\le i\le 2}\left(\max_{1\le j\le 3} P_{ij}\right) = \max\left(\max_{1\le j\le 3} P_{1j}, \max_{1\le j\le 3} P_{2j}\right) = \max(200, 100) = 200,$$

i.e., the optimistic rule implies a gain of at most \$ 200 and the corresponding decision is a_1.

Third, consider the Hurwitz rule

$$\max_{1\le i\le 2} P_i(\alpha) = \max_{1\le i\le 2}\left(\alpha \min_{1\le j\le 3} P_{ij} + (1-\alpha)\max_{1\le j\le 3} P_{ij}\right).$$

From

$$\min_{1\le j\le 3} P_{1j} = \min(P_{11}, P_{12}, P_{13}) = \min(200, 110, -15) = -15,$$

$$\max_{1\le j\le 3} P_{1j} = \max(P_{11}, P_{12}, P_{13}) = \max(200, 110, -15) = 200,$$

$$\min_{1\le j\le 3} P_{2j} = \min(P_{21}, P_{22}, P_{23}) = \min(100, 50, 5) = 5,$$

$$\max_{1\le j\le 3} P_{2j} = \max(P_{21}, P_{22}, P_{23}) = \max(100, 50, 5) = 100,$$

it follows that for $0 \le \alpha \le 1$,

$$\begin{aligned} P_1(\alpha) &= \alpha \min_{1\le j\le 3} P_{1j} + (1-\alpha)\max_{1\le j\le 3} P_{1j} \\ &= -15\alpha + 200(1-\alpha) = 200 - 215\alpha \quad \text{for decision } a_1, \end{aligned}$$

$$\begin{aligned} P_2(\alpha) &= \alpha \min_{1\le j\le 3} P_{2j} + (1-\alpha)\max_{1\le j\le 3} P_{2j} \\ &= 5\alpha + 100(1-\alpha) = 100 - 95\alpha \quad \text{for decision } a_2. \end{aligned}$$

In order to find $\max_{1\le i\le 2} P_i(\alpha) = \max(P_1(\alpha), P_2(\alpha))$, it is only necessary to solve the equation

$$200 - 215\alpha = 100 - 95\alpha.$$

The solution is $\alpha = \frac{5}{6}$. It is clear that

$$\begin{aligned} 200 - 215\alpha &< 100 - 95\alpha \quad \text{if} \quad 0 \le \alpha < \frac{5}{6}; \\ 200 - 215\alpha &> 100 - 95\alpha \quad \text{if} \quad \frac{5}{6} < \alpha \le 1; \\ 200 - 215\alpha &= 100 - 95\alpha \quad \text{if} \quad \alpha = \frac{5}{6}. \end{aligned}$$

Therefore, if $0 \le \alpha \le \frac{5}{6}$, then the optimal decision is a_1; if $\frac{5}{6} \le \alpha \le 1$, then the optimal decision is a_2; if $\alpha = \frac{5}{6}$, both a_1 and a_2 are the optimal decision. □

Example 9.1.2. A computer company knows the sales potential for low-performance, medium-performance, and high-performance computers. The company wishes to decide on the best development plan based on minimizing risk of financial loss and/or maximizing the expected profit. The sales potential ($ 1000) of the company is given by the following table:

	s_1	s_2	s_3
a_1	150	50	−20
a_2	300	175	−100
a_3	450	150	−150

where the decisions a_1, a_2, and a_3 are the low-performance computer, the medium-performance computer, and the high-performance computer, respectively, and the scenarios s_1, s_2, and s_3 are the excellent computer system, the good computer system, and the poor computer system, respectively.

Solution. Consider the Hurwitz rule

$$\max_{1\le i\le 3}\left\{P_i(\alpha)=\alpha\min_{1\le j\le 3}P_{ij}+(1-\alpha)\max_{1\le j\le 3}P_{ij}\right\}\quad(0\le\alpha\le 1).$$

From

$$\min_{1\le j\le 3}P_{1j}=\min(150,50,-20)=-20,\qquad \max_{1\le j\le 3}P_{1j}=\max(150,50,-20)=150,$$
$$\min_{1\le j\le 3}P_{2j}=\min(300,175,-100)=-100,\qquad \max_{1\le j\le 3}P_{2j}=\max(300,175,-100)=300,$$
$$\min_{1\le j\le 3}P_{3j}=\min(450,150,-150)=-150,\qquad \max_{1\le j\le 3}P_{3j}=\max(450,150,-150)=450,$$

it follows that

$$P_1(\alpha)=-20\alpha+150(1-\alpha)=150-170\alpha\quad\text{for decision } a_1,$$
$$P_2(\alpha)=-100\alpha+300(1-\alpha)=300-400\alpha\quad\text{for decision } a_2,$$
$$P_3(\alpha)=-150\alpha+450(1-\alpha)=450-600\alpha\quad\text{for decision } a_3$$

which represent three straight lines. It is clear that the second straight line and the third straight line are parallel and the second straight line is dominated by the third straight line. Therefore, in order to find $\max_{1\le i\le 3}P_i(\alpha)$, it is only necessary to solve the equation

$$450-600\alpha=150-170\alpha.$$

The solution is $\alpha=\frac{30}{43}\approx 0.698$. This indicates that if $\alpha<\frac{30}{43}$, the high-performance computer should be produced and sold; if $\alpha>\frac{30}{43}$, the low-performance computer should be produced and sold. □

9.2 Decision trees

The decision tree is the most commonly used tool in risk-based decision-making. It possesses the ability to represent and analyze multiple stages in the decision-making process. This is an attractive feature of decision trees. The single-objective decision tree consists of the following basic components.

Decision node. Decision nodes are designated by a square (□). Branches emanating from a decision node represent the various decisions to be investigated. Each alternative choice is designated by a_i and identify each branch with that decision choice.

Chance node. Chance nodes are designated by a circle (○). Branches emanating from a chance node represent the various scenarios with their associated probabilities.

Consequences. The value of the consequences is written at the end of each branch. The consequence associated with the i-th decision and j-th scenario is designated by P_{ij}.

To determine the optimal manufacturing scenario, the following three measures are used for calculating the expected value of profits for each of the alternative decision options.

Let a_i $(i = 1, \dots, I)$ be the i-th decision. Denote by $E[a_i]$ the expected value of profits associated with the i-th decision a_i. Let s_j $(j = 1, \dots, J)$ be the j-th scenario. Denote by $p(s_j)$ the probability associated with the j-th scenario s_j and denote by P_{ij} the payoff associated with the decision/scenario pair (a_i, s_j). Assume that $p(s_j)$ and P_{ij} are known.

The *Expected Monetary Value* (EMV) measure is defined as

$$\text{EMV} = \max_{1\le i\le I} \sum_{j=1}^{J} p(s_j)P_{ij} = \max_{1\le i\le I} E[a_i].$$

The *Expected Value of Opportunity Loss* (EOL) measure is defined as

$$\text{EOL} = \min_{1\le i\le I} \sum_{j=1}^{J} p(s_j)(M_j - P_{ij}),$$

where $M_j = \max_{1\le i\le I} P_{ij}$ $(j = 1, \dots, J)$. The EOL measure is essentially a modification of the EMV measure.

The *Most Likely Value* (MLV) measure is not commonly used because the optimal results are very sensitive to the number of scenarios. In the MLV, one selects the outcome with the highest probability for each scenario and then maximizes the corresponding P_{ij}.

Example 9.2.1. Consider the payoff matrix ($ 1000) in Example 9.1.1

$$\begin{pmatrix} P_{11} & P_{12} & P_{13} \\ P_{21} & P_{22} & P_{23} \end{pmatrix} = \begin{pmatrix} 200 & 110 & -15 \\ 100 & 50 & 5 \end{pmatrix}$$

where the payoff P_{ij} is associated with the pair (a_i, s_j) $(i = 1, 2; j = 1, 2, 3)$. The decision tree is represented by

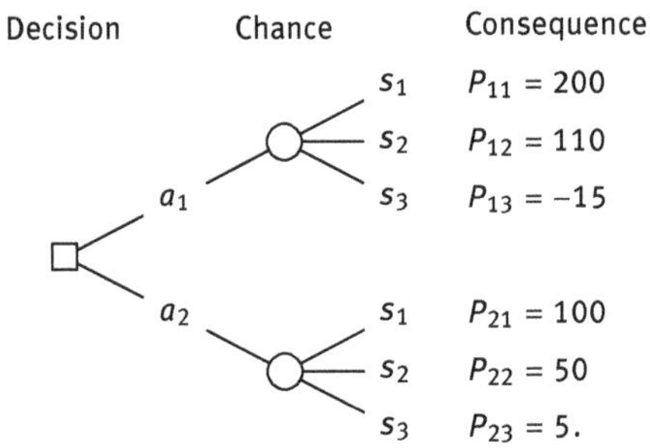

Assume that $p(s_1) = 0.4$, $p(s_2) = 0.5$, and $p(s_3) = 0.1$.

Consider the EMV measure. The expected values of profits for a_1 and a_2 are, respectively,

$$\begin{aligned} E[a_1] &= \sum_{j=1}^{3} p(s_j)P_{1j} \\ &= 0.4 \times 200 + 0.5 \times 110 + 0.1 \times (-15) \\ &= 80 + 55 - 1.5 = 133.5; \\ E[a_2] &= \sum_{j=1}^{3} p(s_j)P_{2j} \\ &= 0.4 \times 100 + 0.5 \times 50 + 0.1 \times 5 \\ &= 40 + 25 + 0.5 = 65.5. \end{aligned}$$

Thus, the EMV measure is

$$\text{EMV} = \max_{1 \le i \le 2} \sum_{j=1}^{3} p(s_j)P_{ij} = \max(E[a_1], E[a_2]) = \max(133,500, 65,000) = 133,500.$$

The decision tree with expected value of profits is represented by

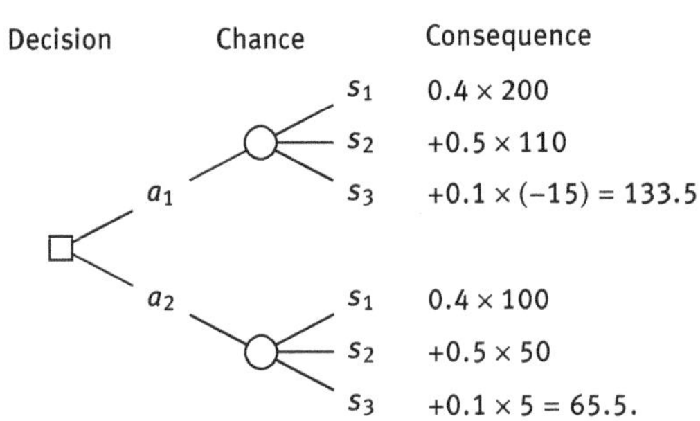

Consider EOL measure. By $M_j = \max_{1\le i\le 2}\{P_{ij}\}$ $(j = 1, 2, 3)$, we get

$$M_1 = \max_{1\le i\le 2}(P_{11}, P_{21}) = \max(200, 100) = 200,$$
$$M_2 = \max_{1\le i\le 2}(P_{12}, P_{22}) = \max(110, 50) = 110,$$
$$M_3 = \max_{1\le i\le 2}(P_{13}, P_{23}) = \max(-15, 5) = 5.$$

By using $\sum_{j=1}^{3} p(s_j)(M_j - P_{ij})$ $(i = 1, 2)$, we get

$$\begin{aligned}\sum_{j=1}^{3} p(s_j)(M_j - P_{1j}) &= 0.4 \times (200 - 200) + 0.5 \times (110 - 110) + 0.1 \times (5 - (-15))\\ &= 0.4 \times 0 + 0.5 \times 0 + 0.1 \times 20\\ &= 0 + 0 + 2 = 2 \quad \text{for } i = 1;\\ \sum_{j=1}^{3} p(s_j)(M_j - P_{2j}) &= 0.4 \times (200 - 100) + 0.5 \times (110 - 50) + 0.1 \times (5 - 5)\\ &= 0.4 \times 100 + 0.5 \times 60 + 0.1 \times 0\\ &= 40 + 30 + 0 = 70 \quad \text{for } i = 2.\end{aligned}$$

Thus, the EOL measure is

$$\text{EOL} = \min_{1\le i\le 2} \sum_{j=1}^{3} p(s_j)(M_j - P_{ij}) = \min(2, 70) = 2.$$

The decision tree with EOL measure is represented by

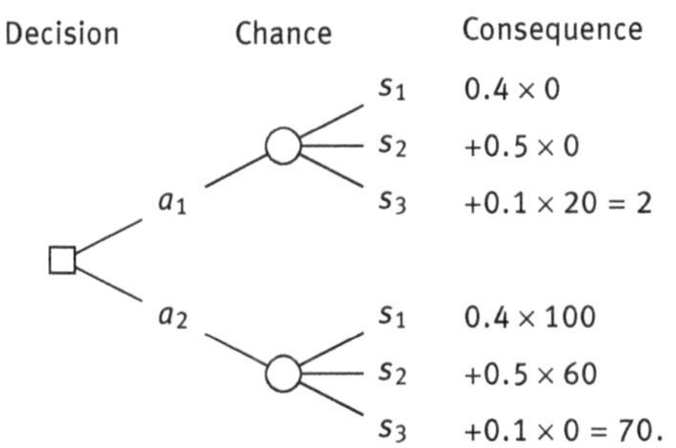

Consider the MLV measure.
For $i = 1$, the highest probability is $\max_{1\le j\le 3} p(s_j) = 0.5$. The corresponding profit is $P_{12} = 110$.
For $i = 2$, the highest probability is $\max_{1\le j\le 3} p(s_j) = 0.5$. The corresponding profit is $P_{22} = 50$.
Thus, the measure is $\text{MLV} = \max(110, 50) = 110$.

9.3 Fractile and triangular methods

The fractile method is used for assessing probability distributions.

A probability distribution can be represented by a cumulative distribution function (cdf) given by

$$F(x) = \int_a^x f(t)\,\mathrm{d}t \quad (a < x < b),$$

where $f(t)$ is a density function and the interval (a, b) is often chosen as one of the three intervals $(0, 1)$, $(0, \infty)$, and $(-\infty, \infty)$. It is seen from this representation that if the density function is continuous on the interval $[a, b]$ and $f(t) > 0$ $(a < t < b)$, then the cdf $F(x)$ is a smooth, positive, strictly increasing function, and the cdf tends to 0 as x tends to a and the cdf tends to 1 as x tends to b. It is well known that the values of the cdf can be interpreted as the probabilities of events

$$F(x) = P(a < \theta < x) \quad (a < x < b).$$

The probabilities on the intervals $a \le \theta < x$, $a < \theta \le x$, $a \le \theta \le x$, and $a < \theta < x$ are the same, i.e.,

$$P(a < \theta < x) = P(a \le \theta < x) = P(a < \theta \le x) = P(a \le \theta \le x).$$

Definition 9.3.1. Given a probability p $(0 \le p \le 1)$, let x_p denote that value of x so that

$$F(x_p) = P(a < \theta < x_p) = \int_a^{x_p} f(t)\,\mathrm{d}t = p.$$

The value x_p is called the *p-th fractile* of θ.

Assume that $f(t) > 0$ $(a < t < b)$. Then

$$F(a) = P(\theta = a) = P(a < \theta < a) = \int_a^a f(t)\,\mathrm{d}t = 0,$$

$$F(b) = P(a < \theta < b) = \int_a^b f(t)\,\mathrm{d}t = 1,$$

and according to the strictly increasing property of the cdf, the value x_p is uniquely defined.

Especially, the 0.5-th fractile $x_{0.5}$ is called the *median* of θ; the 0.25-th fractile $x_{0.25}$ is called the *lower quartile* of θ; and the 0.75-th fractile $x_{0.75}$ is called the *upper quartile* of θ.

For these three fractiles $x_{0.25}$, $x_{0.5}$, $x_{0.75}$, noticing that $\int_a^b f(t)\,dt = 1$, it follows from Definition 9.3.1 that

$$P(a < \theta < x_{0.5}) = \int_a^{x_{0.5}} f(t)\,dt = 0.5,$$

$$P(x_{0.5} < \theta < b) = \int_{x_{0.5}}^{b} f(t)\,dt = \int_a^b f(t)\,dt - \int_a^{x_{0.5}} f(t)\,dt = 1 - 0.5 = 0.5,$$

$$P(a < x < x_{0.25}) = \int_a^{x_{0.25}} f(t)\,dt = 0.25,$$

$$P(x_{0.25} < \theta < x_{0.5}) = \int_{x_{0.25}}^{x_{0.5}} f(t)\,dt = \int_a^{x_{0.5}} f(t)\,dt - \int_a^{x_{0.25}} f(t)\,dt = 0.5 - 0.25 = 0.25,$$

$$P(x_{0.5} < x < x_{0.75}) = \int_{x_{0.5}}^{x_{0.75}} f(t)\,dt = \int_a^{x_{0.75}} f(t)\,dt - \int_a^{x_{0.5}} f(t)\,dt = 0.75 - 0.5 = 0.25,$$

$$P(x_{0.75} < \theta < b) = \int_{x_{0.75}}^{b} f(t)\,dt = \int_a^b f(t)\,dt - \int_a^{x_{0.75}} f(t)\,dt = 1 - 0.75 = 0.25.$$

So the three fractiles $x_{0.25}$, $x_{0.5}$, $x_{0.75}$ have the following probability property:

$$P(a < \theta < x_{0.5}) = P(x_{0.5} < \theta < b),$$
$$P(a < x < x_{0.25}) = P(x_{0.25} < \theta < x_{0.5}) = P(x_{0.5} < x < x_{0.75}) = P(x_{0.75} < \theta < b).$$

These equalities are such that the decision maker often directly assesses these three fractiles and the cdf will have the assessed fractiles, i.e., $F(x_{0.25}) = 0.25$, $F(x_{0.5}) = 0.5$, $F(x_{0.75}) = 0.75$.

The graphical method

Assume that the decision maker has assessed these three fractiles and the points $(x_{0.25}, 0.25)$ and $(x_{0.5}, 0.5)$, and $(x_{0.75}, 0.75)$ are specified. The question is now, how to determine a cdf $F(x)$ whose graph passes through the given points.

Establish an oxp-rectangular coordinate system on a sheet of graph paper, where the vertical op-axis is the probability axis and the horizontal ox-axis is the fractile axis. In the oxp-rectangular coordinate system, plot the three assessed points $(x_{0.25}, 0.25)$, $(x_{0.5}, 0.5)$, $(x_{0.75}, 0.75)$ and two end points $(a, 0)$, $(b, 1)$, and then use the cdf's smooth property and strictly increasing property to draw a graph through the plotted points. The cdf $F(x)$ is obtained graphically.

The analytical method

In this method, the decision maker assumes that the considered probability distribution belongs to some parametric family of distributions, and then the decision maker chooses values for the parameters so that the resulting cdf will have the assessed fractiles.

9.3.1 Assessing a normal distribution

The decision maker first assumes that the considered probability distribution belongs to the family of normal distributions. Denote by $f_N(t)$ a normal density function with parameters μ and σ^2

$$f_N(t) = \frac{1}{\sqrt{2\pi\sigma^2}} e^{-\frac{(t-\mu)^2}{2\sigma^2}},$$

where μ is the expected value and σ^2 is the variance.

Secondly, the decision maker chooses values for parameters μ and σ^2 so that the resulting normal cdf

$$F_N(x) = \int_{-\infty}^{x} \frac{1}{\sqrt{2\pi\sigma^2}} e^{-\frac{(t-\mu)^2}{2\sigma^2}} dt$$

satisfies $F_N(x_{0.25}) = 0.25$, $F_N(x_{0.5}) = 0.5$, and $F_N(x_{0.75}) = 0.75$.

It follows from Definition 9.3.1 that

$$F_N(x_{0.5}) = P(-\infty < \theta < x_{0.5}) = \int_{-\infty}^{x_{0.5}} \frac{1}{\sqrt{2\pi\sigma^2}} e^{-\frac{(t-\mu)^2}{2\sigma^2}} dt = 0.5.$$

Note that

$$\int_{x_{0.5}}^{\infty} \frac{1}{\sqrt{2\pi\sigma^2}} e^{-\frac{(t-\mu)^2}{2\sigma^2}} dt = \int_{-\infty}^{\infty} \frac{1}{\sqrt{2\pi\sigma^2}} e^{-\frac{(t-\mu)^2}{2\sigma^2}} dt - \int_{-\infty}^{x_{0.5}} \frac{1}{\sqrt{2\pi\sigma^2}} e^{-\frac{(t-\mu)^2}{2\sigma^2}} dt.$$

$$= 1 - 0.5 = 0.5.$$

So

$$\int_{-\infty}^{x_{0.5}} \frac{1}{\sqrt{2\pi\sigma^2}} e^{-\frac{(t-\mu)^2}{2\sigma^2}} dt = \int_{x_{0.5}}^{\infty} \frac{1}{\sqrt{2\pi\sigma^2}} e^{-\frac{(t-\mu)^2}{2\sigma^2}} dt.$$

By the change of variables $u = \frac{t-\mu}{\sigma}$, this implies that

$$\int_{-\infty}^{\frac{x_{0.5}-\mu}{\sigma}} \frac{1}{\sqrt{2\pi}} e^{-\frac{u^2}{2}} du = \int_{\frac{x_{0.5}-\mu}{\sigma}}^{\infty} \frac{1}{\sqrt{2\pi}} e^{-\frac{u^2}{2}} du.$$

Since the integrand $e^{-\frac{u^2}{2}}$ is an even function,

$$\frac{x_{0.5} - \mu}{\sigma} = 0 \quad \text{or} \quad \mu = x_{0.5},$$

i.e., the parameter μ (i.e., the expected value) is chosen to be the 0.5-th fractile $x_{0.5}$.

It follows from Definition 9.3.1 and $\mu = x_{0.5}$ that

$$F_N(x_{0.75}) = P(-\infty < \theta < x_{0.75}) = \int_{-\infty}^{x_{0.75}} \frac{1}{\sqrt{2\pi\sigma^2}} e^{-\frac{(t-x_{0.5})^2}{2\sigma^2}} dt = 0.75.$$

By the change of variables $u = \frac{t-x_{0.5}}{\sigma}$, this implies that

$$\int_{-\infty}^{x_{0.75}} \frac{1}{\sqrt{2\pi\sigma^2}} e^{-\frac{(t-x_{0.5})^2}{2\sigma^2}} dt = \int_{-\infty}^{\frac{x_{0.75}-x_{0.5}}{\sigma}} \frac{1}{\sqrt{2\pi}} e^{-\frac{u^2}{2}} du.$$

and so

$$\int_{-\infty}^{\frac{x_{0.75}-x_{0.5}}{\sigma}} \frac{1}{\sqrt{2\pi}} e^{-\frac{u^2}{2}} du = 0.75. \tag{9.3.1}$$

The integrand $e^{-\frac{u^2}{2}}$ is the standard normal density function. However the upper quartile of the standard normal density function can be found in a statistical table to be 0.674. So

$$\frac{x_{0.75} - x_{0.5}}{\sigma} = 0.674$$

or

$$\sigma^2 = \frac{(x_{0.75} - x_{0.5})^2}{(0.674)^2} = 2.20\,(x_{0.75} - x_{0.5})^2.$$

i.e., the parameter σ^2 (i.e., the variance) is chosen to be 2.20 times the square of the difference between the 0.75-th fractile and the 0.5-th fractile.

It follows from Definition 9.3.1 and $\mu = x_{0.5}$ that

$$F_N(x_{0.25}) = P(-\infty < \theta < x_{0.25}) = \int_{-\infty}^{x_{0.25}} \frac{1}{\sqrt{2\pi\sigma^2}} e^{-\frac{(t-x_{0.5})^2}{2\sigma^2}} dt = 0.25.$$

By the change of variables $u = \frac{x_{0.5}-t}{\sigma}$, this implies that

$$\int_{-\infty}^{x_{0.25}} \frac{1}{\sqrt{2\pi\sigma^2}} e^{-\frac{(t-x_{0.5})^2}{2\sigma^2}} dt = \int_{\frac{x_{0.5}-x_{0.25}}{\sigma}}^{\infty} \frac{1}{\sqrt{2\pi}} e^{-\frac{u^2}{2}} du,$$

and so

$$\int_{\frac{x_{0.5}-x_{0.25}}{\sigma}}^{\infty} \frac{1}{\sqrt{2\pi}} e^{-\frac{u^2}{2}} du = 0.25. \tag{9.3.2}$$

On the other hand, by (9.3.1) and $\int_{-\infty}^{\infty} \frac{1}{\sqrt{2\pi}} e^{-\frac{u^2}{2}} du = 1$,

$$\int_{\frac{x_{0.75}-x_{0.5}}{\sigma}}^{\infty} \frac{1}{\sqrt{2\pi}} e^{-\frac{u^2}{2}} du = \int_{-\infty}^{\infty} \frac{1}{\sqrt{2\pi}} e^{-\frac{u^2}{2}} du - \int_{-\infty}^{\frac{x_{0.75}-x_{0.5}}{\sigma}} \frac{1}{\sqrt{2\pi}} e^{-\frac{u^2}{2}} du = 0.25.$$

From this and (9.3.2),

$$\int_{\frac{x_{0.5}-x_{0.25}}{\sigma}}^{\infty} \frac{1}{\sqrt{2\pi}} e^{-\frac{u^2}{2}} du = \int_{\frac{x_{0.75}-x_{0.5}}{\sigma}}^{\infty} \frac{1}{\sqrt{2\pi}} e^{-\frac{u^2}{2}} du.$$

So

$$x_{0.5} - x_{0.25} = x_{0.75} - x_{0.5},$$

i.e., the distance between the 0.5-th fractile and the 0.25-th fractile, $x_{0.5} - x_{0.25}$, is equal to the distance between the 0.75-th fractile and the 0.5-th fractile, $x_{0.75} - x_{0.5}$. So the parameter σ^2 (i.e., the variance) may be also chosen to be 2.20 times the square of the difference between the 0.5-th fractile and the 0.25-th fractile

$$\sigma^2 = 2.20\,(x_{0.5} - x_{0.25})^2.$$

9.3.2 Assessing a beta prior distribution

The decision maker assumes first that the considered probability distribution belongs to the beta family of distributions. Denote by $f_\beta(t)$ the beta density function with parameters r_0 and n_0

$$f_\beta(t) = \frac{1}{B(r_0, n_0)} t^{r_0-1}(1-t)^{n_0-r_0-1} \quad (0 \le t \le 1),$$

where $r_0 > 0$ and $n_0 > 0$ and $B(r_0, n_0)$ is a beta function

$$B(r_0, n_0) = \int_0^1 t^{r_0-1}(1-t)^{n_0-r_0-1}\, dt.$$

Secondly, the decision maker chooses values for parameters r_0 and n_0 so that the resulting beta cdf

$$F_\beta(x) = \int_0^x f_\beta(t)\, dt = B(r_0, n_0)^{-1} \int_0^x t^{r_0-1}(1-t)^{n_0-r_0-1}\, dt \quad (0 \le x \le 1)$$

satisfies $F_\beta(x_{0.25}) = 0.25$, $F_\beta(x_{0.5}) = 0.5$, and $F_\beta(x_{0.75}) = 0.75$. Generally speaking, this will not be possible. To satisfy them approximately, the decision maker considers the error function

$$E(r_0, n_0) = (F_\beta(x_{0.25}) - 0.25)^2 + (F_\beta(x_{0.5}) - 0.5)^2 + (F_\beta(x_{0.75}) - 0.75)^2$$

and uses a computer code to calculate values r_0 and n_0 that minimize $E(r_0, n_0)$.

The *triangular distribution method* is another ideal method. This method qualifies a triangle as a probability density function.

Let the area of the triangle be equal to 1. Denote by a and b the lowest and highest values of the outcome, respectively, and denote by c the most likely value of the outcome. The base of the triangle is equal to $b - a$ and its height is the frequency $p(c)$ of the most likely value. So

$$\frac{1}{2}(b-a)p(c) = 1,$$

and so the frequency of the most likely value of the outcome is given by

$$p(c) = \frac{2}{b-a}.$$

Law and Kelton presented that the cdf $P(x)$ and pdf $p(x)$ of the triangular distribution for a continuous random variable X are as follows:

$$P(x) = P(X \le x) = \begin{cases} 0 & (x < a), \\ \frac{(x-a)^2}{(b-a)(c-a)} & (a \le x \le c), \\ 1 - \frac{(b-x)^2}{(b-a)(b-c)} & (c < x \le b), \\ 1 & (x > b) \end{cases}$$

and

$$p(x) = \begin{cases} 0 & (x < a), \\ \frac{2(x-a)}{(b-a)(c-a)} & (a \le x \le c), \\ \frac{2(b-x)}{(b-a)(b-c)} & (c < x \le b), \\ 0 & (x > b), \end{cases}$$

where $p(x) = \frac{d}{dx}P(x)$.

The expected value of the triangular distribution is

$$E[X] = \frac{a+b+c}{3}.$$

In fact, by the definition, the expected value of the triangular distribution is

$$\begin{aligned} E[x] &= \int_{-\infty}^{\infty} xp(x)\,dx \\ &= \frac{2}{(b-a)(c-a)}\int_a^c (x^2 - ax)\,dx + \frac{2}{(b-a)(b-c)}\int_c^b (bx - x^2)\,dx. \end{aligned}$$

However,

$$\int_a^c (x^2 - ax)\,dx = \frac{c^3 - a^3}{3} - \frac{a(c^2 - a^2)}{2},$$

$$\int_c^b (bx - x^2)\,dx = \frac{b^3 - bc^2}{2} - \frac{b^3 - c^3}{3}.$$

Therefore, the expected value of the triangular distribution is

$$\begin{aligned} E[x] &= \frac{2}{(b-a)(c-a)}\left(\frac{c^3-a^3}{3}-\frac{a(c^2-a^2)}{2}\right)\\ &\quad+\frac{2}{(b-a)(b-c)}\left(\frac{b^3-bc^2}{2}-\frac{b^3-c^3}{3}\right)\\ &=\frac{2c^2-ac-a^2}{3(b-a)}+\frac{b^2+bc-2c^2}{3(b-a)}=\frac{a+b+c}{3}. \end{aligned}$$

The variance of the triangular distribution is

$$\operatorname{Var}(X)=\frac{a^2+b^2+c^2-ab-ac-bc}{18}.$$

In fact, by the definition, the variance of the triangular distribution is

$$\begin{aligned} \operatorname{Var}(X) &= E[(x-E[X])^2]\\ &=\int_{-\infty}^{\infty}(x^2-2xE[X]+(E[X])^2)p(x)\,\mathrm{d}x\\ &=\int_{-\infty}^{\infty}x^2p(x)\,\mathrm{d}x-2E[X]\int_{-\infty}^{\infty}xp(x)\,\mathrm{d}x+(E[X])^2\int_{-\infty}^{\infty}p(x)\,\mathrm{d}x\\ &=I_1-2E[X]I_2+(E[X])^2I_3. \end{aligned} \tag{9.3.3}$$

These three integrals I_1, I_2, and I_3 are computed as follows:

$$\begin{aligned} I_1 &= \int_{-\infty}^{\infty}x^2p(x)\,\mathrm{d}x\\ &=\frac{2}{(b-a)(c-a)}\int_a^c(x^3-ax^2)\,\mathrm{d}x+\frac{2}{(b-a)(b-c)}\int_c^b(bx^2-x^3)\,\mathrm{d}x\\ &=\frac{2}{(b-a)(c-a)}\left(\frac{c^4-a^4}{4}-\frac{a(c^3-a^3)}{3}\right)+\frac{2}{(b-a)(b-c)}\left(\frac{b(b^3-c^3)}{3}-\frac{b^4-c^4}{4}\right)\\ &=\frac{2}{b-a}\left(\frac{a^3-b^3+c(a^2-b^2)+c^2(a-b)}{4}+\frac{b^3-a^3+c(b^2-a^2)+c^2(b-a)}{3}\right)\\ &=\frac{a^2+b^2+c^2+ab+bc+ac}{6}, \end{aligned}$$

$$
\begin{aligned}
I_2 &= \int_{-\infty}^{\infty} xp(x)\,\mathrm{d}x \\
&= \int_a^c \frac{2}{(b-a)(c-a)}(x^2 - ax)\,\mathrm{d}x + \int_c^b \frac{2}{(b-a)(b-c)}(bx - x^2)\,\mathrm{d}x \\
&= \frac{2}{(b-a)(c-a)}\left(\frac{c^3-a^3}{3} - \frac{a(c^2-a^2)}{2}\right) + \frac{2}{(b-a)(b-c)}\left(\frac{b(b^2-c^2)}{2} - \frac{b^3-c^3}{3}\right) \\
&= \frac{2}{b-a}\left(\frac{a^2-b^2+c(a-b)}{3} + \frac{(b^2-a^2)+c(b-a)}{2}\right) \\
&= \frac{a+b+c}{3},
\end{aligned}
$$

and

$$
\begin{aligned}
I_3 &= \int_{-\infty}^{\infty} p(x)\,\mathrm{d}x \\
&= \int_a^c \frac{2}{(b-a)(c-a)}(x-a)\,\mathrm{d}x + \int_c^b \frac{2}{(b-a)(b-c)}(b-x)\,\mathrm{d}x \\
&= \frac{2}{(b-a)(c-a)}\left(\frac{c^2-a^2}{2} - a(c-a)\right) + \frac{2}{(b-a)(b-c)}\left(b(b-c) - \frac{b^2-c^2}{2}\right) \\
&= \frac{2}{b-a}\left(\frac{c+a}{2} - a\right) + \frac{2}{b-a}\left(b - \frac{b+c}{2}\right) \\
&= \frac{2}{b-a}\left(\frac{c-a}{2} + \frac{b-c}{2}\right) = 1.
\end{aligned}
$$

Note that $E[X] = \frac{a+b+c}{3}$. By (9.3.3), the variance of the triangular distribution is

$$
\begin{aligned}
\mathrm{Var}(X) &= I_1 - 2E[X]I_2 + (E[X])^2 I_3 \\
&= \frac{a^2+b^2+c^2+ab+bc+ac}{6} - 2E[X]\left(\frac{a+b+c}{3}\right) + (E[X])^2 \\
&= \frac{a^2+b^2+c^2+ab+bc+ac}{6} - \frac{2(a+b+c)^2}{9} + \frac{(a+b+c)^2}{9} \\
&= \frac{a^2+b^2+c^2-ab-ac-bc}{18}.
\end{aligned}
$$

Example 9.3.2. Given the pdf $p(x)$ of the triangular distribution as above, compute the value of the conditional expected risk functions

$$
g(\cdot) = \frac{\int_\beta^b xp(x)\,\mathrm{d}x}{\int_\beta^b p(x)\,\mathrm{d}x},
$$

where $\beta > c$.

Solution. Since $\beta > c$, $p(x) = \frac{2(b-x)}{(b-a)(b-c)}$, and so

$$g(\cdot) = \frac{\int_\beta^b xp(x)\,dx}{\int_\beta^b p(x)\,dx} = \frac{\int_\beta^b x(b-x)\,dx}{\int_\beta^b (b-x)\,dx}.$$

The numerator is

$$\int_\beta^b x(b-x)\,dx = \frac{b^3 - b\beta^2}{2} - \frac{b^3 - \beta^3}{3}$$

and the denominator is

$$\int_\beta^b (b-x)\,dx = b^2 - b\beta - \frac{b^2 - \beta^2}{2}.$$

Thus, the conditional expected risk functions are

$$g(\cdot) = \frac{\frac{b^3-b\beta^2}{2} - \frac{b^3-\beta^3}{3}}{b^2 - b\beta - \frac{b^2-\beta^2}{2}} = \frac{\frac{b(b+\beta)}{2} - \frac{b^2+b\beta+\beta^2}{3}}{b - \frac{b+\beta}{2}} = \frac{b^2 - \beta^2 + b\beta - \beta^2}{3(b-\beta)} = \frac{b+2\beta}{3}. \qquad \square$$

9.4 The ε-constraint method

Let $f_i : \mathbb{R}^n \to \mathbb{R}$ $(i = 1, \dots, N)$ and $c_i : \mathbb{R}^n \to \mathbb{R}$ $(i = 1, \dots, m)$ all be continuously differentiable functions. The model of the form

$$\min_{\boldsymbol{x}} \{f_1(\boldsymbol{x}), \dots, f_N(\boldsymbol{x})\} \quad \text{subject to} \quad c_i(\boldsymbol{x}) \le 0 \quad (i = 1, \dots, m) \tag{9.4.1}$$

is called an *multiobjective optimization model.* Haimes et al. presented the ε-constraint method and showed that the model (9.4.1) is equivalent to the ε-constraint optimization model of the form

$$\min_{\boldsymbol{x} \in X} f_i(\boldsymbol{x}) \quad \text{subject to} \quad f_j(\boldsymbol{x}) \le \varepsilon_j \quad (j \ne i;\ j = 1, 2, \dots, N), \tag{9.4.2}$$

where ε_j $(j \ne i, j = 1, \dots, N)$ are variables and

$$X = \{\boldsymbol{x} \in \mathbb{R}^n \mid c_i(\boldsymbol{x}) \le 0\ (i = 1, \dots, m)\},$$

$$\varepsilon_j = \min_{\boldsymbol{x} \in X} f_j(\boldsymbol{x}) + \epsilon_j, \quad \epsilon_j > 0 \quad (j \ne i;\ j = 1, \dots, N).$$

From (9.4.2), it is seen that the ε-constraint optimization model replaces $N - 1$ objective functions in (9.4.1) by $N - 1$ constraints, i.e., one objective $f_i(\boldsymbol{x})$ is the principal objective, all others $f_j(\boldsymbol{x})$ $(j \ne i; j = 1, \dots, N)$ are the constraining objectives.

The generalized Lagrange function corresponding to the ε-constraint optimization model is given by

$$L(\boldsymbol{x};\boldsymbol{\lambda}) = f_i(\boldsymbol{x}) + \sum_{\substack{j\neq i\\ i,j=1,\dots,N}} \lambda_{ij}(f_j(\boldsymbol{x}) - \varepsilon_j),$$

where λ_{ij} are called *generalized Lagrange multipliers* and the positive Lagrange multiplies λ_{ij} are called the *trade-off functions* between the principal objective $f_i(\boldsymbol{x})$ and the constraining objective $f_j(\boldsymbol{x})$ $(j \neq i;\ i,j = 1,\dots,N)$. The trade-off function corresponds to the noninferior set of solutions, so this concept is very important.

Example 9.4.1. Let

$$\begin{aligned} f_1(x_1,x_2) &= (x_1-1)^2 + (x_2-2)^2 + 1,\\ f_2(x_1,x_2) &= (x_1-4)^2 + (x_2-5)^2 + 4. \end{aligned}$$

Consider a biobjective optimization model

$$\min_{(x_1,x_2)} \{f_1(x_1,x_2), f_2(x_1,x_2)\}, \quad \text{subject to}$$
$$\text{the restriction:} \quad x_1 \geq 0, x_2 \geq 0.$$

Let $f_1(x_1,x_2)$ be the principal objective and $f_2(x_1,x_2)$ the constraining objective. Then this model is equivalent to the following ε-constraint optimization model:

$$\min_{(x_1,x_2)\in \boldsymbol{X}} f_1(x_1,x_2) \quad \text{subject to}$$
$$f_2(x_1,x_2) \leq \varepsilon_2 \quad ((x_1,x_2) \in X),$$

where $X = \{(x_1,x_2) \mid x_1 \geq 0, x_2 \geq 0\}$.

The corresponding generalized Lagrange function is given by

$$\begin{aligned} L(x_1,x_2;\lambda_{12}) &= f_1(x_1,x_2) + \lambda_{12}(f_2(x_1,x_2) - \varepsilon_2)\\ &= (x_1-1)^2 + (x_2-2)^2 + 1 + \lambda_{12}((x_1-4)^2 + (x_2-5)^2 + 4 - \varepsilon_2). \end{aligned}$$

Differentiating both sides with respect to x_1 and x_2, respectively,

$$\begin{aligned} \frac{\partial L}{\partial x_1} &= 2(x_1-1) + 2\lambda_{12}(x_1-4),\\ \frac{\partial L}{\partial x_2} &= 2(x_2-2) + 2\lambda_{12}(x_2-5). \end{aligned}$$

Let $\frac{\partial L}{\partial x_i} = 0$ $(i = 1,2)$. Then

$$\begin{aligned} 2(x_1-1) + 2\lambda_{12}(x_1-4) &= 0,\\ 2(x_2-2) + 2\lambda_{12}(x_2-5) &= 0. \end{aligned}$$

This implies that the trade-off function λ_{12} between $f_1(x_1,x_2)$ and $f_2(x_1,x_2)$ satisfies

$$\begin{aligned} \lambda_{12} &= \frac{x_1-1}{4-x_1},\\ \lambda_{12} &= \frac{x_2-2}{5-x_2}. \end{aligned} \tag{9.4.3}$$

From (9.4.3) and $\lambda_{12} > 0$, it follows that the decision variables x_1 and x_2 satisfy

$$1 < x_1 < 4,$$
$$2 < x_2 < 5.$$

Several noninferior solutions and trade-off values of the biobjective optimization model are listed in the following table.

x_1	x_2	$f_1(x_1, x_2)$	$f_2(x_1, x_2)$	λ_{12}
1.5	2.5	1.5	16.5	0.2
2	3	3	12	0.5
2.5	3.5	5.5	8.5	1
3	4	9	6	2
3.5	4.5	13.5	4.5	5

From (9.4.3), it follows that $x_2 = x_1 + 1$. So the Pareto-optimal solutions in the decision space are determined by the equation $x_2 = x_1 + 1$ $(1 < x_1 < 4)$, i.e., a line segment jointing two points $(1, 2)$ and $(4, 5)$.

Example 9.4.2. Let

$$f_1(x_1, x_2) = (x_1 - 1)^2 + (x_2 - 2)^2 + 1,$$
$$f_2(x_1, x_2) = (x_1 - 4)^2 + (x_2 - 5)^2 + 4,$$
$$f_3(x_1, x_2) = (x_1 - 6)^2 + (x_2 - 10)^2 + 1.$$

A three-objective optimization model is given as

$$\min_{(x_1,x_2)} \{f_1(x_1, x_2), f_2(x_1, x_2), f_3(x_1, x_2)\},$$
$$\text{subject to} \quad x_1 \geq 0, x_2 \geq 0.$$

Let $f_1(x_1, x_2)$ be the principal objective and $f_2(x_1, x_2), f_3(x_1, x_2)$ the constraining objectives. This model is equivalent to the following ε-constraint optimization model:

$$\min_{(x_1,x_2)\in \boldsymbol{X}} f_1(x_1, x_2) \quad \text{subject to}$$
$$f_2(x_1, x_2) \leq \varepsilon_2,$$
$$f_3(x_1, x_2) \leq \varepsilon_3 \quad ((x_1, x_2) \in X),$$

where $\boldsymbol{X} = \{(x_1, x_2) \mid x_1 \geq 0, x_2 \geq 0\}$. The corresponding generalized Lagrange function is given by

$$\begin{aligned} L(x_1, x_2; \lambda_{12}, \lambda_{13}) &= f_1(x_1, x_2) + \lambda_{12}(f_2(x_1, x_2) - \varepsilon_2) + \lambda_{13}(f_3(x_1, x_2) - \varepsilon_3) \\ &= (x_1 - 1)^2 + (x_2 - 2)^2 + 1 + \lambda_{12}((x_1 - 4)^2 + (x_2 - 5)^2 + 4 - \varepsilon_2) \\ &\quad + \lambda_{13}((x_1 - 6)^2 + (x_2 - 10)^2 + 1 - \varepsilon_3). \end{aligned}$$

Differentiating both sides with respect to x_1 and x_2, respectively,

$$\frac{\partial L}{\partial x_1} = 2(x_1 - 1) + 2\lambda_{12}(x_1 - 4) + 2\lambda_{13}(x_1 - 6),$$
$$\frac{\partial L}{\partial x_2} = 2(x_2 - 2) + 2\lambda_{12}(x_2 - 5) + 2\lambda_{13}(x_2 - 10).$$

Let $\frac{\partial L}{\partial x_i} = 0$ $(i = 1, 2)$. Then

$$2(x_1 - 1) + 2\lambda_{12}(x_1 - 4) + 2\lambda_{13}(x_1 - 6) = 0,$$
$$2(x_2 - 2) + 2\lambda_{12}(x_2 - 5) + 2\lambda_{13}(x_2 - 10) = 0.$$

This implies that the trade-off functions λ_{12} and λ_{13} satisfy

$$\begin{aligned}\lambda_{12} &= \frac{-8x_1 + 5x_2 - 2}{5x_1 - 2x_2 - 10},\\ \lambda_{13} &= \frac{3x_1 - 3x_2 + 3}{5x_1 - 2x_2 - 10}.\end{aligned} \tag{9.4.4}$$

Secondly, let $f_2(x_1, x_2)$ be the principal objective and $f_1(x_1, x_2), f_3(x_1, x_2)$ be the constraining objectives. Then the given three-objective optimization model is also equivalent to the following ε-constraint optimization model:

$$\begin{aligned}\min_{(x_1,x_2)\in \boldsymbol{X}} \; & f_2(x_1, x_2) \quad \text{subject to}\\ & f_1(x_1, x_2) \le \varepsilon_1,\\ & f_3(x_1, x_2) \le \varepsilon_3 \quad ((x_1, x_2) \in X).\end{aligned}$$

The corresponding generalized Lagrange function is given by

$$\begin{aligned}L(x_1, x_2; \lambda_{21}, \lambda_{23}) &= f_2(x_1, x_2) + \lambda_{21}(f_1(x_1, x_2) - \varepsilon_1) + \lambda_{23}(f_3(x_1, x_2) - \varepsilon_3)\\ &= (x_1 - 4)^2 + (x_2 - 5)^2 + 4 + \lambda_{21}((x_1 - 1)^2 + (x_2 - 2)^2 + 1 - \varepsilon_1)\\ &\quad + \lambda_{23}((x_1 - 6)^2 + (x_2 - 10)^2 + 1 - \varepsilon_3).\end{aligned}$$

Let $\frac{\partial L}{\partial x_i} = 0$ $(i = 1, 2)$. Then

$$2(x_1 - 4) + 2\lambda_{21}(x_1 - 1) + 2\lambda_{23}(x_1 - 6) = 0,$$
$$2(x_2 - 5) + 2\lambda_{21}(x_2 - 2) + 2\lambda_{23}(x_2 - 10) = 0.$$

This implies that the trade-off functions λ_{21} and λ_{23} satisfy

$$\begin{aligned}\lambda_{21} &= \frac{5x_1 - 2x_2 - 10}{-8x_1 + 5x_2 - 2},\\ \lambda_{23} &= \frac{3x_1 - 3x_2 + 3}{-8x_1 + 5x_2 - 2}.\end{aligned} \tag{9.4.5}$$

Third, let $f_3(x_1, x_2)$ be the principal objective and $f_1(x_1, x_2), f_2(x_1, x_2)$ the constraining objectives. Then the given three-objective optimization model is also equivalent to the following ε-constraint optimization model:

$$\begin{aligned} \min_{(x_1,x_2)\in X} \; & f_3(x_1, x_2) \quad \text{subject to} \\ & f_1(x_1, x_2) \le \varepsilon_1, \\ & f_2(x_1, x_2) \le \varepsilon_2 \quad ((x_1, x_2) \in X). \end{aligned}$$

The corresponding generalized Lagrange function is given by

$$\begin{aligned} L(x_1, x_2; \lambda_{31}, \lambda_{32}) &= f_3(x_1, x_2) + \lambda_{31}(f_1(x_1, x_2) - \varepsilon_1) + \lambda_{32}(f_2(x_1, x_2) - \varepsilon_2) \\ &= (x_1 - 6)^2 + (x_2 - 10)^2 + 1 + \lambda_{31}((x_1 - 1)^2 + (x_2 - 2)^2 + 1 - \varepsilon_1) \\ &\quad + \lambda_{32}((x_1 - 4)^2 + (x_2 - 5)^2 + 4 - \varepsilon_2). \end{aligned}$$

Let $\frac{\partial L}{\partial x_i} = 0$ $(i = 1, 2)$. Then

$$\begin{aligned} 2(x_1 - 6) + 2\lambda_{31}(x_1 - 1) + 2\lambda_{32}(x_1 - 4) &= 0, \\ 2(x_2 - 10) + 2\lambda_{31}(x_2 - 2) + 2\lambda_{32}(x_2 - 5) &= 0. \end{aligned}$$

This implies that the trade-off functions λ_{31} and λ_{32} satisfy

$$\begin{aligned} \lambda_{31} &= \frac{5x_1 - 2x_2 - 10}{3x_1 - 3x_2 + 3}, \\ \lambda_{32} &= \frac{-8x_1 + 5x_2 - 2}{3x_1 - 3x_2 + 3}. \end{aligned} \tag{9.4.6}$$

The combination of (9.4.4) and (9.4.5), and (9.4.6) gives

$$\begin{aligned} &\lambda_{12} = \frac{1}{\lambda_{21}}, \quad \lambda_{13} = \frac{1}{\lambda_{31}}, \quad \lambda_{23} = \frac{1}{\lambda_{32}}, \\ &\lambda_{13} = \lambda_{12}\lambda_{23}, \qquad \lambda_{23} = \lambda_{21}\lambda_{13}. \end{aligned}$$

Several noninferior solutions and trade-off values of the given three-objective optimization model are listed in the following table.

x_1	x_2	f_1	f_2	f_3	λ_{12}	λ_{13}	λ_{21}	λ_{23}	λ_{31}	λ_{32}
2	3.2	3.44	11.24	63.24	0.31	0.09	3.2	0.3	10.67	3.33
3	5	14	5	35	0.2	0.6	5	3	1.67	0.33
4	6	26	5	21	2	1.5	0.5	0.75	0.67	1.33
5	8	53	14	6	2	6	0.5	3	0.17	0.33

From (9.4.4), (9.4.5), and (9.4.6), the Pareto-optimal solution in the decision space is a triangle determined by the system of inequalities

$$\begin{cases} x_2 > x_1 + 1, \\ x_2 > \frac{5}{2}x_1 - 5, \\ x_2 < \frac{8}{5}x_1 + \frac{2}{5}, \end{cases}$$

i.e., a triangle with three vertices $(1, 2)$, $(4, 5)$, and $(6, 10)$.

Examples 9.4.1 and 9.4.2 indicate that they both have two common objective functions, but by adding the third objective function in Example 9.4.2, a large number of Pareto-optimal solutions have been added.

9.5 The uncertainty sensitivity index method

Uncertainty and sensitivity are two characteristics of systems. *Uncertainty* is defined as the inability to determine the true state of affairs of a system. It is caused by stochastic variability and incomplete knowledge. Taylor shows that the variability uncertainty arises from temporal, spatial, and individually heterogeneous variability. Finkel shows that knowledge about uncertainty comes from model, parameter, and decision uncertainty. *Sensitivity* is defined as the relation between changes in the system's performance index and possible variations in decision variables, constraint levels, and uncontrolled parameters. The *Uncertainty Sensitivity Index Method* (USIM) presented by Li and Haimes is a uncertainty-sensitivity analysis method. This method represents the uncertainty associated with potential variations of the system's parameter by a system's output sensitivity index. The USIM is intrinsic to considering a joint optimality and sensitivity multiobjective model.

Let

$$y = h(\boldsymbol{x}; \alpha)$$

be the system's output response, where $\boldsymbol{x}$ is a vector consisting of n decision control variables and α is a vector consisting of m uncertain random system parameters

$$\boldsymbol{x} = (x_1, \ldots, x_n),$$
$$\alpha = (\alpha_1, \ldots, \alpha_m),$$

and $y \in \mathbb{R}$ is differentiable with respect to $\boldsymbol{x}$ and α. Let

$$f(\boldsymbol{x}; y; \alpha) = f(\boldsymbol{x}; h(\boldsymbol{x}; \alpha); \alpha)$$

be the system's objective function.

Denote the nominal value of α by $\hat{\alpha} = (\hat{\alpha}_1, \ldots, \hat{\alpha}_m)$. The nominal values $\hat{\alpha}_i$ $(i = 1, \ldots, m)$ may be determined by any system's identification procedure. Assume that the m uncertain random system parameters $\alpha_1, \ldots, \alpha_m$ vary in the neighborhood of their nominal value $(\hat{\alpha}_1, \ldots, \hat{\alpha}_m)$. The Taylor theorem shows that

$$h(\boldsymbol{x}; \hat{\alpha}_1 + \Delta\hat{\alpha}_1, \ldots, \hat{\alpha}_m + \Delta\hat{\alpha}_m) \approx h(\boldsymbol{x}; \hat{\alpha}_1, \ldots, \hat{\alpha}_m) + \sum_1^m \frac{\partial h(\boldsymbol{x}; \hat{\alpha}_1, \ldots, \hat{\alpha}_m)}{\partial \hat{\alpha}_i} \Delta\hat{\alpha}_i,$$

where $\Delta\hat{\alpha}_i$ is a small variation in the nominal value $\hat{\alpha}_i$. Thus, associated with variations in the parameters, the variation of the system's output is equal approximately to

$$\sum_1^m \frac{\partial h(\boldsymbol{x}; \hat{\alpha}_1, \dots, \hat{\alpha}_m)}{\partial \hat{\alpha}_i} \Delta\hat{\alpha}_i.$$

and the square of the variation satisfies the Cauchy–Schwarz inequality

$$\left(\sum_1^m \frac{\partial h(\boldsymbol{x}; \hat{\alpha}_1, \dots, \hat{\alpha}_m)}{\partial \hat{\alpha}_i} \Delta\hat{\alpha}_i \right)^2 \leq \sum_1^m \left(\frac{\partial h(\boldsymbol{x}; \hat{\alpha}_1, \dots, \hat{\alpha}_m)}{\partial \hat{\alpha}_i} \right)^2 \cdot \sum_1^m (\Delta\hat{\alpha}_i)^2 .$$

In order to reduce this variation, a control scenario $\boldsymbol{x}$ is chosen such that

$$\sum_1^m \left(\frac{\partial h(\boldsymbol{x}; \hat{\alpha}_1, \dots, \hat{\alpha}_m)}{\partial \hat{\alpha}_i} \right)^2$$

is minimal. Define a *sensitivity index function* as

$$s(\boldsymbol{x}; \hat{\alpha}) = \sum_1^m \left(\frac{\partial h(\boldsymbol{x}; \hat{\alpha}_1, \dots, \hat{\alpha}_m)}{\partial \hat{\alpha}_i} \right)^2 .$$

The multiobjective optimization model of the form

$$\min_{\boldsymbol{x}} \{ f(\boldsymbol{x}; h(\boldsymbol{x}; \hat{\alpha}); \hat{\alpha}), s(\boldsymbol{x}; \hat{\alpha}) \}$$

is called a *joint optimality and sensitivity multiobjective model*. Let $f(\boldsymbol{x}; h(\boldsymbol{x}; \hat{\alpha}); \hat{\alpha})$ be the principal objective and $s(\boldsymbol{x}; \hat{\alpha})$ be the constraining objective. Using the ε-constraint method, the joint optimality and sensitivity multiobjective optimization model is equivalent to the ε-constraint optimization model of the form

$$\begin{aligned} \min_{\boldsymbol{x}} f(\boldsymbol{x}; h(\boldsymbol{x}; \hat{\alpha}); \hat{\alpha}) \quad & \text{subject to} \\ & s(\boldsymbol{x}; \hat{\alpha}) \leq \varepsilon. \end{aligned}$$

The corresponding generalized Lagrangian function is given by

$$L(\boldsymbol{x}; \hat{\alpha}; \lambda_{fs}) = f(\boldsymbol{x}; h(\boldsymbol{x}; \hat{\alpha}); \hat{\alpha}) + \lambda_{fs}(s(\boldsymbol{x}; \hat{\alpha}) - \varepsilon),$$

where λ_{fs} is the generalized Lagrange multiplier. Differentiating both sides with respect to x_i,

$$\frac{\partial L}{\partial x_i} = \frac{\partial f(\boldsymbol{x}; h(\boldsymbol{x}; \hat{\alpha}); \hat{\alpha})}{\partial x_i} + \lambda_{fs} \frac{\partial s(\boldsymbol{x}; \hat{\alpha})}{\partial x_i} \quad (i = 1, \dots, n), \tag{9.5.1}$$

where $\boldsymbol{x} = (x_1, \dots, x_n)$ and λ_{fs} is the trade-off function between the system's objective function $f(\boldsymbol{x}; h(\boldsymbol{x}; \hat{\alpha}); \hat{\alpha})$ and the sensitivity index function $s(\boldsymbol{x}; \hat{\alpha})$. Let $\frac{\partial L}{\partial x_i} = 0$ $(i = 1, \dots, n)$. Then

$$\frac{\partial f(\boldsymbol{x}; h(\boldsymbol{x}; \hat{\alpha}); \hat{\alpha})}{\partial x_i} + \lambda_{fs} \frac{\partial s(\boldsymbol{x}; \hat{\alpha})}{\partial x_i} = 0 \quad (i = 1, \dots, n)$$

or

$$\lambda_{fs} = -\frac{\left(\frac{\partial f(\boldsymbol{x};h(\boldsymbol{x};\hat{\alpha});\hat{\alpha})}{\partial x_i}\right)}{\left(\frac{\partial s(\boldsymbol{x};\hat{\alpha})}{\partial x_i}\right)} \qquad (i = 1, \ldots, n).$$

From this and $\lambda_{fs} > 0$, the set of noninferior solutions can be determined by the systems of inequalities

$$\begin{cases} \frac{\partial f(\boldsymbol{x};h(\boldsymbol{x};\hat{\alpha});\hat{\alpha})}{\partial x_i} > 0, \\ \frac{\partial s(\boldsymbol{x};\hat{\alpha})}{\partial x_i} < 0 \end{cases}$$

or

$$\begin{cases} \frac{\partial f(\boldsymbol{x};h(\boldsymbol{x};\hat{\alpha});\hat{\alpha})}{\partial x_i} < 0, \\ \frac{\partial s(\boldsymbol{x};\hat{\alpha})}{\partial x_i} > 0. \end{cases}$$

Let $\boldsymbol{x}^*$ be a decision variable which minimizes the system's objective function $f(\boldsymbol{x}; h(\boldsymbol{x}; \hat{\alpha}); \hat{\alpha})$, i.e.,

$$f(\boldsymbol{x}^*; h(\boldsymbol{x}^*; \hat{\alpha}); \hat{\alpha}) = \min_{\boldsymbol{x}} f(\boldsymbol{x}; h(\boldsymbol{x}; \hat{\alpha}); \hat{\alpha}),$$

and let $\hat{\boldsymbol{x}}$ be another decision variable which minimizes the sensitivity index function $s(\boldsymbol{x}; \hat{\alpha})$, i.e.,

$$s(\hat{\boldsymbol{x}}; \hat{\alpha}) = \min_{\boldsymbol{x}} s(\boldsymbol{x}; \hat{\alpha}).$$

The most conservative policy $\hat{\boldsymbol{x}}$ provides a very stable solution, whereas the conventional solution $\boldsymbol{x}^*$ suffers the highest deviation. Based on the preference of the decision maker, the best compromise solution may be selected from the noninferior solution set $\{\hat{\boldsymbol{x}} \le \boldsymbol{x} \le \boldsymbol{x}^*\}$.

Example 9.5.1. Consider a system that has the following output response and objective function:

$$\begin{aligned} h(x; \alpha) &= (\alpha_1^2 + \alpha_2^2 + \alpha_3^2 + \alpha_4^2)x, \\ f(x; h(x; \alpha); \alpha) &= x^2 - 2(\alpha_1\alpha_2^2 + \alpha_1\alpha_3^2 + \alpha_1\alpha_4^2 + \alpha_2\alpha_3^2 + \alpha_2\alpha_4^2 + \alpha_3\alpha_4^2)x \\ &\quad + \alpha_1^2\alpha_2 + \alpha_1^2\alpha_3 + \alpha_1^2\alpha_4 + \alpha_2^2\alpha_3 + \alpha_2^2\alpha_4 + \alpha_3^2\alpha_4, \end{aligned}$$

where x is the one-dimensional decision variable and α is a vector consisting of four uncertain random system parameters $\alpha_1, \alpha_2, \alpha_3, \alpha_4$. The system's sensitivity index function is

$$s(x; \alpha) = \sum_1^4 \left(\frac{\partial h(x; \alpha_1, \alpha_2, \alpha_3, \alpha_4)}{\partial \alpha_i}\right)^2 = 4\,(\alpha_1^2 + \alpha_2^2 + \alpha_3^2 + \alpha_4^2)x.$$

Assume that the nominal values $\hat{\alpha}_1, \hat{\alpha}_2, \hat{\alpha}_3, \hat{\alpha}_4$ of the system's parameters $\alpha_1, \alpha_2, \alpha_3, \alpha_4$ are, respectively,

$$\hat{\alpha}_1 = 1, \quad \hat{\alpha}_2 = 2,$$
$$\hat{\alpha}_3 = 1, \quad \hat{\alpha}_4 = 2.$$

Then

$$\begin{aligned} h(x;\hat{\alpha}_1,\hat{\alpha}_2,\hat{\alpha}_3,\hat{\alpha}_4) &= h(x;1,2,1,2) = 10x,\\ f(x;h(x;\hat{\alpha}_1,\hat{\alpha}_2,\hat{\alpha}_3,\hat{\alpha}_4);\hat{\alpha}_1,\hat{\alpha}_2,\hat{\alpha}_3,\hat{\alpha}_4) &= f(x;h(x;1,2,1,2);1,2,1,2)\\ &= x^2-23x+19\\ s(x;\hat{\alpha}_1,\hat{\alpha}_2,\hat{\alpha}_3,\hat{\alpha}_4) &= s(x;1,2,1,2) = 40x^2. \end{aligned} \tag{9.5.2}$$

When the nominal values $\hat{\alpha}_1,\hat{\alpha}_2,\hat{\alpha}_3,\hat{\alpha}_4$ are perturbed by $\Delta\hat{\alpha}_1,\Delta\hat{\alpha}_2,\Delta\hat{\alpha}_3,\Delta\hat{\alpha}_4$ and

$$\Delta\hat{\alpha}_1 = \Delta\hat{\alpha}_2 = \Delta\hat{\alpha}_3 = \Delta\hat{\alpha}_4 = 0.1,$$

the variation of the system's output response is

$$\Delta h = \left(\sum_1^4(\hat{\alpha}_i+\Delta\hat{\alpha}_i)^2\right)x - \left(\sum_1^4 \alpha_i^2\right)x = \sum_1^4(2\hat{\alpha}_i\Delta\hat{\alpha}_i + (\Delta\hat{\alpha}_i)^2)\,x = 1.24\,x. \tag{9.5.3}$$

The joint optimality and sensitivity multiobjective model is the form

$$\min_x \begin{cases} f(x;h(x;\hat{\alpha}_1,\hat{\alpha}_2,\hat{\alpha}_3,\hat{\alpha}_4);\hat{\alpha}_1,\hat{\alpha}_2,\hat{\alpha}_3,\hat{\alpha}_4) = x^2-23x+19\\ s(x;\hat{\alpha}_1,\hat{\alpha}_2,\hat{\alpha}_3,\hat{\alpha}_4) = 40x^2 \end{cases}$$

Using the ε-constraint method, this model is equivalent to the following ε-constraint optimization model:

$$\begin{aligned} &\min_x\{f(x;h(x;\hat{\alpha}_1,\hat{\alpha}_2,\hat{\alpha}_3,\hat{\alpha}_4);\hat{\alpha}_1,\hat{\alpha}_2,\hat{\alpha}_3,\hat{\alpha}_4) = x^2-23x+19\}\\ &\text{subject to the } \varepsilon-\text{constraint:}\quad s(x;\hat{\alpha}_1,\hat{\alpha}_2,\hat{\alpha}_3,\hat{\alpha}_4) = 40x^2 \le \varepsilon. \end{aligned}$$

The corresponding generalized Lagrangian function is given by

$$\begin{aligned} L(x,\hat{\alpha}_1,\hat{\alpha}_2,\hat{\alpha}_3,\hat{\alpha}_4,\lambda_{fs}) &= f(x;h(x;\hat{\alpha}_1,\hat{\alpha}_2,\hat{\alpha}_3,\hat{\alpha}_4);\hat{\alpha}_1,\hat{\alpha}_2,\hat{\alpha}_3,\hat{\alpha}_4)\\ &\quad + \lambda_{fs}(s(x;\hat{\alpha}_1,\hat{\alpha}_2,\hat{\alpha}_3,\hat{\alpha}_4)-\varepsilon)\\ &= (x^2-23x+19) + \lambda_{12}(40x^2-\varepsilon), \end{aligned}$$

where λ_{fs} is the trade-off between the system's objective function and system's sensitivity index function. Let $\frac{\partial L}{\partial x} = 0$. Then $2x - 23 + \lambda_{fs}80x = 0$, i.e.,

$$\lambda_{fs} = -\frac{2x-23}{80x}.$$

From this and $\lambda_{fs} > 0$, it follows that

$$\begin{cases} 2x-23 > 0,\\ 80x < 0 \end{cases}$$

or

$$\begin{cases} 2x-23 < 0,\\ 80x > 0. \end{cases}$$

So the set of noninferior solutions is $0 \le x \le 11.5$.

By (9.5.2) and (9.5.3), a sample of noninferior solutions of the joint optimality and sensitivity model, the corresponding values of the system's objective function f, the system's sensitivity index function s, the trade-off λ_{fs}, the system's output response h, and its variation Δh are listed in the following table.

x	0	4	8	10	11.5
f	19	−57	−101	−111	−113.25
s	0	640	2560	4000	5290
λ_{fs}	∞	0.0456	0.0109	0.00375	0
h	0	40	80	100	115
Δh	0	4.96	9.92	12.4	14.26

Let x^* be a decision variable which minimizes $f(x, h(x; \hat{\alpha}_1, \dots, \hat{\alpha}_n); \hat{\alpha}_1, \dots, \hat{\alpha}_n)$, and let $\hat{x}$ be another decision variable which minimizes $s(x; \hat{\alpha}_1, \dots, \hat{\alpha}_n)$, i.e.,

$$f(x^*; h(x^*; \hat{\alpha}_1, \hat{\alpha}_2, \hat{\alpha}_3, \hat{\alpha}_4); \hat{\alpha}_1, \hat{\alpha}_2, \hat{\alpha}_3, \hat{\alpha}_4) = \min_x f(x; h(x; \hat{\alpha}_1, \dots, \hat{\alpha}_n); \hat{\alpha}_1, \hat{\alpha}_2, \hat{\alpha}_3, \hat{\alpha}_4),$$
$$s(\hat{x}; \hat{\alpha}_1, \hat{\alpha}_2, \hat{\alpha}_3, \hat{\alpha}_4) = \min_x s(x; \hat{\alpha}_1, \hat{\alpha}_2, \hat{\alpha}_3, \hat{\alpha}_4). \tag{9.5.4}$$

From (9.5.2), it follows that

$$f(x, h(x, \hat{\alpha}_1, \hat{\alpha}_2, \hat{\alpha}_3, \hat{\alpha}_4); \hat{\alpha}_1, \hat{\alpha}_2, \hat{\alpha}_3, \hat{\alpha}_4) = x^2 - 23x + 19 = (x - 11.5)^2 - 113.25,$$
$$s(x; \hat{\alpha}_1, \hat{\alpha}_2, \hat{\alpha}_3, \hat{\alpha}_4) = 40x^2.$$

The combination of this with (9.5.4) gives

$$x^* = 11.5,$$
$$\hat{x} = 0.$$

Thus, the best compromise solution can be selected from the noninferior solution set $\{0 \le x \le 11.5\}$.

9.6 The partitioned multiobjective risk method

The *Partitioned Multiobjective Risk* (PMR) method is a risk analysis method. In the PMR method, the concept of the traditional expected value of damage is extended to generate a number of conditional expected value functions (or *conditional expected risk functions*), each associated with a particular range of exceedance probabilities (or the corresponding range of damage severities). The resulting conditional expected risk functions, together with the traditional expected value of damage, provide a family of risk measures associated with a particular policy.

Let a damage severity associated with a particular policy s_j ($j = 1, \dots, m$) be represented by the continuous random variable X_j ($j = 1, \dots, m$), and $P_j(x)$ and $p_j(x)$ denote the cdf and the pdf of damage, respectively. Let $1 - \alpha_i$ ($i = 1, \dots, n$) denote n exceedance probabilities, where $0 < \alpha_1 < \dots < \alpha_n < 1$. For the policy s_j, there is a unique damage β_i^j ($i = 1, \dots, n$) on the damage axis that corresponds to the exceedance probability $1 - \alpha_i$ ($i = 1, \dots, n$) on the probability axis. Thus, the partition of the probability axis

$$[0, 1 - \alpha_n], \quad [1 - \alpha_n, 1 - \alpha_{n-1}], \quad \dots, \quad [1 - \alpha_1, 1]$$

corresponds the partition of the damage axis

$$[\beta_0^j, \beta_1^j], \quad [\beta_1^j, \beta_2^j], \quad \dots, \quad [\beta_n^j, \beta_{n+1}^j],$$

where β_0^j and β_{n+1}^j are the lower bound and upper bound of damage, respectively. The policies s_j, the exceedance probabilities $1 - \alpha_i$, and the bounds β_i^j of damage ranges satisfy the relationship

$$\begin{aligned} P_j(\beta_1^j) &= 1 - \alpha_1, \\ &\vdots \\ P_j(\beta_n^j) &= 1 - \alpha_n. \end{aligned}$$

If the inverse $P_j^{-1}(x)$ exists, then

$$\begin{aligned} P_j^{-1}(1 - \alpha_1) &= \beta_1^j, \\ &\vdots \\ P_j^{-1}(1 - \alpha_n) &= \beta_n^j. \end{aligned}$$

The conditional expected risk functions $f_i(s_j)$ ($i = 2, \dots, n + 2$) of the damage are given by

$$f_2(s_j) = E[\, X_j \mid p_j(x), x \in [\beta_0^j, \beta_1^j]\,] = \frac{\int_{\beta_0^j}^{\beta_1^j} x p_j(x)\,\mathrm{d}x}{\int_{\beta_0^j}^{\beta_1^j} p_j(x)\,\mathrm{d}x},$$

$$f_3(s_j) = E[\, X_j \mid p_j(x), x \in [\beta_3^j, \beta_4^j]\,] = \frac{\int_{\beta_1^j}^{\beta_2^j} x p_j(x)\,\mathrm{d}x}{\int_{\beta_1^j}^{\beta_2^j} p_j(x)\,\mathrm{d}x},$$

$$\vdots$$

$$f_{n+2}(s_j) = E[\, X_j \mid p_j(x), x \in [\beta_n^j, \beta_{n+1}^j]\,] = \frac{\int_{\beta_n^j}^{\beta_{n+1}^j} x p_j(x)\,\mathrm{d}x}{\int_{\beta_n^j}^{\beta_{n+1}^j} p_j(x)\,\mathrm{d}x}.$$

The unconditional excepted risk function $f_{n+3}(s_j)$ of the damage is given by

$$f_{n+3}(s_j) = E[X_j] = \frac{\int_{\beta_0^j}^{\beta_{n+1}^j} x p_j(x)\,\mathrm{d}x}{\int_{\beta_0^j}^{\beta_{n+1}^j} p_j(x)\,\mathrm{d}x}.$$

Since the total probability of the damage is equal to 1, i.e.,

$$\int_{\beta_0^j}^{\beta_{n+1}^j} p_j(x)\,\mathrm{d}x = 1,$$

the unconditional excepted risk function of the damage is

$$f_{n+3}(s_j) = \int_{\beta_0^j}^{\beta_{n+1}^j} x p_j(x)\,\mathrm{d}x.$$

Denote their denominators by q_i $(i = 2, \ldots, n+2)$, respectively, i.e.,

$$\begin{aligned} q_2 &= \int_{\beta_{0j}}^{\beta_1^j} p_j(x)\,\mathrm{d}x, \\ q_3 &= \int_{\beta_1^j}^{\beta_2^j} p_j(x)\,\mathrm{d}x, \\ &\vdots \\ q_{n+2} &= \int_{\beta_n^j}^{\beta_{n+1}^j} p_j(x)\,\mathrm{d}x, \end{aligned}$$

where $q_i > 0$ $(i = 2, \ldots, n+2)$ and $q_2 + \cdots + q_{n+2} = 1$. So the conditional expected risk functions f_i $(i = 2, \ldots, n+2)$ and the unconditional expected risk function f_{n+3} satisfy the balance relationship

$$f_{n+3}(s_j) = q_2 f_2(s_j) + q_3 f_3(s_j) + \cdots + q_{n+2} f_{n+2}(s_j) = \sum_2^{n+2} q_i f_i(s_j).$$

Let $f_1(s_j)$ denote the cost objective function.

Combining any one of the conditional expected risk functions or the unconditional excepted risk function with the cost objective function constitutes a set of biobjective optimization models, i.e.,

$$\begin{cases} \min\{f_1(s_j), f_2(s_j)\}, \\ \quad\vdots \\ \min\{f_1(s_j), f_{n+2}(s_j)\}, \\ \min\{f_1(s_j), f_{n+3}(s_j)\}. \end{cases}$$

These $n + 2$ biobjective optimization models can offer more information about the probabilistic behavior than the single biobjective optimization model $\min\{f_1(s_j), f_{n+3}(s_j)\}$.

Let $f_1(s_j)$ be the principal objective and $f_i(s_j)$ $(i = 2, \ldots, n + 2)$ the constraining objective. The above $n + 2$ biobjective optimization models can be transformed into the $n + 2$ equivalent ε-constraint optimization models and the trade-offs λ_{1i} $(i = 2, \ldots, n + 3)$ between the cost function f_1 and risk functions f_i $(i = 2, \ldots, n + 3)$ satisfy

$$\frac{1}{\lambda_{1,n+3}} = \frac{q_2}{\lambda_{12}} + \frac{q_3}{\lambda_{13}} + \cdots + \frac{q_{n+2}}{\lambda_{1,n+2}} = \sum_{2}^{n+2} \frac{q_i}{\lambda_{1i}},$$

where $\lambda_{1i} = -\frac{\partial f_1}{\partial f_i}$ $(i = 2, \ldots, n + 3)$ and q_i $(i = 2, \ldots, n + 2)$ are stated as above.

9.7 The multiobjective multistage impact analysis method

A general definition of impact is the effect of one thing upon another. For impact analysis in a multiobjective framework, Gomide developed a theoretical basis. He formulated a multiobjective multistage optimization model and presented the *Multiobjective Multistage Impact Analysis* (MMIA) method.

Let $g\colon \mathbb{R}^n \times \mathbb{R}^r \to \mathbb{R}^n$ be a continuously differentiable function. Assume that the evolution of a system in stages can be represented by a multistage process

$$x(0) = x_0, \quad x(k+1) = g(x(k), u(k); k) \qquad (k = 0, \ldots, T-1),$$

where $x(k) \in \mathbb{R}^n$ is the system's state at stage k and $u(k) \in \mathbb{R}^r$ is the decision at stage k, and T is the horizon of interest to the system's decision maker. Denote by F the *universe* of objectives of interest to the system

$$F = \{f_i^k(x(k), u(k); k), \quad i = 1, 2, \ldots, N_k;\ k = 0, \ldots, T-1\},$$

where N_k is the number of objectives considered as important for stage k, and f_i^k are continuously differentiable functions. Assume further that Ω_k $(k = 0, \ldots, T)$ are specified by a system of inequality constraints

$$\begin{aligned} \Omega_k &= \{(x(k), u(k))\colon h(x(k), u(k); k) \le 0\} \quad (k = 0, \ldots, T-1), \\ \Omega_T &= \{x(T)\colon h(x(T); T) \le 0\}, \end{aligned}$$

where $h(x(k), u(k); k) : \mathbb{R}^n \times \mathbb{R}^r \to \mathbb{R}^{m_k}$ $(k = 0, \ldots, T-1)$ and $h(x(k); T) : \mathbb{R}^n \to \mathbb{R}^{m_T}$ are both continuously differentiable functions. The Multiobjective Multistage Impact Analysis (MMIA) model is the form

$$\min_u \begin{Bmatrix} f_1^0(x,u;0) \\ \vdots \\ f_{N_0}^k(x,u;0) \end{Bmatrix}, \quad \dots, \quad \min_u \begin{Bmatrix} f_1^{T-1}(x,u;T-1) \\ \vdots \\ f_{N_{T-1}}^k(x,u;T-1) \end{Bmatrix} \quad \text{subject to}$$

$$x(0) = x_0,$$
$$x(k+1) = g(x(k), u(k); k) \quad (k = 0, \dots, T-1),$$

where

$$(x(k), u(k)) \in \Omega_k \subset \mathbb{R}^n \times \mathbb{R}^r \quad (k = 0, \dots, T-1),$$
$$x(T) \in \Omega_T \subset \mathbb{R}^n.$$

From this, it is seen that the meaning of the MMIA model is to solve a sequence of static or single-stage multiobjective optimization models where decisions made at stage k affect stages $k+1, k+2, \dots, T-1$.

Gomide and Haimes characterized noninferior policy decisions for the MMIA model. To obtain noninferior solutions, the MMIA model is converted into a series of single-objective models using the ε-constraint method as follows.

Let

$$X = (x(0), \dots, x(T)),$$
$$U = (u(0), \dots, u(T-1))$$

and

$$\mathbb{K} = \{0, \dots, T-1\},$$
$$\mathbb{N}_k = \{1, \dots, N_k\} \quad (k = 0, \dots, T-1)$$

and let $\Omega = \bigcup_{k=0}^{T} \Omega_k$. Using the ε-constraint method, the MMIA model is represented as

$$\min_U f_s^t(x(t), u(t); t) \quad \text{subject to}$$
$$(X, U) \in \left(\Omega \bigcap (\Gamma_s^t(\varepsilon) \bigcup \Gamma_{s,t}^\tau(\varepsilon)) \bigcap V\right),$$

where

$$\Gamma_s^t(\varepsilon) = \{(X,U) : f_i^t(x(t), u(t); t) \le \varepsilon_i^t \quad \text{for all } i \in \mathbb{N}_t \ (i \ne s), \ t \in \mathbb{K}\},$$
$$\Gamma_{s,t}^\tau(\varepsilon) = \{(X,U) : f_i^\tau(x(\tau), u(\tau); \tau) \le \varepsilon_i^\tau \quad \text{for all } i \in \mathbb{N}_\tau, \quad \text{for all } \tau \in \mathbb{K} \ (\tau \ne t)\},$$
$$V = \{(X,U) : x(k+1) = g(x(k), u(k); k) \ (k = 0, \dots, T-1), \ x(0) = x_0\}$$

and $\varepsilon \in Y_s$ with

$$\varepsilon = \{\varepsilon_1^0, \dots, \varepsilon_{N_0}^0, \dots, \varepsilon_1^{t-1}, \dots, \varepsilon_{N_{t-1}}^{t-1}, \varepsilon_1^t, \dots, \varepsilon_{s-1}^t \varepsilon_{s+1}^t, \dots, \varepsilon_{N_t}^t, \dots,$$
$$\varepsilon_1^{t+1}, \dots, \varepsilon_{N_{t+1}}^{t+1}, \dots, \varepsilon_1^{T-1}, \dots, \varepsilon_{N_{T-1}}^{T-1}\},$$
$$Y_s = \{\varepsilon : (\Gamma_s^t(\varepsilon) \bigcup \Gamma_{s,t}^\tau(\varepsilon)) \ne \emptyset\}.$$

Gomide and Haimes proved that if a policy decision is the unique solution of the ε-constraint problem, then it is a noninferior policy decision. Conversely, any noninferior policy decision solves the ε-constraint problem.

9.8 Multiobjective risk impact analysis method

The PMR method is a risk analysis method. The MMIA method is an impact analysis method. The *Multiobjective Risk Impact Analysis* (MRIA) method is a combination of the MMIA method with the PMR method.

Let $x(k)$ and $y(k)$ be two normally distributed random variables. Assume that $x(k)$ represents the system's state and $y(k)$ represents the system's output, both at stage k, and

$$\begin{aligned} x(k+1) &= Ax(k) + Bu(k) + w(k), \quad x(0) = x_0, \\ y(k) &= Cx(k) + \upsilon(k) \quad (k = 0, \dots, T-1), \end{aligned} \tag{9.8.1}$$

where A, B, C are system parameters, $u(k)$ is the decision, and $w(k), \upsilon(k)$ are two normally distributed, purely random sequences. Assume further that the system satisfies the following statistical properties:

$$\text{for } 0 \le k,\ l \le T-1,$$

$$\begin{cases} E[x(0)] = x_0, \\ E[(x(0) - x_0)^2] = X_0, \\ E[(x(0) - x_0)\upsilon(k)] = 0, \\ E[(x(0) - x_0)w(k)] = 0 \end{cases}$$

and

$$\begin{cases} E[\upsilon(k)] = 0, \\ E[w(k)] = 0, \\ E[\upsilon^2(k)] = Q(k) = Q, \\ E[w^2(k)] = P(k) = P, \\ E[\upsilon(k)\upsilon(l)] = 0 & (k \ne l), \\ E[w(k)w(l)] = 0 & (k \ne l), \\ E[w(k)\upsilon(l)] = 0. \end{cases}$$

Proposition 9.8.1. *Under the above assumption, the mean of $y(k)$ satisfies that*

$$E[y(k)] = CA^k x_0 + \sum_{i=0}^{k-1} CA^i Bu(k-1-i).$$

Proof. Note that $E(w(k)) = 0$ and $E(\upsilon(k)) = 0$. Since $u(k)$ is constant, it follows by (9.8.1) that

$$\begin{aligned} E[x(k+1)] &= E[Ax(k) + Bu(k) + w(k)] \\ &= AE[x(k)] + BE[u(k)] + E[w(k)] = AE[x(k)] + Bu(k), \\ E[y(k)] &= E[Cx(k) + \upsilon(k)] = CE[x(k)] + E[\upsilon(k)] = CE[x(k)]. \end{aligned} \tag{9.8.2}$$

Now we prove by induction that the mean of $x(k)$ satisfies that

$$E[x(k)] = A^k x_0 + \sum_{i=0}^{k-1} A^i Bu(k-1-i). \tag{9.8.3}$$

In the case $k = 0$. By (9.8.2) and $E[x(0)] = x_0$, it follows that

$$E[x(1)] = AE[x(0)] + Bu(0) = Ax_0 + Bu(0),$$

i.e., (9.8.3) holds clearly for $k = 0$.

Assume for k that

$$E[x(k)] = A^k x_0 + \sum_{i=0}^{k-1} A^i Bu(k-1-i). \tag{9.8.4}$$

Prove that (9.8.3) holds for $k + 1$.

By (9.8.2) and (9.8.4), it follows that

$$\begin{aligned} E[x(k+1)] &= AE[x(k)] + Bu(k) \\ &= A\left(A^k x_0 + \sum_{i=0}^{k-1} A^i Bu(k-1-i)\right) + Bu(k) \\ &= A^{k+1} x_0 + \sum_{i=0}^{k-1} A^{i+1} Bu(k-1-i) + Bu(k). \end{aligned}$$

Let $j = 1 + i$. Note that $Bu(k) = A^0 Bu(k-0)$ since $A^0 = 1$ and $B(k) = B(k-0)$. Then

$$E[x(k+1)] = A^{k+1} x_0 + \sum_{j=1}^{k} A^j Bu(k-j) + Bu(k) = A^{k+1} x_0 + \sum_{j=0}^{k} A^j Bu(k-j),$$

i.e., (9.8.3) holds for $k + 1$.

Finally, by (9.8.2) and (9.8.3), a direct computation deduces that

$$E[y(k)] = CE[x(k)] = CA^k x_0 + \sum_{i=0}^{k-1} CA^i Bu(k-1-i). \qquad \square$$

Proposition 9.8.2. *Under the above assumption, the variances of $y(k)$ satisfy that*

$$\mathrm{Var}(y(k)) = C^2 A^{2k} X_0 + \sum_{i=0}^{k-1} C^2 A^{2i} P + Q.$$

Proof. It is clear by (9.8.1) that

$$\mathrm{Var}(y(k)) = \mathrm{Var}(Cx(k)) + \mathrm{Var}(\upsilon(k)).$$

A direct computation shows that

$$\mathrm{Var}(Cx(k)) = E[C^2 x^2(k)] - (E[Cx(k)])^2 = C^2\left(E[x^2(k)] - (E[x(k)])^2\right) = C^2\,\mathrm{Var}(x(k)).$$

By $E[v^2(k)] = Q$ and $E[v(k)] = 0$, it follows that $\mathrm{Var}(v(k)) = E(v^2(k)) - (E[v(k)])^2 = Q$. Thus,

$$\mathrm{Var}(y(k)) = C^2 \mathrm{Var}(x(k)) + Q. \tag{9.8.5}$$

Similarly, by (9.8.1), it is clear that

$$\mathrm{Var}(x(k+1)) = \mathrm{Var}(Ax(k)) + \mathrm{Var}(Bu(k)) + \mathrm{Var}(w(k)).$$

A direct computation shows that

$$\mathrm{Var}(Ax(k)) = E[A^2x^2(k)] - (E[Ax(k)])^2 = A^2(E[x^2(k)] - (E[x(k)])^2) = A^2 \mathrm{Var}(x(k)).$$

Since B and $u(k)$ are constants,

$$\mathrm{Var}(Bu(k)) = E[B^2u^2(k)] - (E[Bu(k)])^2 = B^2u^2(k) - (Bu(k))^2 = 0.$$

By $E[w(k)] = 0$ and $E[w^2(k)] = P$, it follows that $\mathrm{Var}(w(k)) = E[w^2(k)] - (E[w(k)])^2 = P$. Thus,

$$\mathrm{Var}(x(k+1)) = A^2 \mathrm{Var}(x(k)) + P. \tag{9.8.6}$$

Now we prove by induction that the variance of $x(k)$ satisfies that

$$\mathrm{Var}(x(k)) = A^{2k}X_0 + \sum_{i=0}^{k-1} A^{2i}P. \tag{9.8.7}$$

In the case $k = 0$. From (9.8.6), it follows that

$$\mathrm{Var}(x(1)) = A^2 \mathrm{Var}(x(0)) + P = A^2\left(E[x^2(0)] - (E[x(0)])^2\right) + P.$$

By $E[(x(0) - x_0)^2] = X_0$ and $E[x(0)] = x_0$, and $E[x_0^2] = x_0^2$, it follows that

$$\begin{aligned} E[x^2(0)] &= E[((x(0) - x_0)^2 + 2x(0)x_0 - x_0^2)] \\ &= E[(x(0) - x_0)^2] + 2x_0E[x(0)] - E[x_0^2] = X_0 + x_0^2, \\ (E[x(0)])^2 &= x_0^2. \end{aligned}$$

Thus,

$$\mathrm{Var}(x(1)) = A^2(X_0 + x_0^2 - x_0^2) + P = A^2X_0 + P,$$

i.e., (9.8.7) holds for $k = 0$.

Assume for k that

$$\mathrm{Var}(x(k)) = A^{2k}X_0 + \sum_{i=0}^{k-1} A^{2i}P. \tag{9.8.8}$$

Prove that (9.8.7) holds for $k + 1$.

By (9.8.6) and (9.8.8), it follows that

$$\begin{aligned} \mathrm{Var}(x(k+1)) &= A^2 \mathrm{Var}(x(k)) + P \\ &= A^2\left(A^{2k}X_0 + \sum_{i=0}^{k-1} A^{2i}P\right) + P \\ &= A^{2(k+1)}X_0 + \sum_{i=0}^{k-1} A^{2(i+1)}P + P. \end{aligned}$$

Let $j = 1 + i$. Then

$$\mathrm{Var}(x(k+1)) = A^{2(k+1)}X_0 + \sum_{j=1}^{k} A^{2j}P + P = A^{2(k+1)}X_0 + \sum_{j=0}^{k} A^{2j}P,$$

i.e., (9.8.7) holds for $k+1$.

Finally, by (9.8.5) and (9.8.8), a direct computation shows that

$$\mathrm{Var}(y(k)) = C^2A^{2k}X_0 + \sum_{i=0}^{k-1} C^2A^{2i}P + Q. \qquad \square$$

From Propositions 9.8.1 and 9.8.2, it is seen that $E[y(k)]$ is dependent on k and u, while $\mathrm{Var}(y(k))$ is only dependent on k. Thus, $E[y(k)]$ and $\mathrm{Var}(y(k))$ may be denoted by $\mu_y(k,u)$ and $\sigma_y^2(k)$, respectively, i.e.,

$$\mu_y(k,u) := E[y(k)],$$
$$\sigma_y^2(k) := \mathrm{Var}(y(k)).$$

Let $y(k)$ be the damage and its pdf $p(y)$ be the normal distribution

$$p(y) := p(y(k)) = \frac{1}{\sqrt{2\pi}\,\sigma_y(k)} \exp\left(-\frac{1}{2}\left(\frac{y(k)-\mu_y(k,u)}{\sigma_y(k)}\right)^2\right),$$

where $\sigma_y(k)$ is the positive square root of $\mathrm{Var}(y(k))$. Correspondingly,

- $f_1^k(u)$ represents the cost objective function of $y(k)$ at stage k;
- $f_j^k(u)$ $(j = 2,\dots,N_k+2)$ represents the j-th conditional expected risk function of $y(k)$ at stage k and

$$f_j^k(u) = \frac{\int_{\beta_{j-2}^k}^{\beta_{j-1}^k} y(k)p(y(k))\,\mathrm{d}y(k)}{\int_{\beta_{j-2}^k}^{\beta_{j-1}^k} p(y(k))\,\mathrm{d}y(k)} = \frac{\int_{\beta_{j-2}^k}^{\beta_{j-1}^k} \frac{y(k)}{\sqrt{2\pi}\sigma_y(k)} \exp\left(-\frac{1}{2}\left(\frac{y(k)-\mu_y(k,u)}{\sigma_y(k)}\right)^2\right)\mathrm{d}y(k)}{\int_{\beta_{j-2}^k}^{\beta_{j-1}^k} \frac{1}{\sqrt{2\pi}\sigma_y(k)} \exp\left(-\frac{1}{2}\left(\frac{y(k)-\mu_y(k,u)}{\sigma_y(k)}\right)^2\right)\mathrm{d}y(k)}; \tag{9.8.9}$$

- $f_{N_k+3}^k(u)$ represents the unconditional excepted risk function of $y(k)$ at stage k and

$$f_{N_k+3}^k(u) = \frac{\int_{\beta_0^k}^{\beta_{N_k+1}^k} y(k)p(y(k))\,\mathrm{d}y(k)}{\int_{\beta_0^k}^{\beta_{N_k+1}^k} p(y(k))\,\mathrm{d}y(k)} = \frac{\int_{\beta_0^k}^{\beta_{N_k+1}^k} \frac{y(k)}{\sqrt{2\pi}\sigma_y(k)} \exp\left(-\frac{1}{2}\left(\frac{y(k)-\mu_y(k,u)}{\sigma_y(k)}\right)^2\right)\mathrm{d}y(k)}{\int_{\beta_0^k}^{\beta_{N_k+1}^k} \frac{1}{\sqrt{2\pi}\sigma_y(k)} \exp\left(-\frac{1}{2}\left(\frac{y(k)-\mu_y(k,u)}{\sigma_y(k)}\right)^2\right)\mathrm{d}y(k)}, \tag{9.8.10}$$

where N_k is the number of partitions for the probability exceedance axis at stage k, $\beta_0^k = -\infty$ and $\beta_{N_k+1}^k = \infty$ are the lower and upper bounds of the damage at stage k, respectively, and β_j^k $(j = 1,\dots,N_k)$ are the partitioned points of the damage at stage k. Denote the dominator of (9.8.9) by q_j^k, i.e.,

$$q_j^k = \int_{\beta_{j-2}^k}^{\beta_{j-1}^k} \frac{1}{\sqrt{2\pi}\,\sigma_y(k)} \exp\left(-\frac{1}{2}\left(\frac{y(k)-\mu_y(k,u)}{\sigma_y(k)}\right)^2\right)\mathrm{d}y(k) \quad (j = 2,\dots,N_k+2).$$

By (9.8.9) and (9.8.10), it follows that

$$f_{N_k+3}^k(u) = q_2^k f_2^k(u) + \cdots + q_{N_k+2}^k f_{N_k+2}^k(u) = \sum_{j=2}^{N_k+2} q_j^k f_j^k(u),$$

where $q_j^k > 0$ and $q_2^k + \cdots + q_{N_k+2}^k = \sum_{j=2}^{N_k+2} q_j^k = 1$.

The $f_j^k(u)$ $(j = 2, \ldots, N_k + 2)$ is computed as follows.

Note that $y(k) = \mu_y(k,u) + (y(k) - \mu_y(k,u))$. The numerator of (9.8.9) becomes

$$\begin{aligned}&\int_{\beta_{j-2}^k}^{\beta_{j-1}^k} \frac{y(k)}{\sqrt{2\pi}\,\sigma_y(k)} \exp\left(-\frac{1}{2}\left(\frac{y(k)-\mu_y(k,u)}{\sigma_y(k)}\right)^2\right) \mathrm{d}y(k)\\ &\quad = \int_{\beta_{j-2}^k}^{\beta_{j-1}^k} \frac{\mu_y(k,u)}{\sqrt{2\pi}\,\sigma_y(k)} \exp\left(-\frac{1}{2}\left(\frac{y(k)-\mu_y(k,u)}{\sigma_y(k)}\right)^2\right) \mathrm{d}y(k)\\ &\qquad + \int_{\beta_{j-2}^k}^{\beta_{j-1}^k} \frac{y(k)-\mu_y(k,u)}{\sqrt{2\pi}\,\sigma_y(k)} \exp\left(-\frac{1}{2}\left(\frac{y(k)-\mu_y(k,u)}{\sigma_y(k)}\right)^2\right) \mathrm{d}y(k).\end{aligned}$$

Let $\tau = \frac{y(k)-\mu_y(k,u)}{\sigma_y(k)}$ and $\mathrm{d}\tau = \frac{\mathrm{d}y(k)}{\sigma_y(k)}$. Then

$$\int_{\beta_{j-2}^k}^{\beta_{j-1}^k} \frac{\mu_y(k,u)}{\sqrt{2\pi}\,\sigma_y(k)} \exp\left(-\frac{1}{2}\left(\frac{y(k)-\mu_y(k,u)}{\sigma_y(k)}\right)^2\right) \mathrm{d}y(k) = \mu_y(k,u) \int_{\alpha_{j-2}^k}^{\alpha_{j-1}^k} \frac{1}{\sqrt{2\pi}} \mathrm{e}^{-\frac{\tau^2}{2}} \mathrm{d}\tau,$$

$$\int_{\beta_{j-2}^k}^{\beta_{j-1}^k} \frac{(y(k)-\mu_y(k,u))}{\sqrt{2\pi}\,\sigma_y(k)} \exp\left(-\frac{1}{2}\left(\frac{y(k)-\mu_y(k,u)}{\sigma_y(k)}\right)^2\right) \mathrm{d}y(k) = \sigma_y(k) \int_{\alpha_{j-2}^k}^{\alpha_{j-1}^k} \frac{\tau}{\sqrt{2\pi}} \mathrm{e}^{-\frac{\tau^2}{2}} \mathrm{d}\tau,$$

where

$$\alpha_{j-i}^k = \frac{\beta_{j-i}^k - \mu_y(k,u)}{\sigma_y(k)} \quad (j = 2, \ldots, N_k + 2;\ i = 1, 2), \tag{9.8.11}$$

and so the numerator of (9.8.9) becomes

$$\begin{aligned}&\int_{\beta_{j-2}^k}^{\beta_{j-1}^k} \frac{y(k)}{\sqrt{2\pi}\,\sigma_y(k)} \exp\left(-\frac{1}{2}\left(\frac{y(k)-\mu_y(k,u)}{\sigma_y(k)}\right)^2\right) \mathrm{d}y(k)\\ &\quad = \mu_y(k,u) \int_{\alpha_{j-2}^k}^{\alpha_{j-1}^k} \frac{1}{\sqrt{2\pi}} \mathrm{e}^{-\frac{\tau^2}{2}} \mathrm{d}\tau + \sigma_y(k) \int_{\alpha_{j-2}^k}^{\alpha_{j-1}^k} \frac{\tau}{\sqrt{2\pi}} \mathrm{e}^{-\frac{\tau^2}{2}} \mathrm{d}\tau.\end{aligned}$$

Similarly, the denominator of (9.8.9) becomes

$$\int_{\beta_{j-2}^k}^{\beta_{j-1}^k} \frac{1}{\sqrt{2\pi}\,\sigma_y(k)} \exp\left(-\frac{1}{2}\left(\frac{y(k)-\mu_y(k,u)}{\sigma_y(k)}\right)^2\right) \mathrm{d}y(k) = \int_{\alpha_{j-2}^k}^{\alpha_{j-1}^k} \frac{1}{\sqrt{2\pi}} \mathrm{e}^{-\frac{\tau^2}{2}} \mathrm{d}\tau.$$

Therefore, the j-th conditional expected risk function of the damage at stage k becomes

$$f_j^k(u) = \mu_y(k,u) + \sigma_y(k) b_j^k \quad (j = 2, \ldots, N_k + 2), \tag{9.8.12}$$

where

$$b_j^k = \frac{\int_{\alpha_{j-2}^k}^{\alpha_{j-1}^k} \tau e^{-\frac{\tau^2}{2}} \mathrm{d}\tau}{\int_{\alpha_{j-2}^k}^{\alpha_{j-1}^k} e^{-\frac{\tau^2}{2}} \mathrm{d}\tau} \quad (j = 2, \ldots, N_k + 2) \tag{9.8.13}$$

and α_{j-i}^k is stated in (9.8.11).

Similarly, the $f_{N_k+3}^k(u)$ is computed as follows:

$$f_{N_k+3}^k(u) = \frac{\mu_y(k,u) \int_{\beta_0^k}^{\beta_{N_k+1}^k} \frac{1}{\sqrt{2\pi}} e^{-\frac{\tau^2}{2}} \mathrm{d}\tau + \sigma_y(k) \int_{\beta_0^k}^{\beta_{N_k+1}^k} \frac{\tau}{\sqrt{2\pi}} e^{-\frac{\tau^2}{2}} \mathrm{d}\tau}{\int_{\beta_0^k}^{\beta_{N_k+1}^k} \frac{1}{\sqrt{2\pi}} e^{-\frac{\tau^2}{2}} \mathrm{d}\tau}.$$

Note that $\beta_0^k = -\infty$ and $\beta_{N_k+1}^k = \infty$. Since $e^{-\frac{\tau^2}{2}}$ is an even function and $\tau e^{-\frac{\tau^2}{2}}$ is an odd function,

$$\int_{-\infty}^{\infty} \frac{1}{\sqrt{2\pi}} e^{-\frac{\tau^2}{2}} \mathrm{d}\tau = 1,$$

$$\int_{-\infty}^{\infty} \frac{\tau}{\sqrt{2\pi}} e^{-\frac{\tau^2}{2}} \mathrm{d}\tau = 0,$$

and so

$$f_{N_k+3}^k(u) = \frac{\mu_y(k,u) \int_{-\infty}^{\infty} \frac{1}{\sqrt{2\pi}} e^{-\frac{\tau^2}{2}} \mathrm{d}\tau + \sigma_y(k) \int_{-\infty}^{\infty} \frac{\tau}{\sqrt{2\pi}} e^{-\frac{\tau^2}{2}} \mathrm{d}\tau}{\int_{-\infty}^{\infty} \frac{1}{\sqrt{2\pi}} e^{-\frac{\tau^2}{2}} \mathrm{d}\tau} = \mu_y(k,u), \tag{9.8.14}$$

i.e., the unconditional excepted risk function of the damage at stage k is equal to the mean of the damage.

Note that $\sigma_y(k)$ is independent of u and b_j^k $(j = 2, \ldots, N_k + 2)$ is constant. Then

$$\begin{aligned} \min_u\{f_j^k(u)\} &= \min_u\{\mu_y(k,u) + b_j^k \sigma_y(k)\} \\ &= b_j^k \sigma_y(k) + \min_u\{\mu_y(k,u)\} \quad (j = 2, \ldots, N_k + 2), \end{aligned}$$

i.e., the minimizing of any one of the conditional excepted risk functions is reduced to minimizing the mean of the damage.

Combining the cost objective function with any one of the risk functions (the conditional or the unconditional) constitutes a set of biobjective optimization problems at stage k, i.e.,

$$\begin{cases} \min\{f_1^k(u), f_2^k(u)\}, \\ \quad \vdots \\ \min\{f_1^k(u), f_{N_k+2}^k(u)\}, \\ \min\{f_1^k(u), f_{N_k+3}^k(u)\}. \end{cases}$$

These biobjective optimization problems can be solved using the ε-constraint method. The trade-off functions λ_{1j}^k between the cost objective function $f_1^k(u)$ and the risk function $f_j^k(u)$ both at stage k are given by

$$\lambda_{1j}^k = -\frac{\partial f_1^k}{\partial f_j^k} \quad (j = 2, \ldots, N_k + 3).$$

For $j = 2, \ldots, N_k + 2$, since $\sigma_y(k)$ is independent of u and b_j^k is constant, by (9.8.12), it follows that

$$\partial f_j^k(u) = \partial\left(\mu_y(k,u) + \sigma_y(k)\, b_j^k\right) = \partial \mu_y(k,u),$$

and for $j = N_k + 3$, by (9.8.14), it is clear that

$$\partial f_{N_k+3}^k(u) = \partial \mu_y(k,u).$$

Thus, the trade-off functions λ_{1j}^k between the cost function $f_1^k(u)$ and the risk function $f_j^k(u)$ at stage k are

$$\lambda_{1j}^k = -\frac{\partial f_1^k}{\partial \mu_y} \quad (j = 2, \ldots, N_k + 3),$$

i.e., all trade-offs for the given stage k are equal. Therefore, for the normal distribution, the risk functions of the damage at stage k are parallel curves. This greatly simplifies the multiobjective optimization.

Example 9.8.3. Let $N_k = 2$ and $\alpha_0^k = -\infty$, $\alpha_1^k = -1$, $\alpha_2^k = 1$, $\alpha_3^k = \infty$ in (9.8.11). Denote

$$\Phi(x) := \frac{1}{\sqrt{2\pi}} \int_{-\infty}^{x} \mathrm{e}^{-\frac{\tau^2}{2}}\, \mathrm{d}\tau.$$

By (9.8.13),

$$\begin{aligned}
b_2^k &= \frac{\int_{-\infty}^{-1} \tau \mathrm{e}^{-\frac{\tau^2}{2}}\, \mathrm{d}\tau}{\sqrt{2\pi}\, \Phi(-1)} = -\frac{\mathrm{e}^{-\frac{1}{2}}}{\sqrt{2\pi}\, \Phi(-1)} = -\frac{0.24197}{0.158655} = -1.52514, \\
b_3^k &= \frac{\int_{-1}^{1} \tau \mathrm{e}^{-\frac{\tau^2}{2}}\, \mathrm{d}\tau}{\sqrt{2\pi}\, (\Phi(1) - \Phi(-1))} = 0, \\
b_4^k &= \frac{\int_{1}^{\infty} \tau \mathrm{e}^{-\frac{\tau^2}{2}}\, \mathrm{d}\tau}{\sqrt{2\pi}(1 - \Phi(1))} = \frac{\mathrm{e}^{-\frac{1}{2}}}{\sqrt{2\pi}\Phi(-1)} = \frac{0.24197}{0.158655} = 1.52514.
\end{aligned}$$

Then, by (9.8.12), the conditional expected risk functions at stage k are

$$\begin{aligned}
f_2^k(u) &= \mu_y(k,u) + \sigma_y(k) b_2^k = \mu_y(k,u) - 1.52514\, \sigma_y(k), \\
f_3^k(u) &= \mu_y(k,u) + \sigma_y(k) b_3^k = \mu_y(k,u), \\
f_4^k(u) &= \mu_y(k,u) + \sigma_y(k) b_4^k = \mu_y(k,u) + 1.52514\, \sigma_y(k),
\end{aligned} \tag{9.8.15}$$

where $\sigma_y(k)$ is the positive square root of $\sigma_y^2(k)$. By Propositions 9.8.1 and 9.8.2, $\mu_y(k,u)$ and $\sigma_y^2(k)$ satisfy the following:

$$\begin{aligned} \mu_y(k,u) &= CA^k x_0 + \sum_{i=0}^{k-1} CA^i Bu(k-1-i), \\ \sigma_y^2(k) &= C^2 A^{2k} X_0 + \sum_{i=0}^{k-1} C^2 A^{2i} P + R. \end{aligned} \tag{9.8.16}$$

Assume that for one scenario,

$$\begin{aligned} &A = 1.02, \quad B = -0.01, \quad C = 1, \quad P = 200, \quad R = 0, \\ &x_0 = 100, \quad X_0 = 100, \quad u(0) = 200, \quad u(1) = 150. \end{aligned}$$

We will find the values of conditional expected risk functions at stage k $(k = 1, 2)$. For $k = 1$, by (9.8.16),

$$\begin{aligned} \mu_y(1,u) &= (1.02)^1(100) - (1.02)^0(0.01)(200) = 100, \\ \sigma_y^2(1) &= (1.02)^2(100) + (1.02)^0(200) = 304.04, \\ \sigma_y(1) &= \sqrt{304.04} = 17.4. \end{aligned}$$

From this and (9.8.15), we get

$$\begin{aligned} f_2^1(u) &= 100 - (1.52514)(17.4) = 73.46 \\ f_3^1(u) &= 100, \\ f_4^1(u) &= 100 + (1.52514)(17.4) = 126.54. \end{aligned}$$

For $k = 2$, by (9.8.16),

$$\begin{aligned} \mu_y(2,u) &= (1.02)^2(100) - (1.02)^0(0.01)(150) - (1.02)(0.01)(200) = 100.5, \\ \sigma_y^2(2) &= (1.02)^4(100) + (1.02)^0(200) + (1.02)^2(200) = 516.32, \\ \sigma_y(2) &= \sqrt{516.32} = 22.7. \end{aligned}$$

From this and (9.8.15), we get

$$\begin{aligned} f_2^2(u) &= 100.5 - (1.52514)(22.7) = 65.88 \\ f_3^2(u) &= 100.5, \\ f_4^2(u) &= 100.5 + (1.52514)(22.7) = 135.12. \end{aligned}$$

These values are summarized in the following table.

Scenario	$k = 1$	$k = 2$
f_2^k	73.46	65.88
f_3^k	100	100.5
f_4^k	126.54	135.12

9.9 The Leslie model

In the simplest *population dynamics model* of a state variable, let

- $p(t)$ represent the level of population at time t $(t = 0, 1, \dots)$;
- B and D be the numbers of births and deaths in any one year, respectively;
- $b(t)$ and $d(t)$ represent birth and death rates for the time between t and $t+1$ $(t = 0, 1, \dots)$, respectively.

Clearly,

$$B = b(t)p(t),$$
$$D = d(t)p(t).$$

Assume that the population level $p(0)$ is known, and that the birth and death rates do not change with time, i.e., $b(t) = b$ and $d(t) = d$. Then population growth from t to $t+1$ is balanced as follows:

$$p(t+1) = p(t) + B - D = p(t)r, \tag{9.9.1}$$

where $r = 1 + b - d$ is the overall growth rate. Meyer calls it the *Malthusian parameter*.

For any t $(t \in \mathbb{Z}_+)$, it follows from (9.9.1) that

$$p(t) = p(t-1)r = p(t-2)r^2 = \cdots = p(0)r^t, \tag{9.9.2}$$

i.e., the growth rate of the population can be represented by an exponential function, thus the simplest population dynamics model of a state variable is called an *exponential model.*

Example 9.9.1. Assume that the current worker population in a small factory is 70 workers, and that the rate of increasing due to new hiring has been 0.04 and the rate of leaving the factory has been 0.02. How many workers will be at that factory in 5 years? How many years will it take for the number of workers to double?

Solution. By the assumption, $p(0) = 70$, $b = 0.04$, $d = 0.02$, and $r = 1 + b - d = 1.02$. By (9.9.2),

$$p(5) = p(0)r^5 = (70)(1.02)^5 \approx (70)(1.104) \approx 77.$$

Thus the number of workers in 5 years is about 77. Let t be the number of years until doubling. Then $p(t) = 2p(0)$. From this and (9.9.2),

$$p(0)r^t = 2p(0) \quad \text{or} \quad 2 = (1.02)^t.$$

The solution is $t = \ln 2/\ln(1.02) \approx 35$, i.e., the number of years of doubling is about 35. □

The simplified version of the *Leslie model* adapted by Meyer is as follows. In the Leslie model, only the female population is considered and the female population is divided into n age categories

$$[0, \Delta), \quad [\Delta, 2\Delta), \quad \dots, \quad [(n-1)\Delta, n\Delta),$$

where Δ is the width of each age interval of the population, and define

- $F_i(t)$ is the number of females in the age interval $[i\Delta, (i+1)\Delta)$,
- m_i is the Δ-year maternity rate for the age interval $[i\Delta, (i+1)\Delta)$,
- p_i is the survival rate in the age interval $[i\Delta, (i+1)\Delta)$,

  where $i = 0, 1, \dots, n-1$.

The female population of the i-th age group at the next $t + \Delta$ period is given by

$$F_{i+1}(t+\Delta) = p_i F_i(t) \quad (t = 0, \Delta, 2\Delta, \dots).$$

The number of newborns at the lowest age group at time $(t + \Delta)$ is given by

$$F_0(t+\Delta) = \sum_0^{n-1} m_i F_i(t).$$

The combination of these two equations gives

$$\begin{pmatrix} F_0(t+\Delta) \\ F_1(t+\Delta) \\ F_2(t+\Delta) \\ \vdots \\ F_{n-1}(t+\Delta) \end{pmatrix} = \begin{pmatrix} m_0 & m_1 & m_2 & \cdots & \cdots & m_{n-1} \\ p_0 & 0 & 0 & \cdots & \cdots & 0 \\ 0 & p_1 & 0 & \cdots & \cdots & 0 \\ \vdots & \vdots & \vdots & \cdots & \cdots & \vdots \\ 0 & 0 & 0 & \cdots & p_{n-2} & 0 \end{pmatrix} \begin{pmatrix} F_0(t) \\ F_1(t) \\ F_2(t) \\ \vdots \\ F_{n-1}(t) \end{pmatrix}.$$

Its matrix form is

$$\boldsymbol{F}(t+\Delta) = M\boldsymbol{F}(t), \tag{9.9.3}$$

where

$$\boldsymbol{F}(t) = \begin{pmatrix} F_0(t) \\ F_1(t) \\ F_2(t) \\ \vdots \\ F_{n-1}(t) \end{pmatrix} \qquad M = \begin{pmatrix} m_0 & m_1 & m_2 & \cdots & \cdots & m_{n-1} \\ p_0 & 0 & 0 & \cdots & \cdots & 0 \\ 0 & p_1 & 0 & \cdots & \cdots & 0 \\ \vdots & \vdots & \vdots & \cdots & \cdots & \vdots \\ 0 & 0 & 0 & \cdots & p_{n-2} & 0 \end{pmatrix}.$$

The vector $\boldsymbol{F}$ is called the *age distribution vector* and the $n \times n$ matrix M is called the *Leslie matrix*.

For any k ($k \in \mathbb{Z}_+$), it follows from (9.9.3) that

$$\boldsymbol{F}(k\Delta) = M\boldsymbol{F}((k-1)\Delta) = \cdots = M^k \boldsymbol{F}(0) \quad (k \in \mathbb{Z}_+).$$

Example 9.9.2. Based on an observation, the population, the maternity rate, and the survival rate of female birds are given by the following:

$$F_0(0) = 100, \quad F_1(0) = 60, \quad F_2(0) = 40,$$
$$m_0 = 0, \quad m_1 = 3, \quad m_2 = 1,$$
$$p_0 = \frac{1}{2}, \quad p_1 = \frac{3}{4}, \quad p_2 = 0.$$

Find the age distribution vectors $\boldsymbol{F}(\Delta)$ and $\boldsymbol{F}(2\Delta)$.

Solution. The initial vector and the Leslie matrix are, respectively,

$$\boldsymbol{F}(0) = \begin{pmatrix} F_0(0) \\ F_1(0) \\ F_2(0) \end{pmatrix} = \begin{pmatrix} 100 \\ 60 \\ 40 \end{pmatrix}, \quad M = \begin{pmatrix} 0 & 3 & 1 \\ \frac{1}{2} & 0 & 0 \\ 0 & \frac{3}{4} & 0 \end{pmatrix}.$$

To find $\boldsymbol{F}(\Delta)$, by $\boldsymbol{F}(\Delta) = M\boldsymbol{F}(0)$, we get

$$\boldsymbol{F}(\Delta) = M \begin{pmatrix} F_0(0) \\ F_1(0) \\ F_2(0) \end{pmatrix} = \begin{pmatrix} 0 & 3 & 1 \\ \frac{1}{2} & 0 & 0 \\ 0 & \frac{3}{4} & 0 \end{pmatrix} \begin{pmatrix} 100 \\ 60 \\ 40 \end{pmatrix} = \begin{pmatrix} 220 \\ 50 \\ 45 \end{pmatrix},$$

To find $\boldsymbol{F}(2\Delta)$, by $\boldsymbol{F}(2\Delta) = M\boldsymbol{F}(\Delta)$, we get

$$\boldsymbol{F}(2\Delta) = \begin{pmatrix} 0 & 3 & 1 \\ \frac{1}{2} & 0 & 0 \\ 0 & \frac{3}{4} & 0 \end{pmatrix} \begin{pmatrix} 220 \\ 50 \\ 45 \end{pmatrix} = \begin{pmatrix} 195 \\ 110 \\ 37.5 \end{pmatrix}$$

or by $\boldsymbol{F}(2\Delta) = M^2\boldsymbol{F}(0)$, noticing that

$$M^2 = \begin{pmatrix} 0 & 3 & 1 \\ \frac{1}{2} & 0 & 0 \\ 0 & \frac{3}{4} & 0 \end{pmatrix}^2 = \begin{pmatrix} \frac{3}{2} & \frac{3}{4} & 0 \\ 0 & \frac{3}{2} & \frac{1}{2} \\ \frac{3}{8} & 0 & 0 \end{pmatrix},$$

we get

$$\boldsymbol{F}(2\Delta) = \begin{pmatrix} \frac{3}{2} & \frac{3}{4} & 0 \\ 0 & \frac{3}{2} & \frac{1}{2} \\ \frac{3}{8} & 0 & 0 \end{pmatrix} \begin{pmatrix} 100 \\ 60 \\ 40 \end{pmatrix} = \begin{pmatrix} 195 \\ 110 \\ 37.5 \end{pmatrix}.$$ □

Example 9.9.3. Suppose that the numbers of females, the maternity rates, and the survival rates in the age intervals are given as follows:

$$F_0(0) = 50, \quad F_1(0) = 30, \quad F_2(0) = 20, \quad F_3(0) = 10,$$
$$m_0 = 0, \quad m_1 = 1, \quad m_2 = 2, \quad m_3 = 3$$
$$p_0 = \frac{3}{4}, \quad p_1 = \frac{1}{2}, \quad p_2 = \frac{1}{2}.$$

Find the age distribution vectors $\boldsymbol{F}(\Delta)$ and $\boldsymbol{F}(2\Delta)$.

Solution. The initial vector and the Leslie matrix are, respectively,

$$\boldsymbol{F}(0) = \begin{pmatrix} F_0(0) \\ F_1(0) \\ F_2(0) \\ F_3(0) \end{pmatrix} = \begin{pmatrix} 50 \\ 30 \\ 20 \\ 10 \end{pmatrix}, \quad M = \begin{pmatrix} 0 & 1 & 2 & 3 \\ \frac{3}{4} & 0 & 0 & 0 \\ 0 & \frac{1}{2} & 0 & 0 \\ 0 & 0 & \frac{1}{2} & 0 \end{pmatrix}.$$

To find $\boldsymbol{F}(\Delta)$, by $\boldsymbol{F}(\Delta) = M\boldsymbol{F}(0)$, we get

$$\boldsymbol{F}(\Delta) = M \begin{pmatrix} F_0(0) \\ F_1(0) \\ F_2(0) \\ F_3(0) \end{pmatrix} = \begin{pmatrix} 0 & 1 & 2 & 3 \\ \frac{3}{4} & 0 & 0 & 0 \\ 0 & \frac{1}{2} & 0 & 0 \\ 0 & 0 & \frac{1}{2} & 0 \end{pmatrix} \begin{pmatrix} 50 \\ 30 \\ 20 \\ 10 \end{pmatrix} = \begin{pmatrix} 100 \\ 37.5 \\ 15 \\ 10 \end{pmatrix},$$

To find $\boldsymbol{F}(2\Delta)$, by $\boldsymbol{F}(2\Delta) = M\boldsymbol{F}(\Delta)$, we get

$$\boldsymbol{F}(2\Delta) = \begin{pmatrix} 0 & 1 & 2 & 3 \\ \frac{3}{4} & 0 & 0 & 0 \\ 0 & \frac{1}{2} & 0 & 0 \\ 0 & 0 & \frac{1}{2} & 0 \end{pmatrix} \begin{pmatrix} 100 \\ 37.5 \\ 15 \\ 10 \end{pmatrix} = \begin{pmatrix} 97.5 \\ 75 \\ 18.75 \\ 7.5 \end{pmatrix},$$

or by $\boldsymbol{F}(2\Delta) = M^2\boldsymbol{F}(0)$, noticing that

$$M^2 = \begin{pmatrix} 0 & 1 & 2 & 3 \\ \frac{3}{4} & 0 & 0 & 0 \\ 0 & \frac{1}{2} & 0 & 0 \\ 0 & 0 & \frac{1}{2} & 0 \end{pmatrix}^2 = \begin{pmatrix} \frac{3}{4} & 1 & \frac{3}{2} & 0 \\ 0 & \frac{3}{4} & \frac{3}{2} & \frac{9}{4} \\ \frac{3}{8} & 0 & 0 & 0 \\ 0 & \frac{1}{4} & 0 & 0 \end{pmatrix},$$

we get

$$\boldsymbol{F}(2\Delta) = \begin{pmatrix} \frac{3}{4} & 1 & \frac{3}{2} & 0 \\ 0 & \frac{3}{4} & \frac{3}{2} & \frac{9}{4} \\ \frac{3}{8} & 0 & 0 & 0 \\ 0 & \frac{1}{4} & 0 & 0 \end{pmatrix} \begin{pmatrix} 50 \\ 30 \\ 20 \\ 10 \end{pmatrix} = \begin{pmatrix} 97.5 \\ 75 \\ 18.75 \\ 7.5 \end{pmatrix}. \qquad \square$$

9.10 Leontief's and inoperability input-output models

Leontief's input-output model studies the equilibrium state of an economy consisting of a number of individual economic sectors. The *inoperability input-output model* presented by Haimes and Jiang studies the equilibrium state of a system consisting of n critical complex intra- and interconnected infrastructures. Although the equations of these two models are the same, they connote different meanings.

In Leontief's input-output model, define the following notations:

- x_i is the output (for the total economy) of the i-th goods, $i = 1, \ldots, n$;
- r_k is the input (for the total economy) of the k-th resource, $k = 1, \ldots, m$;

- x_{ij} is the amount of the i-th goods used in the production of the j-th goods;
- r_{kj} is the amount of the k-th resource input used in the production of the j-th goods;
- c_i is the i-th input for the production of other commodities, $i = 1, \dots, n$;

Assume that the inputs of both goods and resources required to produce any commodity are proportional to the output of that commodity, i.e.,

$$x_{ij} = a_{ij}x_j \quad (i, j = 1, \dots, n),$$
$$r_{kj} = b_{kj}x_j \quad (k = 1, \dots, m;\ j = 1, \dots, n).$$

These two equations are called *Leontief's proportionality equations* of goods and resource, respectively. The equation

$$x_i = \sum_{j=1}^{n} x_{ij} + c_i \quad (i = 1, \dots, n).$$

is called the *Leontief-based equation*. The combination of the proportionality equation of goods with the Leontief-based equation gives the following *Leontief equation*:

$$x_i = \sum_{j=1}^{n} a_{ij}x_j + c_i \quad (i = 1, \dots, n)$$

which is written in a matrix form $\boldsymbol{x} = A\boldsymbol{x} + \boldsymbol{c}$, where $\boldsymbol{x} = (x_1, \dots, x_n)^{\mathrm{T}}$ and $\boldsymbol{c} = (c_1, \dots, c_n)^{\mathrm{T}}$, and

$$A = \begin{pmatrix} a_{11} & \cdots & a_{1n} \\ \vdots & \ddots & \vdots \\ a_{n1} & \cdots & a_{nn} \end{pmatrix}.$$

Similarly, applying the proportionality assumption of the resources gives

$$\sum_{j=1}^{n} r_{kj} = \sum_{j=1}^{n} b_{kj}x_j \quad (k = 1, \dots, m).$$

Since the demand for the k-th resource cannot exceed its supply, then

$$\sum_{j=1}^{n} b_{kj}x_j \le r_k, \quad r_k \ge 0 \quad (k = 1, \dots, m).$$

In the Inoperability Input-Output Model (IIM), define the following notations:

- x_i is the overall risk of inoperability of the i-th intra- and interconnected infrastructure that can be triggered by one or multiple failures caused by accidents or acts of terrorism, $i = 1, \dots, n$;
- x_{ij} is the degree of inoperability triggered by one or multiple failures that the j-th infrastructure can contribute to the i-th infrastructure due to their complex intra- and interconnectedness, $i, j = 1, \dots, n$;

- a_{ij} is the probability of inoperability that the j-th infrastructure contributes to the i-th infrastructure;
- c_i is the natural or man-made perturbation into the i-th critical infrastructure, $i = 1, \dots, n$;

and assume that

$$x_{ij} = a_{ij} x_j \quad (i, j = 1, \dots, n).$$

This equation is called the *proportionality equation* of IIM. The equation

$$x_i = \sum_{j=1}^{n} x_{ij} + c_i \quad (i = 1, \dots, n).$$

is called the *balance equation* of IIM. The combination of the proportionality equation with the balance equation gives that

$$x_i = \sum_{j=1}^{n} a_{ij} x_j + c_i \quad (i = 1, \dots, n)$$

which is called the *inoperability equation* of IIM. Its matrix form is $\mathbf{x} = A\mathbf{x} + \mathbf{c}$, where $\mathbf{x} = (x_1, \dots, x_n)^{\mathrm{T}}$, $\mathbf{c} = (c_1, \dots, c_n)^{\mathrm{T}}$, and the A-matrix is

$$A = \begin{pmatrix} a_{11} & \cdots & a_{1n} \\ \vdots & \ddots & \vdots \\ a_{n1} & \cdots & a_{nn} \end{pmatrix}.$$

Example 9.10.1. Assume that a system consists of three subsystems. Denote by x_1, x_2, x_3 the inoperability of these three subsystems. If Subsystem 1 fails completely, then Subsystems 2 and 3 can perform to only 40 % and 75 %, respectively, of their functionality. If Subsystem 2 fails completely, then Subsystems 1 and 3 can perform to only 20 % and 10 %, respectively, of their functionality. Assume that the inoperability of three subsystems has no impact on each other. Thus, the A-matrix for the system is

$$A = \begin{pmatrix} 0 & 0.8 & 0 \\ 0.6 & 0 & 0 \\ 0.25 & 0.9 & 0 \end{pmatrix}.$$

Assume that Subsystem 2 loses 40 % of its functionality due to an external perturbation. The inoperability equation of IIM is

$$\begin{pmatrix} x_1 \\ x_2 \\ x_3 \end{pmatrix} = \begin{pmatrix} 0 & 0.8 & 0 \\ 0.6 & 0 & 0 \\ 0.25 & 0.9 & 0 \end{pmatrix} \begin{pmatrix} x_1 \\ x_2 \\ x_3 \end{pmatrix} + \begin{pmatrix} 0 \\ 0.4 \\ 0 \end{pmatrix} = \begin{pmatrix} 0.8x_2 \\ 0.6x_1 + 0.4 \\ 0.25x_1 + 0.9x_2 \end{pmatrix}$$

or

$$\begin{cases} x_1 = 0.8x_2, \\ x_2 = 0.6x_1 + 0.4, \\ x_3 = 0.25x_1 + 0.9x_2. \end{cases}$$

Solving this set of equations gives

$$x_1 = \frac{8}{13} \approx 0.615,$$
$$x_2 = \frac{10}{13} \approx 0.769,$$
$$x_3 = \frac{11}{13} \approx 0.846.$$

This means that the inoperability of Subsystems 1, 2, and 3 is 0.615, 0.769, and 0.846, respectively.

Assume $h \times 100\,\%$ of the operability of Subsystem 2 is lost due to the attack alone. Then the inoperability equation of IIM is

$$\begin{pmatrix} x_1 \\ x_2 \\ x_3 \end{pmatrix} = \begin{pmatrix} 0 & 0.8 & 0 \\ 0.6 & 0 & 0 \\ 0.25 & 0.9 & 0 \end{pmatrix} \begin{pmatrix} x_1 \\ x_2 \\ x_3 \end{pmatrix} + \begin{pmatrix} 0 \\ h \\ 0 \end{pmatrix} = \begin{pmatrix} 0.8x_2 \\ 0.6x_1 + h \\ 0.25x_1 + 0.9x_2 \end{pmatrix}$$

or

$$\begin{cases} x_1 = 0.8x_2, \\ x_2 = 0.6x_1 + h, \\ x_3 = 0.25x_1 + 0.9x_2. \end{cases}$$

Note that the constraint $0 \le x_1, x_2, x_2 \le 1$. Solving the set of equations, the solution is

$$\begin{array}{llll} x_1 = \frac{20}{13}h, & x_2 = \frac{25}{13}h, & x_3 = \frac{55}{26}h & \text{for } 0 \le h \le 0.47, \\ x_1 = \frac{8}{11} \approx 0.727, & x_2 = \frac{10}{11} \approx 0.909, & x_3 = 1 & \text{for } 0.47 < h \le 1. \end{array}$$

Subsystem 3 fails completely when the external attack brings down 47 % of its operability. The remaining 53 % is taken away by its dependency on Subsystems 1 and 2.

Further reading

[1] Asbeck E, Haimes YY. The partitioned multiobjective risk method. *Large Scale Systems*. 1984(6):13–38.

[2] Aven T. Risk assessment and risk management: Review of recent advances on their foundation. *European Journal of Operational Research*. 2016(253):1–13.

[3] Browning J, Thomas N. An assessment of the tsunami risk in Muscat and Salalah, Oman, based on estimations of probable maximum loss. *International Journal of Disaster Risk Reduction*. 2016(16):75–87.

[4] Chankong V. Multiobjective decision making analysis: The interactive surrogate worth trade-off method, PhD dissertation, Systems Engineering Department, Case Western Reserve, Cleveland, OH, 1977.

[5] de Carvalho AL, Antunes CH, Freire F, Henriques CO. A multi-objective interactive approach to assess economic-energy-environment trade-offs in Brazil. *Renewable and Sustainable Energy Reviews*. 2016(54):1429–1442.

[6] Finkel A. *Confronting Uncertainty in Risk Management: A guide for Decision-Makers*. Resources for the Future, Center for Risk Management, Washington, DC., 1990.

[7] Gomide F, Haimes YY. The multiobjective, multistage impact analysis method, theoretical basis. *IEEE Transactions on System, Man, and Cybernetics SMC*. 1984(14):88–98.

[8] Haimes YY, Lasdon LS, Wismer DA. On the bicriterion formulation of the integrated system identification and systems optimization. *IEEE Transactions on Systems, Man, and Cybernetics SMC*. 1971(1):296–297.

[9] Haimes YY, Lambert JH, Li D. Risk of extreme events in a multiobjective framework. *Water Resources Research*. 1992(28):201–209.

[10] Hamdy M, Nguyen AT, Hensen JLM. A performance comparison of multi-objective optimization algorithms for solving nearly-zero-energy-building design problems. *Energy and Buildings*. 2016(121):57–71.

[11] He W, He Q, Zhou J. Soil weathering-water environment-ecological risks in Hanjiang River Basin, China. *Quaternary International*. 2015(380–381):297–304.

[12] Kankara RS, Arockiaraj S, Prabhu K. Environmental sensitivity mapping and risk assessment for oil spill along the Chennai Coast in India. *Marine Pollution Bulletin*. 2016(106):95–103.

[13] Kennedy WV. The directive on environmental impact assessment. *Environmental Policy and Law*. 1982(8):84–95.

[14] Law AM, Kelton WD. *Simulation Modeling and Analysis*. McGraw-Hill, New York, 1991.

[15] Leach MR. Risk and impact analysis in a multiobjective framework, MSc thesis, Systems Engineering Department, Case Western Reserve University, Cleveland, OH, 1984.

[16] Leontief WW. *The Structure of the American Economy, 1919–1939*. Second edition, Oxford University Press, New York, 1951.

[17] Leontief WW. *Input-Output Economics*. Second edition, Oxford University Press, New York, 1986.

[18] Li C, Sun L, Jia J, Cai Y, Wang X. Risk assessment of water pollution sources based on an integrated k-means clustering and set pair analysis method in the region of Shiyan, China. *Science of the Total Environment*. 2016(557–558):307–316.

[19] Li D, Haimes YY. The uncertainty sensitivity index method and its extension. *Naval Research Logistic*. 1988(35):655–672.

[20] Mohamad N, Latif MT, Khan MF. Source apportionment and health risk assessment of PM10 in a naturally ventilated school in a tropical environment. *Ecotoxicology and Environmental Safety*. 2016(124):351–362.

[21] Morgan G, Henrion M. *Uncertainty*. Cambridge University Press, Cambridge, 1990.

[22] Naidu R, Espana VAA, Liu Y, Jit J. Emerging contaminants in the environment: Risk-based analysis for better management. *Chemosphere*. 2016(154):350–357.

[23] Oxley T, Simon HM. Space, time and nesting Integrated Assessment Models. *Environmental Modelling & Software*. 2007(22):1732–1749.

[24] Senapati N, Jansson PE, Smith P, Chabbi A. Modelling heat, water and carbon fluxes in mown grassland under multi-objective and multi-criteria constraints. *Environmental Modelling & Software*. 2016(80):201–224.

[25] Shrestha S, Semkuyu DJ, Pandey VP. Assessment of groundwater vulnerability and risk to pollution in Kathmandu Valley, Nepal. *Science of the Total Environment*. 2016(556):23–35.

10 Life cycle assessments

Life cycle means the entire life cycle of a product system from the raw material acquisition, manufacture, use/reuse/maintenance, and recycle/waste management, to its final disposal of a product, process or system. The concept of life cycle was mentioned first by Novick in 1959. At that time, a main application was the cost of weapon systems including the cost of purchase, development, and end-of-life operations. When environmental policy became a major issue in all industrialized societies, the life cycle concept was fully developed. Currently, life cycle assessment (LCA) has become a major instrument to assess the ecological burdens and impacts throughout the consecutive and interlinked stages of a product system, from raw material acquisition or generation from natural resources, through production and use to final disposal.

10.1 Classic life cycle assessment

Life Cycle Assessment (LCA) is a cradle-to-grave approach for assessing industrial systems. It begins with the initial gathering of raw materials from the earth to create the product and ends at the point when all residuals are returned to the earth. LCA is used for identifying, quantifying, and decreasing the overall environmental impact of a product, process, or system. The International Standards Organization (ISO) documents on LCA include the following several parts:
- ISO-14040, Life Cycle Assessment – Principles and Framework (ISO, 1997)
- ISO-14041, Life Cycle Assessment – Goal and Scope Definition and Inventory Analysis (ISO, 1998)
- ISO-14042, Life Cycle Assessment – Life Cycle Impact Assessment (ISO, 2000)
- ISO-14043, Life Cycle Assessment – Life Cycle Interpretation (ISO, 2000)
- ISO-14044, Life Cycle Assessment – Requirements and Guidelines (ISO, 2006)

The LCA process is a systematic, phased approach. An LCA consists of the following four interrelated phases:
- goal and scope definition,
- life cycle inventory analysis (or inventory analysis for short),
- life cycle impact assessment (or impact assessment for short),
- life cycle interpretation (or interpretation for short).

The framework of LCA is shown in Figure 10.1. Arrows indicate that all phases are linked to each other, the life cycle interpretation is also linked to all phases, and the LCA has direct applications including product development and improvement, strategic planning, public policy making, marketing, and so on.

DOI 10.1515/9783110424904-011

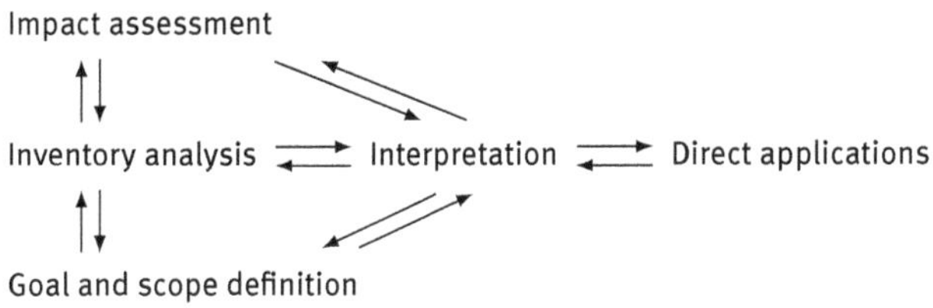

Fig. 10.1: The framework of LCA.

10.1.1 Goal and scope definition

There is no explicit ISO definition of this phase of LCA. But it obviously centers around formulating the question and stating the context of answering this question. In this phase, no data is collected and no results are calculated. Rather, it is a place where the plan of the LCA study is defined as clearly and unambiguously as possible such that one can quickly find out the precise question addressed and main principles chosen.

The goal of LCAs deals with the following topics:

- the intended application;
- reasons for carrying out the study;
- the intended audience;
- whether the results will be used as the basis for public comparative assertions.

The scope definition makes a number of major choices. For instance, the product system or systems to be studied, the function the system delivers, and the functional unit. The scope definition further sets the main outline on a number of subjects that are discussed and further refined in more detail in the later phases. For instance, system boundaries, impact categories, and treatment of uncertainty.

10.1.2 Life cycle inventory analysis

Life Cycle Inventory Analysis (LCI) is the second phase of LCA. ISO defines it as the phase of LCA involving the compilation and quantification of inputs and outputs for a product throughout its life cycle. In this phase, quantification is an important aspect and numbers (data and calculations) are of central concern.

Unit processes are the central element of inventory analysis. A unit process in ISO 14040 is defined as the smallest element considered in the LCI. Examples of unit process are coal mining, steel production, refining of oil, recycling of waste paper, and transport by lorry. In LCA a unit process is considered commonly as a black box that converts a bundle of inputs into a bundle of outputs. LCA studies can connect different unit process into a system in simple upstream-downstream connections or more complicated connections. In the present era of digital databases, LCA studies can easily connect several thousands of unit processes.

In the LCI, all unit processes have to be quantified. This means that the sizes of the inflows and outflows per unit process have to be specified. In scaling the unit processes, the web-like nature of the system quickly creates complications. Two complications are mentioned as follows:
- upstream or downstream processes of some products may be difficult to quantify;
- some unit processes produce several co-products such that the balance equations become impossible.

The first issue can be solved by a procedure known as cut-off. The second issue can be solved by co-product allocation. After appropriate cut-off and allocation, the final inventory results can be calculated.

The key steps of LCI are to develop a flow diagram and an LCI data collection plan, collect data of inputs and outputs, and evaluate and document the LCI results.

10.1.3 Life cycle impact assessment

Life Cycle Impact Assessment (LCIA) is the third phase of LCA aimed at understanding and evaluating the magnitude and significance of the potential environmental impacts for a product system throughout the life cycle of the product. According to ISO 14042, LCIA is divided into mandatory and optional steps.

The mandatory steps are as follows.
- *Selection of impact categories.* The impact category is the central element in life cycle impact assessment. ISO defines the impact category as a class representing environmental issues of concern to which LCI results may be assigned. Impact categories and the corresponding category indicators can be organized at a midpoint level and at an endpoint level along the cause-effect chain. Important environmental impact categories are Acidification Potential (AP), Eutrophication Potential (EP), Global Warming Potential (GWP), and Ozone Depletion Potential (ODP).
- *Classification.* ISO defines it as the assignment of LCI results to the selected impact categories.
- *Characterization.* This step converts the results of LCI into a common metric and aggregates the converted LCI results. The result from characterization is a list of numbers. ISO calls these numbers *category indicator results* (*score*).
  Many LCIA studies stop at the characterization.

The optional steps are as follows.
- *Normalization.* This refers to calculating the magnitude of category indicator results relative to reference information. It fulfills several functions that it provides insight into the meaning of the impact indicator results, it helps to check for errors, and it prepares for a possible weighting step.

- *Grouping.* This is seldom seen in LCA studies. ISO defines it as the assignment of impact categories into one or more sets. ISO mentions two ways: sorting on a nominal basis and ranking on an ordinal basis.
- *Weighting.* This starts with the characterization (or normalization) results. Weighting factors are applied to the characterization indicator results or their normalized version. Weighting produces one final number by

$$W = \sum_{c} WF_c \times I_c,$$

  where I_c is the impact score and WF_c is the weighting factor, both for the impact category c, and W is the weighted result.
- *Data quality analysis.* This develops a better understanding of the reliability of the indicator results in the LCIA profile.

10.1.4 Life cycle interpretation

Life cycle interpretation is the final phase of LCA. It integrates the LCI and LCIA results to develop conclusions and recommendations that relate to the goal and scope of the study. ISO 2006a defines two objectives of interpretation:
- Analyze results, reach conclusion, explain limitations, and provide recommendations based on the findings of the preceding phases, and report the results of the interpretation in a transparent manner.
- Provide a readily understandable, complete, and consistent presentation of the results of an LCA study in accordance with the goal and scope of the study.

LCA is best used as an iterative approach. It is especially important to determine that if the result of the impact assessment or the underlying inventory data are incomplete or unacceptable for drawing conclusions and making recommendations, then the previous steps must be repeated until the results can support the original goals of the study.

10.2 Exergetic life cycle assessment

Exergy is a thermodynamic quantity. The exergy of a system is defined as the maximum shaft work that can be attained when it is in a reference environment. Exergy is conserved only when all processes occurring in a system and its surroundings are reversible. Exergy is destroyed whenever an irreversible process occurs. Like energy, exergy can also be transferred across the boundary of a system. There is an exergy transfer corresponding to each type of energy transfer. The exergy transfer associated with shaft work is equal to the shaft work, while the exergy transfer associated with heat transfer depends on the temperature of the reference environment. A system in complete equilibrium with its environment has no exergy.

Exergy analysis is based on the second law of thermodynamics. An exergy balance for a process or system is the following:

Accumulated exergy = Input exergy – Output exergy – Destroyed exergy.

The exergy quantities in an exergy balance include the exergy of a matter flow, a thermal energy transfer, and electricity. The exergy of a matter flow is equal to the sum of physical, chemical, kinetic, and potential components. The exergy associated with a thermal energy transfer relates to system and reference environment temperatures. The exergy associated with electricity is equal to the energy.

Reducing exergy losses or increasing exergy efficiencies can often decrease environmental impacts associated with systems or processes. Exergy losses occur during the lifetime of a product or process. Reducing exergy losses helps improve sustainability. As exergy efficiency approaches ideality, environmental impacts approach zero. As exergy efficiency approaches zero, sustainability approaches zero. The exergy and environmental impact occur generally during all phases of the life cycle. Their main connections are waste exergy emissions, resource degradation, and order destruction and chaos creation.

Exergetic life cycle assessment (ExLCA) is a different approach from LCA. Like LCA, it is also a useful analytical tool to identify, quantify, and decrease the overall environmental impact of a process or a system. The general methodological framework of ExLCA is similar to that of LCA. The main differences are the following:

(a) The inventory analysis of ExLCA is more detailed than that of LCA. All inputs and outputs must be identified and quantified. The material and energy balances have to be closed.
(b) The impact assessment of ExLCA focuses on the determination of the exergies of the flow, the exergy destructions, and exergy efficiencies of the overall process and its subprocesses.
(c) The improvement analysis in ExLCA is intended to reduce its life cycle irreversibilities. The summation of all exergy destructions in the life cycle identifies the life cycle irreversibility of the product or process.

Throughout ExLCA, the calculation of exergy values requires that the conditions and composition of the reference environment are specified.

10.3 Ecologically-based life cycle assessment

Ecosystem goods and services, such as fresh water, soil, and pollination, are essential to all human activity. These goods and services can be divided into the following categories:

- provisioning services that supply goods;
- regulating services that provide benefits by controlling ecosystem processes;
- cultural services that are all of the intangible benefits;
- supporting services that are required for all of the other ecosystem services to take place.

Twenty-four different ecosystem services relating to provisioning, regulating, and cultural services are included in the Millennium Ecosystem Assessment (MEA). Of these twenty-four services, scientists have found that in the last fifty years, fifteen ecosystem services have globally degraded, another five ecosystem services have mixed results, and the remaining four ecosystem services have enhanced performance. If these trends continue, the earth may no longer be able to sustain human life. This motivates scientists to put these ecosystem services into the life cycle assessment methodology.

Ecologically-based Life Cycle Assessment (Eco-LCA) is a hybrid LCA method. In a hybrid study, the most important parts of the process are modeled using Process-based LCA (Process LCA), the less important parts are modeled using Economic Input-Output LCA (EIO-LCA). Process LCA and EIO-LCA are two methods used routinely to perform the assessment of products and processes by LCA practitioners. Eco-LCA has both the precision of Process LCA and the completeness of EIO-LCA.

Eco-LCA contains twenty different ecosystem services in which many provisioning and regulating services were assessed in the MEA, some supporting services were not included in the MEA.

Many engineering analyses undervalue or completely ignore the role of ecosystems, i.e., energy consumption and emissions. It is well known that the second law of thermodynamics has profound implications on the capability of man-made technology in meeting sustainability goals. The second law says that decreasing entropy in an open system must result in a greater increasing entropy in the surroundings. This increase manifests as environmental impact since the surrounding environment must dissipate the local increasing entropy. Eco-LCA can quantify the role of ecosystem goods and services in the life cycle. Different goods and services have different units that cannot be added together. To compare various ecosystem services Eco-LCA converts many different units of ecosystem goods and services to a common basis of thermodynamic work using exergy and emergy. Exergy allows the comparison of the different resources on a common basis of Joules of work. Emergy is also a basis of comparison. Emergy's unit is the *solar emergy Joule* (seJ). The seJ allows the difference in energy quality to be accounted for in different resources. The factors that are used to convert from the original energy unit to seJ are called *transformities* (units: seJ/J) or, more generally, *Unit Emergy Value* (UEV, units: seJ/unit). Transformity indicates how many seJ are required to get a single Joule of the product.

10.4 Case studies

Life cycle assessment (LCA) has become a major instrument to assess ecological burdens and impacts throughout the consecutive and interlinked stages of a product system, from raw material acquisition or generation from natural resources, through production and use to final disposal. Here we give some case studies as examples.

10.4.1 Energy crops

The need to tackle climate change has pushed the cultivation of energy crops for the production of bio-fuels to the top of the global agenda. Christoforou et al. (2016) use LCA to quantify the environmental impact of first generation energy crops in Cyprus which include maize, sweet sorghum, winter wheat, sugar-beets, potato and winter barley. The goal of the study was to evaluate the environmental impact of different energy crop systems. Life cycle inventory are classified as pre-farm activities (e.g. fuel consumption, electricity consumption, seed, fertilizers and pesticides transportation) and on-farm activities (e.g. land preparation, fertilizer and pesticides application, irrigation and harvesting). Main impact assessment includes global warming potential, acidification potential, eutrophication potential, ozone depletion potential and abiotic depletion potential. The results show that barley, potato, and wheat crop presented the highest environment impact in Cyprus.

10.4.2 Wastewater treatment

Wastewater treatment plays a key role in assuring the continued utility of ecosystems. The three main inputs needed to treat wastewater are electricity, chlorine, and transportation services for the solid and sludge waste products. Morrison et al. (2016) investigated a wastewater treatment plant (WWTP) at a university campus. Their analysis began at the point where raw wastewater from individual buildings on the campus deposits into the central wastewater collection system and is gravity fed to the wastewater inlet port at the WWTP and extends to the point that it has been treated to discharge standards. Based on LCA, Morrison et al. (2016) predicted the total energy consumption and CO_2 emissions over the lifetime of the WWTP.

10.4.3 Livestock production

Commercial livestock production has significant impacts on the environment. Pig production is a highly complex global system which involves the production of fertilizers and pesticides for crop production, land transformation, transportation to and from farms, energy for light and heat, water for animal consumption and farmyard washing, and waste management. LCA can be used to measure the potential environmental performance of pig production, including acidification potential, eutrophication potential and global warming potential (Graham et al. 2016).

Further reading

[1] Bare JC. *Developing a Consistent Decision-Making Framework by using U.S. EPA's TRACI.* National Risk Management Research Laboratory, US Environmental Protection Agency, Cincinnati, OH, 2002.

[2] Bare JC, Gloria TP. Critical analysis of the mathematical relationships and comprehensiveness of life cycle impact assessment approaches. *Environmental Science & Technology*. 2006(40): 1104–1113.

[3] Brand G, Braunschweig A et al. *Weighting in Ecobalances with the Ecoscarcity Method – Ecofactors*. Environment Series, No.297, Bern, Switzerland, Swiss Agency for the Environment, Forests, and Landscape (SAEFL), 1997.

[4] BUS. *Ökobilanzen von Packstoffen*. Schriftenreihe Umweltschutz Nr. 24. Bern, Switzerland, Bundesamt für Umweltschutz, 1984.

[5] Christoforou E, Fokaides PA, Koroneos CJ, Recchia L. Life Cycle Assessment of first generation energy crops in arid isolated island states: The case of Cyprus. *Sustainable Energy Technologies and Assessments*. 2016(14):1–8.

[6] Daniel JJ, Rosen MA. Exergetic environmental assessment of life cycle emissions for various automobiles and fuels. *Exergy, an International Journal*. 2002(2):283–294.

[7] Dincer I. Thermodynamics, exergy and environmental impact. *Energy Sources*. 2000(22) 723–732.

[8] Dincer I, Rosen MA. *Exergy: Energy, Environment and Sustainable Development*. Elsevier, UK, 2007.

[9] EC-JRC. Framework and Requirements for Life Cycle Impact Assessment (LCIA) Models and Indicators. ILCD Handbook-International Reference Life Cycle Data System, European Commission-Joint Research Center, 2010.

[10] EC-JRC. An Analysis of existing Environmental Impact Assessment Methodologies for use in Life Cycle Assessment-Background Document. ILCD Handbook-International Reference Life Cycle Data System, European Commission-Joint Research Center, 2010.

[11] EC-JRC. General Guide for Life Cycle Assessment-Detailed Practice. ILCD Handbook-International Reference Life Cycle Data System, European Commission-Joint Research Center, 2010.

[12] Goedkoop M, Demmers M, Collignon M. The Eco-indicator 95 Manual for Designers. National Reuse of Waste Research Programme, The Netherlands, 1996.

[13] Guinée JB, Gorrée M et al. *Handbook on Life Cycle Assessment, Operational guide to the ISO Standards*. I: LCA in perspective, IIa: Guide, IIb: Operational annex, III: Scientific background. Kluwer Academic Publishers, Dordrecht, 2002.

[14] Heijungs R, Guinee J et al. *Environment Life Cycle Assessment of Products: Guide and Background*. CML, Leiden, The Netherlands, 1992.

[15] Heijungs R, Kleijn R. Numerical approaches to life-cycle interpretation, five examples. *International Journal of Life Cycle Assessment*. 2001(6):141–148.

[16] Heijungs R, Suh S. *The Computational Structure of Life Cycle Assessment*. Kluwer Academic Publishers, Dordrecht, 2002.

[17] Hendrickson CT, Lave LB, Matthews HS. *Environmental Life Cycle Assessment of Goods and Services*. Washington DC, REF Press, 2006.
[18] Hermann WA. Quantifying global exergy resources. *Energy*. 2006(31):1349–1366.
[19] Huesemann MH. The limits of technological solutions to sustainable development. *Clean Technologies and Environmental Policy*. 2003(5):21–34.
[20] Huijbregts MA, Norris G, Bretz R, Ciroth A, Maurice B, von Bahr B et al. Framework for modelling data uncertainty in life cycle inventories. *International Journal of Life Cycle Assessment*. 2001(6):127–132.
[21] ISO. Environmental Management – Life Cycle Assessment – Requirements and Guidelines (ISO 14044). International Organization for Standardization, Geneva, 2006.
[22] Jolliet O, Margni M, Charles R, Humbert S, Payet J, Rebitzer G, Rosenbaum R. IMPACT 2002+: A new life cycle impact assessment methodology. *International Journal of Life Cycle Assessment*. 2003(8):324–330.
[23] Kotas TJ. *The Exergy Method of Thermal Plant Analysis*. Kriger: Malabar, Florida, 1995.
[24] McAuliffe GA, Chapman DV, Sage CL. A thematic review of life cycle assessment (LCA) applied to pig production. *Environmental Impact Assessment Review*. 2016(56):12–22.
[25] Millennium Ecosystem Assessment Board. Living Beyond Our Means: Natural Assets and Human Well-being, 2005.
[26] Moran MJ. *Availability Analysis: A Guide to Efficient Energy Use*. American Society of Mechanical Engineers, New York, 1989.
[27] Morrison M, Srinivasan RS, Ries R. Complementary life cycle assessment of wastewater treatment plants: An integrated approach to comprehensive upstream and downstream impact assessments and its extension to building-level wastewater generation. *Sustainable Cities and Society*. 2016(23):37–49.
[28] Novick D. *The Federal Budget as an Indicator of Government Intentions and the Implications of Intentions*. Santa Monica, CA: Rand Corporation, 1959, P-1803.
[29] Parry ML, Canziani OF, Palutikof JP, van der Linden PJ, Hanson CE. *Contribution of Working Group II to the Fourth Assessment Report of the Intergovernmental Panel on Climate Change*. Cambridge Univ. Press, Cambridge, UK, 2007.
[30] Rosen MA, Dincer I. On exergy and environmental impact. *International Journal of Energy Research*. 1997(21):643–654.
[31] Rosen MA, Dincer I. Exergy as the confluence of energy, environment and sustainable development. *Exergy, an International Journal*. 2001(1) 3–13.
[32] Steen B. A systematic approach to environmental priority strategies in product development (EPS), Version 2000-General system characteristics. CPM report 1999: 4, Center for Environmental Assessment, Chalmers Univ. of Technology, Gothenburg, Sweden, 1999.
[33] Ultiati S. Energy quality, emergy, and transformity: H.T. Odum's contributions to quantifying and understanding systems. *Ecological Modelling*. 2004(178):201–213.
[34] Zhang Y, Baral A, Bakshi BR. Accounting for ecosystem services in life cycle assessment, Part II, Toward an ecologically based LCA. *Environmental Science & Technology*. 2010(44): 2624–2631.
[35] Zhang Y, Singh S, Bakshi BR. Accounting for ecosystem services in life cycle assessment, Part I: A critical review. *Environmental Science & Technology*. 2010(44):2232–2242.

Index

www.ingramcontent.com/pod-product-compliance
Ingram Content Group UK Ltd.
Pitfield, Milton Keynes, MK11 3LW, UK
UKHW050944290926
15470UKWH00005B/1619

9 783110 430011